Multi-Cloud Strategy for Cloud Architects

Second Edition

Learn how to adopt and manage public clouds by leveraging BaseOps, FinOps, and DevSecOps

Jeroen Mulder

BIRMINGHAM—MUMBAI

Multi-Cloud Strategy for Cloud Architects

Second Edition

Senior Publishing Product Manager: Rahul Nair

Acquisition Editor – Peer Reviews: Gaurav Gavas

Project Editor: Rianna Rodrigues

Content Development Editor: Grey Murtagh

Copy Editor: Safis Editing

Technical Editor: Srishty Bhardwaj

Proofreader: Safis Editing

Indexer: Pratik Shirodkar

Presentation Designer: Ganesh Bhadwalkar

Developer Relations Marketing Executive: Monika Sangwan

First published: December 2020

Second edition: April 2023

Production reference: 1210423

Published by Packt Publishing Ltd.
Livery Place
35 Livery Street
Birmingham
B3 2PB, UK.

ISBN 978-1-80461-673-4

www.packt.com

To my wife: I owe you big time. And to my inspirational sources, Eckart Wintzen and Wubbo Ockels. You left too soon.

–Jeroen Mulder

Contributors

About the author

After his study in Journalism, **Jeroen Mulder** (born 1970) started his career as editor for the economic pages of Dutch newspapers. In 1998 he got involved in internet projects for the British publisher Reed Business Information, creating websites and digital platforms. Highly attracted by the possibilities of the new digital era, Jeroen decided to pursue a career in digital technologies. In 2000 he joined the IT company Origin, which later became Atos Origin and eventually Atos. Within Atos he has fulfilled many roles, but always in the heart of technology.

Jeroen is a certified enterprise and security architect. From 2014 onwards he started concentrating more and more on cloud technology. This included architecture for cloud infrastructure, serverless and container technology, DevOps, security, and AI.

In March 2017 he joined the Japanese technology company Fujitsu, focusing on cloud solutions. In 2021 he was appointed principal cloud architect at Philips Precision Diagnosis, but decided to return to Fujitsu in 2022, where he currently holds the position of principal consultant for the company's global multi-cloud and security solutions.

Jeroen regularly publishes articles about cloud developments, AI, and emerging tech, and is frequently invited to perform as a speaker at tech events. With Packt, he has previously published books about multi-cloud, DevOps for enterprise architects, and the digital transformation of healthcare.

Once again, I must thank my wonderful wife, Judith, and my two girls for granting me the space and time to write. A big thank you goes out to the whole Packt editing team for making this another epic journey. Lastly, thank you, Fujitsu, for rehiring me. I'm having a blast.

About the reviewers

Juan Ramón Cabrera is a Sr. Cloud Solutions Architect at Microsoft with 25+ years of IT experience. He has worked in the army, as an entrepreneur, and in various IT roles, from developer to team lead and IT manager. He has expertise in software development, DevOps, Agile, security, infrastructure, and cloud architecture. He holds several certifications from Microsoft (such as Azure Architect Expert), AWS (CSA Associate), and the Linux Foundation (CKA, CKAD).

Thanks to the author and the publisher for giving me the opportunity to participate and contribute my knowledge and experience to this book. I enjoyed it a lot. Congrats for your work.

Kamesh Ganesan is a seasoned technology professional, an author, and a leader with over 25 years of IT experience in all major cloud technologies, including AWS, Azure, GCP, Oracle and Alibaba. He has over 55 IT and cloud certifications. He has played many IT roles and architected and delivered mission-critical, innovative technology solutions that have helped commercial enterprise and government clients to be very successful. He has written AWS and Azure books and has reviewed many IT/cloud technology books and courses.

I am extremely thankful for all the Gods' blessings in my life. A special thanks to my wife, Hemalatha, for her motivation and continuous support in all my pursuits, and many thanks to my kids, Sachin and Arjun, for their unconditional love. I am very grateful to my father, Ganesan, and mother, Kasthuri, for their unwavering encouragement throughout my life.

Table of Contents

Chapter 3: Starting the Multi-Cloud Journey 49

Chapter 4: Service Designs for Multi-Cloud 75

Chapter 6: Controlling the Foundation Using Well-Architected Frameworks 131

Chapter 7: Designing Applications for Multi-Cloud 161

Chapter 9: Creating a Foundation for IoT 207

Chapter 10: Managing Costs with FinOps 221

Chapter 11: Maturing FinOps 249

Chapter 12: Cost Modeling in the Cloud 265

Chapter 13: Implementing DevSecOps 277

Chapter 14: Defining Security Policies 303

Chapter 15: Implementing Identity and Access Management 327

Chapter 16: Defining Security Policies for Data 343

Preface

Enterprises are increasingly adopting a multi-cloud strategy, using a mix of **Software as a Service (SaaS)**, **Platform as a Service (PaaS)**, and **Infrastructure as a Service (IaaS)**, hosted on platforms such as AWS, Azure, and other technology stacks. This leaves the architects and lead engineers with the challenge of how to integrate architectures and manage the enterprise cloud. Architects and engineers will learn how to design, implement, and integrate cloud solutions and set up controls for governance.

The first edition of this book was released in 2020, but developments in the cloud are rapidly evolving. This edition is extended with two cloud platforms that have grown significantly over the past years, Oracle Cloud Infrastructure and Alibaba Cloud. New methodologies have also been adopted by companies to improve cloud management. This includes the financial controls of FinOps and embedded security in DevSecOps.

After the introduction of the concept of multi-cloud, this book covers all of the topics that architects should consider when designing systems for multi-cloud platforms. That starts with designing connectivity to and between the various platforms and creating landing zones in Azure, AWS, **Google Cloud Platform (GCP)**, **Oracle Cloud Infrastructure (OCI)**, and Alibaba Cloud. These clouds will be discussed in this book.

The book is divided into four main sections, covering the following:

1. Introduction to cloud and multi-cloud architecture and governance
2. Operations, including setting up and managing the landing zones that provide the infrastructure for cloud environments
3. Financial operations using the principles of FinOps
4. Continuous delivery and deployment using DevSecOps, covering identity and access management, securing data, security information, and new concepts such as AIOps and **Site Reliability Engineering (SRE)**

The book contains best practices for the major providers, discusses common pitfalls and how to avoid them, and gives recommendations for methodologies and tools. Of course, a book about multi-cloud could never be complete, but this book will provide you with good guidelines to get started with architecting for multi-cloud.

Who this book is for

This book targets architects and lead engineers who are involved in architecting multi-cloud environments using Azure, AWS, GCP, OCI, and Alibaba Cloud. A basic understanding of cloud platforms and overall Cloud Adoption Frameworks is required.

What this book covers

Chapter 1, Introduction to Multi-Cloud, provides the definition of multi-cloud and why companies have a multi-cloud strategy.

Chapter 2, Collecting Business Requirements, discusses how enterprises could collect requirements using various enterprise architecture methodologies and how they can accelerate business results by implementing a multi-cloud strategy.

Chapter 3, Starting the Multi-Cloud Journey, explains how businesses can start developing and implementing cloud platforms, describing the steps in transition and transformation.

Chapter 4, Service Designs for Multi-Cloud, discusses governance in multi-cloud using the Cloud Adoption Frameworks of cloud providers.

Chapter 5, Managing the Enterprise Cloud Architecture, covers the architecture principles of various domains, such as security, data, and applications. You will learn how to create an enterprise architecture for multi-cloud.

Chapter 6, Controlling the Foundation Using Well-Architected Frameworks, explains how to define policies to manage the landing zone and get a deeper understanding of handling accounts in landing zones. The Well-Architectured Frameworks of cloud providers are used as guidance in setting up landing zones in various clouds.

Chapter 7, Designing Applications for Multi-Cloud, covers how to gather and validate business requirements for the resilience and performance of applications in the cloud.

Chapter 8, Creating a Foundation for Data Platforms, discusses the basic architecture of data lakes and considers the various solutions that cloud providers offer. You will also learn about the challenges that come with collecting and analyzing vast amounts of data.

Chapter 9, Creating a Foundation for IoT, explores the architecture principles of an IoT ecosystem and discusses how the cloud can help in managing IoT devices. We will explore some of these cloud solutions and also look at crucial elements in IoT, such as connectivity and security.

Chapter 10, Managing Costs with FinOps, focuses on the basics of financial operations in the cloud – for instance, the provisioning of resources and the costs that come with the deployment of resources.

Chapter 11, Maturing FinOps, talks about the transformation to managed FinOps in an organization by setting up a FinOps team, which has a major task in the adoption of the FinOps principles that we discussed in *chapter 10*.

Chapter 12, Cost Modeling in the Cloud, teaches how to develop and implement a cost model that allows organizations to identify cloud costs (showback) and allocate (chargeback) costs to the budgets of teams or units.

Chapter 13, Implementing DevSecOps, discusses setting up DevOps practices to develop and deploy applications to the cloud, but always with security as a priority, making sure that code, pipelines, applications, and infrastructure remain secure at every stage of the release cycle.

Chapter 14, Defining Security Policies, introduces the security frameworks of cloud providers and overall frameworks such as the **Center for Internet Security** (**CIS**) controls. You will learn how to define policies using these frameworks.

Chapter 15, Implementing Identity and Access Management, covers authenticating and authorizing identities. It also provides a good understanding of how to deal with least privileged accounts and the use of eligible accounts.

Chapter 16, Defining Security Policies for Data, starts with explaining data models and data classification. Next, you will learn how to protect data using cloud technologies such as encryption.

Chapter 17, Implementing and Integrating Security Monitoring, discusses the function and the need for integrated security monitoring, using **SIEM** (**Security Information and Event Management**) and **SOAR** (**Security Orchestration, Automation, and Response**).

Chapter 18, Developing for Cloud with DevOps and DevSecOps, studies the principles of DevOps, how CI/CD pipelines work with push and pull mechanisms, and how pipelines are designed so that they fit multi-cloud environments. Next, you will learn how to secure DevOps processes using the principles of the DevSecOps maturity model and the most common security frameworks.

Chapter 19, Introducing AIOps and GreenOps, introduces the concept of **Artificial Intelligence Operations** (**AIOps**) and how enterprises can optimize their cloud environments using AIOps. You will also learn about achieving sustainability in the cloud using GreenOps.

Chapter 20, Conclusion: The Future of Multi-Cloud, provides a peek into the future of emerging clouds and how enterprises can manage the growth of cloud technology within their organizations. The chapter contains sections about SRE as a method to ensure the stability of systems, while development is done at high speed.

To get the most out of this book

- It's recommended to have a basic understanding of IT architecture and more specific cloud architecture. Architects are advised to study the foundation of enterprise architecture, for instance, TOGAF (The Open Group Architecture Framework).
- Since this book also covers aspects of service management as part of governance, it's also recommended to have knowledge about IT service management (ITSM). Common basic knowledge about cloud patterns in public and private clouds is assumed.
- All chapters contain a *Further reading* section that provides information on more in-depth literature about topics discussed in the chapters.

Download the color images

We also provide a PDF file that has color images of the screenshots/diagrams used in this book. You can download it here: `https://packt.link/pDhXa`.

Conventions used

There are a number of text conventions used throughout this book.

`CodeInText`: Indicates code words in text, database table names, folder names, filenames, file extensions, pathnames, dummy URLs, user input, and Twitter handles. For example: "Mount the downloaded `WebStorm-10*.dmg` disk image file as another disk in your system."

Bold: Indicates a new term, an important word, or words that you see on the screen. For instance, words in menus or dialog boxes appear in the text like this. For example: "Select **System info** from the **Administration** panel."

Warnings or important notes appear like this.

Tips and tricks appear like this.

Get in touch

Feedback from our readers is always welcome.

General feedback: Email `feedback@packtpub.com` and mention the book's title in the subject of your message. If you have questions about any aspect of this book, please email us at `questions@packtpub.com`.

Errata: Although we have taken every care to ensure the accuracy of our content, mistakes do happen. If you have found a mistake in this book, we would be grateful if you reported this to us. Please visit `http://www.packtpub.com/submit-errata`, click **Submit Errata**, and fill in the form.

Piracy: If you come across any illegal copies of our works in any form on the internet, we would be grateful if you would provide us with the location address or website name. Please contact us at `copyright@packtpub.com` with a link to the material.

If you are interested in becoming an author: If there is a topic that you have expertise in and you are interested in either writing or contributing to a book, please visit `http://authors.packtpub.com`.

Share your thoughts

Once you've read *Multi-Cloud Strategy for Cloud Architects, Second Edition*, we'd love to hear your thoughts! Scan the QR code below to go straight to the Amazon review page for this book and share your feedback.

`https://packt.link/r/1804616737`

Your review is important to us and the tech community and will help us make sure we're delivering excellent quality content.

Join us on Discord!

Read this book alongside other users, cloud experts, authors, and like-minded professionals. Ask questions, provide solutions to other readers, chat with the authors via. Ask Me Anything sessions and much more.

Scan the QR code or visit the link to join the community now.

`https://packt.link/cloudanddevops`

Download a free PDF copy of this book

Thanks for purchasing this book!

Do you like to read on the go but are unable to carry your print books everywhere? Is your eBook purchase not compatible with the device of your choice?

Don't worry, now with every Packt book you get a DRM-free PDF version of that book at no cost.

Read anywhere, any place, on any device. Search, copy, and paste code from your favorite technical books directly into your application.

The perks don't stop there, you can get exclusive access to discounts, newsletters, and great free content in your inbox daily

Follow these simple steps to get the benefits:

1. Scan the QR code or visit the link below

`https://packt.link/free-ebook/9781804616734`

2. Submit your proof of purchase
3. That's it! We'll send your free PDF and other benefits to your email directly

1

Introduction to Multi-Cloud

Multi-cloud is a hot topic with companies. Most companies are already multi-cloud, sometimes even without realizing it. They have **Software as a Service** (**SaaS**) such as Office 365 from Microsoft and Salesforce, for instance, next to applications that they host in a public cloud such as **Amazon Web Services** (**AWS**) or **Google Cloud Platform** (**GCP**). It's all part of the digital transformation that companies are going through, that is, creating business agility by adopting cloud services where companies develop a best-of-breed strategy: picking the right cloud service for specific business functions. The answer might be multi-cloud, rather than going for a single cloud provider.

The main goal of this chapter is to develop a foundational understanding of what multi-cloud is and why companies have a multi-cloud strategy. We will focus on the main public cloud platforms of Microsoft Azure, **AWS, and GCP**, next to the different on-premises variants of these platforms, such as Azure Stack, AWS Outposts, Google Anthos, and some emerging players.

The most important thing before starting the transformation to multi-cloud is gathering requirements, making sure a company is doing the right thing and making the right choices. Concepts such as **The Open Group Architecture Framework** (**TOGAF**) and **Quality Function Deployment** (**QFD**) will be discussed as tools to capture the **voice of the customer** (**VOC**). Lastly, you will learn that any transformation starts with people. The final section discusses the changes to the organization itself needed to execute the digital transformation.

In this chapter, we're going to cover the following main topics:

- Understanding multi-cloud concepts
- Multi-cloud—more than just public and private
- Setting out a real strategy for multi-cloud

- Introducing the main players in the field
- Evaluating cloud service models
- Gathering requirements for multi-cloud
- Understanding the business challenges of multi-cloud

Understanding multi-cloud concepts

This book aims to take you on a journey along the different major cloud platforms and will try to answer one crucial question: *if my organization deploys IT systems on various cloud platforms, how do I keep control?* We want to avoid cases where costs in multi-cloud environments grow over our heads, where we don't have a clear overview of who's managing the systems, and, most importantly, where system sprawl introduces severe security risks. But before we start our deep-dive, we need to agree on a common understanding of *multi-cloud* and multi-cloud concepts.

There are multiple definitions of multi-cloud, but we're using the one stated at `https://www.techopedia.com/definition/33511/multi-cloud-strategy`:

> *Multi-cloud refers to the use of two or more cloud computing systems at the same time. The deployment might use public clouds, private clouds, or some combination of the two. Multi-cloud deployments aim to offer redundancy in case of hardware/software failures and avoid vendor lock-in.*

Let's focus on some topics in that definition. First of all, we need to realize where most organizations come from: traditional datacenters with physical and virtual systems, hosting a variety of functions and business applications. If you want to call this *legacy*, that's OK. But do realize that the cutting edge of today is the legacy of tomorrow. Hence, in this book, we will refer to "traditional" IT when we're discussing the traditional systems, typically hosted in physical, privately owned datacenters. And with that, we've already introduced the first problem in the definition that we just gave for multi-cloud.

A lot of enterprises call their virtualized environments private clouds, whether these are hosted in external datacenters or in self-owned, on-premises datacenters. What they usually mean is that these environments host several business units that get billed for consumption on a centrally managed platform. You can have long debates on whether this is really using the cloud, but the fact is that there is a broad description that sort of fits the concept of private clouds.

Of course, when talking about the cloud, most of us will think of the major public cloud offerings that we have today: AWS, Microsoft Azure, and GCP. These are public clouds: providers that offer IT services on demand from centralized platforms using the public internet. They are centralized platforms that provide IT services such as compute, storage, and networking but distributed across datacenters around the globe. The cloud provider is responsible for managing these datacenters and, with that, the cloud. Companies "rent" the services, without the need to invest in datacenters themselves.

By another definition, multi-cloud is a best-of-breed solution from these different platforms, creating added value for the business in combination with this solution and/or service. So, using the *cloud* can mean either a combination of solutions and services in the public cloud or combined with private cloud solutions.

But the simple feature of combining solutions and services from different cloud providers and/or private clouds does not make up the multi-cloud concept alone. There's more to it.

Maybe the best way to explain this is by using the analogy of the smartphone. Let's assume you are buying a new phone. You take it out of the box and switch it on. Now, what can you do with that phone? First of all, if there's no subscription with a telecom provider attached to the phone, you will discover that the functionality of the device is probably very limited. There will be no connection from the phone to the outside world, at least not on a mobile network. An option would be to connect it through a Wi-Fi device, if Wi-Fi is available. In short, one of the first actions, in order to actually use the phone, would be making sure that it has connectivity.

Now you have a brand-new smartphone set to its factory defaults and you have it connected to the outside world. Ready to go? Probably not. You probably want to have all sorts of services delivered to your phone, usually through the use of apps, delivered through online catalogs such as an app store. The apps themselves come from different providers and companies, including banks and retailers, and might even be coded in different languages. Yet, they will work on different phones with different versions of mobile operating systems such as iOS or Android.

You will also very likely want to configure these apps according to your personal needs and wishes. Lastly, you need to be able to access the data on your phone. All in all, the phone has turned into a landing platform for all sorts of personalized services and data.

The best part is that in principle, you, the user of the phone, don't have to worry about updates. Every now and then the operating system will automatically be updated and most of the installed apps will still work perfectly. It might take a day or two for some apps to adapt to the new settings, but in the end, they will work. And the data that is stored on the phone or accessed via some cloud directory will also still be available. The whole ecosystem around that smartphone is designed in such a way that from the end user's perspective, the technology is completely transparent:

Figure 1.1: Analogy of the smartphone—a true multi-cloud concept

Well, this mirrors the concept of the cloud, where the smartphone in our analogy is the actual integrated landing zone, where literally everything comes together, providing a seamless user experience.

How is this an analogy for multi-cloud? The first time we enter a portal for any public cloud, we will notice that there's not much to see. We have a platform—the cloud itself—and we probably also have connectivity through the internet, so we can reach the portal. But we don't want everyone to be able to see our applications and data on this platform, so we need to configure it for our specific usage. After we've done that, we can load our applications and the data on to the platform. Only authorized people can access those applications and that data. However, just like the user of a smartphone, a company might choose to have applications and data on other platforms. They will be able to connect to applications on a different platform.

The company might even decide to migrate applications to a different platform. Think of the possibility of having Facebook on both an iPhone and an Android phone; with just one Facebook account, the user will see the same data, even when the platforms—the phones—use different operating systems.

Multi-cloud—more than just public and private

There's a difference between hybrid IT and multi-cloud, and there are different opinions on the definitions. One is that hybrid platforms are homogeneous and multi-cloud platforms are heterogeneous. *Homogeneous* here means that the cloud solutions belong to one stack, for instance, the Azure public cloud with Azure Stack on-premises. *Heterogeneous,* then, would mean combining Azure and AWS, for instance.

Key definitions are:

- **Hybrid**: Combines on-premises and cloud.
- **Multi-cloud**: Two or more cloud providers.
- **Private**: Resources dedicated to one company or user.
- **Public**: Resources are shared (note, this doesn't mean anyone has access to your data. In the public cloud, we will have separate tenants, but these tenants will share resources, for instance, in networking).

For now, we will keep it very simple: a hybrid environment combines an on-premises stack—a private cloud—with a public cloud. It is a very common deployment model within enterprises and most consultancy firms have concluded that these hybrid deployments will be the most implemented future model of the cloud.

Two obvious reasons for hybrid—a mixture between the public and private clouds—are security and latency, besides the fact that a lot of companies already had on-premises environments before the cloud entered the market.

To start with security: this is all about sensitive data and privacy, especially concerning data that may not be hosted outside a country, or outside certain regional borders, such as the **European Union** (**EU**). Data may not be accessible in whatever way to—as an example—US-based companies, which in itself is already quite a challenge in the cloud domain. Regulations, laws, guidelines, and compliance rules often prevent companies from moving their data off-premises, even though public clouds offer frameworks and technologies to protect data at the very highest level. We will discuss this later on in *Part 4* of this book in *Chapters 13* to *18*, where we talk about security, since security and data privacy are of the utmost importance in the cloud.

Latency is the second reason to keep systems on-premises. One example that probably everyone can relate to is that of print servers. Print servers in the public cloud might not be a good idea. The problem with print servers is the spooling process. The spooling software accepts the print jobs and controls the printer to which the print assignment has to be sent. It then schedules the order in which print jobs are actually sent to that printer. Although print spoolers have improved massively in recent years, it still takes some time to execute the process. Print servers in the public cloud might cause delays in that process. Fair enough: it can be done, and it will work if configured in the right way, in a cloud region close to the sending PC and receiving printer device, plus accessed through a proper connection.

You get the idea, in any case: there are functions and applications that are highly sensitive to latency. One more example: retail companies have warehouses where they store their goods. When items are purchased, the process of order picking starts. Items are labeled in a supply system so that the company can track how many of a specific item are still in stock, where the items originate from, and where they have to be sent. For this functionality, items have a barcode or QR code that can be scanned with RFID or the like. These systems have to be close to the production floor in the warehouse or—if you do host them in the cloud—accessible through really high-speed, dedicated connections on fast, responsive systems.

These are pretty simple and easy-to-understand examples, but the issue really comes to life if you start thinking about the medical systems used in operating theatres, or the systems controlling power plants. It is not that useful to have an all-public-cloud, cloud-first, or cloud-only strategy for quite a number of companies and institutions. That goes for hospitals, utility companies, and also for companies in less critical environments.

Yet, all of these companies discovered that the development of applications was way more agile in the public cloud. Usually, that's where cloud adoption starts: with developers creating environments and apps in public clouds. It's where hybrid IT is born: the use of private systems in private datacenters for critical production systems that host applications with sensitive data that need to be on-premises for latency reasons, while the public cloud is used to enable the fast, agile development of new applications. That's where new cloud service models come into the picture. These models are explored in the next section.

The terms multi-cloud and hybrid get mixed up a lot and the truth is that a solution can be a mix. You can have, as an example, dedicated private hosts in Azure and AWS, hence running private servers in a public cloud. Or, run cloud services on a private host that sits in a private datacenter, for instance, with Azure Stack or AWS Outposts. That can lead to confusion. Still, when we discuss hybrid in this book, we refer to an on-premises environment combined with a public cloud. Multi-cloud is when we have two or more cloud providers.

Introducing the main players in the field

We have been talking about public and private clouds. Although it's probably clear what we commonly understand by these terms, it's still a good idea to have a very clear definition of both. We adhere to the definition as presented on the Microsoft website (`https://azure.microsoft.com/en-us/resources/cloud-computing-dictionary/what-is-cloud-computing`): the public cloud is defined as *computing services offered by third-party providers over the public internet, making them available to anyone who wants to use or purchase them*. The private cloud is defined as *computing services offered either over the internet or a private internal network and only to select users instead of the general public*. There are many more definitions, but these serve our purpose very well.

Public clouds

In the public cloud, the best-known providers are AWS, Microsoft Azure, GCP, Oracle Cloud Infrastructure, and Alibaba Cloud, next to a number of public clouds that have OpenStack as their technological foundation. An example of OpenStack is Rackspace. These are all public clouds that fit the definition that we just gave, but they also have some major differences.

AWS, Azure, and GCP all offer a wide variety of managed services to build environments, but they all differ very much in the way you apply the technology. In short: the concepts are more or less alike, but under the hood, these are completely different beasts. It's exactly this that makes managing multi-cloud solutions complex.

In this book, we will mainly focus on the major players in the multi-cloud portfolio.

Private clouds

Most companies are planning to move, or are actually in the midst of moving, their workloads to the cloud. In general, they have a selected number of major platforms that they choose to host the workloads: Azure, AWS, GCP, and that's about it. Fair enough, there are more platforms, but the three mentioned are the most dominant ones, and will continue to be throughout the forthcoming decades, if we look at analysts' reports. Yet, we will also address **Oracle Cloud Infrastructure** (**OCI**) and Alibaba Cloud in this book when appropriate and when adding valuable extra information, since both clouds have gained quite some market growth over the recent years.

As we already found out in the previous paragraphs, in planning for and migrating workloads to these platforms, organizations also discover that it gets complex. Even more important, there are more and more regulations in terms of compliance, security, and privacy that force these companies to think twice before they bring our data onto these platforms. And it's all about the data, in the end. It's the most valuable asset in any company—next to people.

In the private cloud, VMware seems to be the dominant platform, next to environments that have Microsoft with Hyper-V technology as their basis. Yet, Microsoft is pushing customers more and more toward consumption in Azure, and where systems need to be kept on-premises, they have a broad portfolio available with Azure Stack and Azure Arc, which we will discuss in a bit more detail later in this chapter.

Especially in European governmental environments, OpenStack still seems to do very well, to avoid having data controlled or even viewed by non-European companies. However, the adoption and usage of OpenStack seem to be declining.

The following diagram provides an example of a multi-cloud stack, dividing private from public clouds.

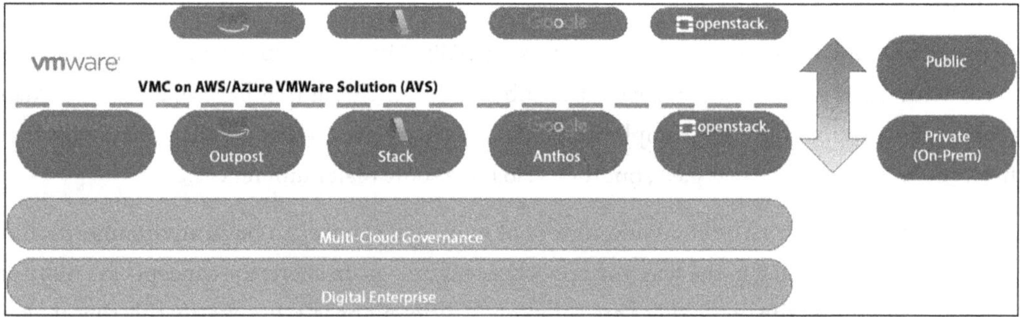

Figure 1.2: An example multi-cloud portfolio: the main players

In this section, we will look briefly at both VMware and OpenStack as private stack foundations. After that, we'll have a deeper look at AWS Outposts and Google Anthos. Basically, both propositions extend the public clouds of AWS and GCP into a privately owned datacenter. Next to this, we have to mention Azure Arc, which extends Azure to anywhere, either on-premises onto other clouds.

Outposts is an appliance that comes as a preconfigured rack with compute, storage, and network facilities. Anthos by Google is more a set of components that can be utilized to specifically host container platforms in on-premises environments using **Google Kubernetes Engine** (**GKE**). Finally, in this section, we will have a look at the Azure Stack portfolio.

VMware

In essence, VMware is still a virtualization technology. It started off with the virtualization of x86-based physical servers, enabling multiple virtual machines **Virtual Machines** (**VMs**) on one physical host. Later, VMware introduced the same concept to storage with **virtualized SAN** (**vSAN**) and **network virtualization and security** (**NSX**), which virtualizes the network, making it possible to adopt micro-segmentation in private clouds.

The company has been able to constantly find ways to move along with the shift to the cloud—as an example, by developing a proposition together with AWS where VMware private clouds can be seamlessly extended to the public cloud. The same applies to Azure: the joint offering is **Azure VMware Solution** (**AVS**).

VMware Cloud on AWS (**VMConAWS**) was a jointly developed proposition by AWS and VMware, but today Azure and VMware also supply migration services to migrate VMware workloads to Azure. VMware, acquired by Broadcom in 2022, has developed new services to stay relevant in the cloud. It has become a strong player in the field of containerization with the Tanzu portfolio, for instance. Over the last few years, the company has also strengthened its position in the security domain, again targeting the multi-cloud stack.

OpenStack

There absolutely are benefits to OpenStack. It's a free and open-source software platform for cloud computing, mostly used as **Infrastructure as a Service** (**IaaS**). OpenStack uses KVM as its main hypervisor, although there are more hypervisors available for OpenStack. It was—and still is, with a group of companies and institutions—popular since it offers a stable, scalable solution while avoiding vendor lock-in on the major cloud and technology providers. Major integrators and system providers such as IBM and Fujitsu adopted OpenStack in their respective cloud platforms, Bluemix and K5 (K5 was decommissioned internationally in 2018).

However, although OpenStack is open source and can be completely tweaked and tuned to specific business needs, it is also complex, and companies find it cumbersome to manage. Most of these OpenStack platforms do not have the richness of solutions that, for example, Azure, AWS, and GCP offer to their clients. Over the last few years, OpenStack seems to have lost its foothold in the enterprise world, yet it still has a somewhat relevant position and certain aspects are therefore considered in this book.

AWS Outposts

Everything you run on the AWS public cloud, you can now run on an appliance, including **Elastic Compute Cloud** (**EC2**), **Elastic Block Store** (**EBS**), databases, and even Kubernetes clusters with **Elastic Kubernetes Service** (**EKS**). It all seamlessly integrates with the **virtual private cloud** (**VPC**) that you would have deployed in the public cloud, using the same APIs and controls. That is, in a nutshell, AWS Outposts: the AWS public cloud on-premises.

One question might be what this means for the **VMConAWS** proposition that both VMware and AWS have in their portfolio. VMConAWS actually extends the private cloud to the public cloud, based on HCX by VMware. VMware uses bare-metal instances in AWS to which it deploys vSphere, vSAN storage, and NSX for software-defined networking.

You can also use AWS services on top of the configuration of VMConAWS through integration with AWS. Outposts works exactly the other way around: bringing AWS to the private cloud. The portfolio for Outposts is growing rapidly. Customers can buy small appliances with single servers and also so-called rack solutions. In both cases, the infrastructure is completely managed by AWS.

Google Anthos

Anthos brings Google Cloud—or more accurately, GKE—to the on-premises datacenter, just as Azure Stack does for Azure and Outposts for AWS, but it focuses on the use of Kubernetes as a landing platform, moving and converting workloads directly into containers using GKE. It's not a standalone box like Azure Stack or Outposts. The solution runs on top of virtualized machines using vSphere and is more of a **Platform of a Service (PaaS)** solution. Anthos really accelerates the transformation of applications to more cloud-native environments, using open-source technology including Istio for microservices and Knative for the scaling and deployment of cloud-native apps on Kubernetes.

More information on the specifics of Anthos can be found at `https://cloud.google.com/anthos/gke/docs/on-prem/how-to/vsphere-requirements-basic`.

Azure Stack

The Azure Stack portfolio contains Stack **Hyperconverged Infrastructure (HCI)**, Stack Hub, and Stack Edge.

The most important feature of Azure Stack **HCI** is that it can run "disconnected" from Azure, running offline without internet connectivity. Stack HCI is delivered as a service, providing the latest security and feature updates.

To put it very simply: HCI works like the commonly known branch office server. Basically, HCI is a box that contains compute power, storage, and network connections. The box holds Hyper-V-based virtualized workloads that you can manage with Windows Admin Center. So, why would you want to run this as Azure Stack then? Well, Azure Stack HCI also has the option to connect to Azure services, such as Azure Site Recovery, Azure Backup, Microsoft Defender (formerly Azure Security Center), and Azure Monitor.

It's a very simple solution that only requires Microsoft-validated hardware, the installation of the Azure Stack operating system plus Windows Admin Center, and optionally an Azure account to connect to specific Azure cloud services.

Pre-warning: it might get a bit complicated from this point onward. Azure Stack HCI is also the foundation of Azure Stack Hub. Yet, Hub is a different solution. Whereas you can run Stack HCI *standalone*, Hub as a solution is integrated with the Azure public cloud—and that's really a different ballgame. It's not possible to upgrade HCI to Hub.

Azure Stack Hub is an extension of Azure that brings the agility and innovation of cloud computing to your on-premises environment. Almost everything you can do in the public cloud of Microsoft, you could also deploy on Hub: from VMs to apps, all managed through the Azure portal or even PowerShell. It all really works like Azure, including things such as configuring and updating fault domains. Hub also supports having an availability set with a maximum of three fault domains to be consistent with Azure. This way, you can create high availability on Hub just as you would in Azure.

The perfect use case for Hub and the Azure public cloud would be to do development on the public cloud and move production to Hub, should apps or VMs need to be hosted on-premises for compliance reasons. The good news is that you can configure your pipeline in such a manner that development and testing can be executed on the public cloud and run deployment of the validated production systems, including the desired state configuration, on Hub. This will work fine since both *entities* of the Azure platform use the Azure resource providers in a consistent way.

There are a few things to be aware of, though. The compute resource provider will create its own VMs on Hub. In other words: it does not *copy* the VM from the public cloud to Hub. The same applies to network resources. Hub will create its own network features such as load balancers, vNets, and **network security groups (NSGs)**. As for storage, Hub allows you to deploy all storage forms that you would have available on the Azure public cloud, such as blobs, queues, and tables. Obviously, we will discuss all of this in much more detail in this book, so don't worry if a number of terms don't sound familiar at this time.

One last Stack product is Stack Edge. Edge makes it easy to send data to Azure. But Edge does more: it runs containers to enable data analyses, perform queries, and filter data at edge locations. Therefore, Edge supports Azure VMs and **Azure Kubernetes Service (AKS)** clusters, which you can run containers on.

Edge, for that matter, is quite a sophisticated solution since it also integrates with **Azure Machine Learning** (**AML**). You can build and train machine learning models in Azure, run them in Azure Stack Edge, and send the datasets back to Azure. For this, the Edge solution is equipped with the **Field-Programmable Gate Arrays** (**FPGAs**) and **Graphics Processing Units** (**GPUs**) required to speed up building and (re)training the models.

Having said this, the obvious use case comes with the implementation of data analytics and machine learning where you don't want raw data to be uploaded to the public cloud straight away.

Azure Arc

There's one more service that needs to be discussed at this point and that's Azure Arc, launched at Ignite 2019. Azure Arc allows you to manage and govern at scale the following resource types hosted outside of Azure: servers, Kubernetes clusters, and SQL Server instances. In addition, Azure Arc allows you to run Azure data services anywhere using Kubernetes as clusters for containers, use GitOps to deploy configuration across the Kubernetes clusters from Git repositories, and manage these non-Azure workloads as if they were fully deployed on Azure itself.

If you want to connect a machine to Arc, you need to install an agent on that machine. It will then get a resource ID and become part of a resource group in your Azure tenant. However, this won't happen until you've configured some settings in the network, such as a proxy allowing for traffic from and to Arc-controlled servers, and registered the appropriate resource providers. The `Microsoft.HybridCompute`, `Microsoft.GuestConfiguration`, and `Microsoft.HybridConnectivity` resource providers must be registered on your subscription. This only has to be done once.

If you perform the actions successfully, then you can have non-Azure machines managed through Azure. In practice, this means that you perform many operational functions, just as you would with native Azure virtual machines. That sort of defines the use case: managing the non-Azure machines in line with the same policies as the Azure machines. These do not necessarily have to be on-premises. That's likely the best part of Arc: Azure Arc-enabled servers let you manage Windows and Linux physical servers and virtual machines hosted outside of Azure, on your corporate network, or on another cloud provider (such as AWS or GCP, but not exclusively).

With that last remark on Arc, we've come to the core of the multi-cloud discussion, and that's integration. All of the platforms that we've studied in this chapter have advantages, disadvantages, dependencies, and even specific use cases. Hence, we see enterprises experimenting with and deploying workloads in more than one cloud. That's not just to avoid cloud vendor lock-in: it's mainly because there's not a "one size fits all" solution.

In short, it should be clear that it's really not about *cloud-first*. It's about getting *cloud-fit*, that is, getting the best out of an ever-increasing variety of cloud solutions. This book will hopefully help you to master working with a mix of these solutions.

Emerging players

Looking at the cloud market, it's clear that it is dominated by a few major players, that is, the ones that were mentioned before: AWS, Microsoft Azure, and GCP. However, a number of players are emerging in both the public and private clouds, for a variety of reasons. The most common reason is geographical and that finds its cause in compliance rules. Some industries or companies in specific countries are not allowed to use, for instance, American cloud providers. Or the provider must have a local presence in a specific country.

From China, two major players have emerged to the rest of the world: Alibaba Cloud and Tencent. Both have been leading providers in China for many years, but are also globally available, but they focus on the Chinese market. Alibaba Cloud, especially, can certainly compete with the major American providers, offering a wide variety of services.

In Europe, a new initiative has recently started with Gaia-X, providing a pure European cloud, based in the EU. Gaia-X seems to concentrate mainly on the healthcare industry to allow European healthcare institutions to use a public cloud and still have privacy-sensitive patient data hosted within the EU.

Finally, big system integrators have stepped into the cloud market as well. A few have found niches in the market, such as IBM Cloud, which collaborates with Red Hat. Japanese technology provider Fujitsu did offer global cloud services with K5 for a while, offering a fully OpenStack public cloud, but found itself not being able to compete with Azure or AWS without enormous investments.

For specific use cases, a number of these clouds will offer good solutions, but the size and breadth of the services typically don't match those of the major public providers.

Where appropriate, new players will be discussed in this book. In the next section, we will first study the various cloud service models.

Evaluating cloud service models

In the early days, the cloud was merely another datacenter that hosted a multitude of customers, sharing resources such as compute power, network, and storage. The cloud has evolved over the years, now offering a variety of service models. In this section, you will learn the fundamentals of these models.

IaaS

IaaS is likely still the best-known service model of the cloud. Typically, enterprises still start with IaaS when they initiate the migration to cloud providers. In practice, this means that enterprises perform a lift and shift of their (virtual) machines to resources that are hosted in the cloud. The cloud provider will manage only the infrastructure for the customer: network, storage, compute, and the virtualization layer. The latter is important, since customers will share physical resources in the cloud. These resources—for instance, servers—are virtualized so they can host multiple instances.

PaaS

With **PaaS** cloud providers take more responsibility over resources, now including operating systems and middleware. A good example is a database platform. Customers don't need to take care of the database platform, but simply run a database instance on a database platform. The database software, for example, MySQL or PostgreSQL, is taken care of by the cloud provider, including the underlying operating systems.

SaaS

SaaS is generally perceived as the future model for cloud services. It basically means that the cloud provider manages everything in the software stack, from the infrastructure to the actual application with all its components, including data. Software updates, bug fixes, and infrastructure maintenance are all handled by the cloud provider. The user, who typically uses the application through some form of subscription, connects to the app through a portal or API without installing software on local machines.

FaaS

FaaS refers to a cloud service that enables the development and management of serverless computing applications. Serverless does not mean that there are no services involved, but developers can program services without having to worry about setting up and maintaining a server: that's taken care of by the cloud provider. The big advantage is that the programmed service only uses the exact amount of, for instance, CPU and memory, instead of an entire virtual machine.

CaaS

A growing number of enterprises are adopting container technology to host, run, and scale their applications. To run containers, developers must set up a runtime environment for these containers. Typically, this is done with Kubernetes, which has developed as the industry standard to host, orchestrate, and run containers. Setting up Kubernetes clusters can be complex and time-consuming. **Container as a Service** (**CaaS**) is the solution. CaaS provides an easy way to set up container clusters.

XaaS

Anything as a Service (**XaaS**) is a term used to express the idea that users can have everything as a service. The concept is widely spread with, for instance, **Hardware as a Service** (**HaaS**), **Desktop as a Service** (**DaaS**), or **Database as a Service** (**DBaaS**). This is not limited to IT, though. The general idea is that companies will offer services and products in an *as a service* model, using the cloud as the digital enabler. Examples are food delivery to homes, ordering taxis, or consulting a doctor using apps.

Although we will touch upon SaaS and containers, we will focus mainly on IaaS and PaaS as starting points to adopt multi-cloud. With that in mind, we can start by setting out our multi-cloud strategy.

Setting out a real strategy for multi-cloud

A cloud strategy emerges from the business and the business goals. Business goals, for example, could include the following:

- Creating more brand awareness
- Releasing products to the market faster
- Improving profit margins

Business strategies often start with increasing revenue as a business goal. In all honesty: that should indeed be a goal; otherwise, you'll be out of business before you know it. The strategy should focus on *how* to generate and increase revenue. We will explore more on this in *Chapter 2, Business Acceleration Using a Multi-Cloud Strategy*.

How do you get from business goals to defining an IT strategy? That is where enterprise architecture comes into play. The most used framework for enterprise architecture is **TOGAF**, although there are many more frameworks that can be used for this. Companies will not only work with TOGAF but also adopt practices from IT4IT and ITIL to manage their IT, as examples.

The core of TOGAF is the **ADM** cycle, short for **Architecture Development Method.** Also, in architecting multi-cloud environments, ADM is applicable. The ground principle of ADM is **B-D-A-T**: the cycle of **business, data, applications, and technology**. This perfectly matches the principle of multi-cloud, where the technology should be transparent. Businesses have to look at their needs, define what data is related to those needs, and how this data is processed in applications. This is translated into technological requirements and finally drives the choice of technology, integrated into the architecture vision as follows:

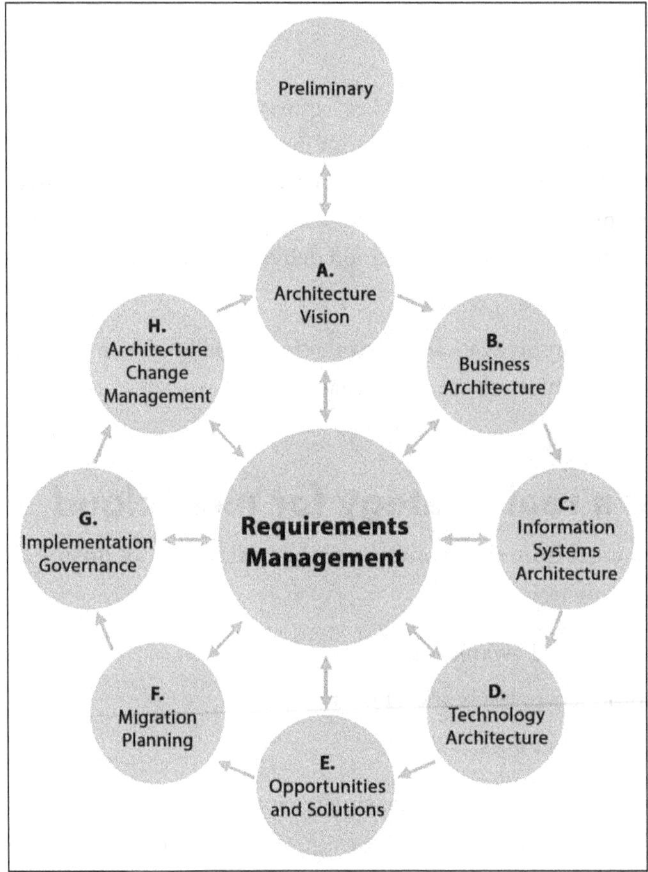

Figure 1.3: The ADM cycle in the TOGAF

This book is not about TOGAF, but it does make sense to have knowledge of enterprise architecture and, for that matter, TOGAF is the leading framework. TOGAF is published and maintained by The Open Group. More information can be found at `https://www.opengroup.org/togaf`.

The good news is that multi-cloud offers organizations flexibility and freedom of choice. That also brings a risk: lack of focus, along with the complexity of managing multi-cloud. Therefore, we need a strategy. Most companies adopt cloud and multi-cloud strategies since they are going through a process of transformation from a more-or-less traditional environment into a digital future. Is that relevant for all businesses? The answer is yes. In fact, more and more businesses are coming to the conclusion that IT is one of their core activities.

Times have changed over the last few years in that respect. At the end of the nineties and even at the beginning of the new millennium, a lot of companies outsourced their IT since it was not considered to be a core activity. That has changed dramatically over the last 10 years or so. Every company is a software company—a message that was rightfully quoted by Microsoft CEO Satya Nadella, following an earlier statement by the father of software quality, Watts S. Humphrey, who already claimed at the beginning of the millennium that every business is a software business.

Both Humphrey and Nadella are right. Take banks as an example: they have transformed to become more and more like IT companies. They deal with a lot of data streams, execute data analytics, and develop apps for their clients. A single provider might not be able to deliver all of the required services, hence these companies look for multi-cloud, best-of-breed solutions to fulfill these requirements.

These best-of-breed solutions might contain traditional workloads with a classic server-application topology, but will more and more shift to the use of PaaS, SaaS, container, and serverless solutions in an architecture that is more focused on microservices and cloud-native options. This has to be considered when defining a multi-cloud strategy: a good strategy would not be "cloud first" but "cloud-fit," meaning that a good strategy would not be about multi-cloud or single cloud, but having the right cloud.

Gathering requirements for multi-cloud

One of the first things that companies need to do is gather requirements before they start designing and building environments on cloud platforms. The most important question is probably: what will be the added value of using cloud technology to the business? Enterprises don't move to the cloud because they can, but because cloud technology can provide them with benefits. Think about not only agility, speed, and flexibility in the development of new services and products but also financial aspects: paying only for resources that they actually use or because of automation being able to cut costs in operations.

Using TOGAF for requirements management

Design and build start with requirements. TOGAF provides good guidance for collecting business or enterprise requirements. As you have learned in the previous section, TOGAF starts with the business' vision and mission. From there, an enterprise architect will define the business requirements, before diving into architectures that define the use of data, the need for specific applications, and lastly, the underlying technology. Indeed, technology comes in last. The architect will first have to determine what goals the business wants to achieve.

Requirements management is at the heart of the ADM of TOGAF. This means that from every phase of developing the architecture, requirements can be derived. Every requirement might have an impact on various stages of developing the architecture.

- TOGAF lists the following activities in gathering and managing requirements:
- Identify requirements
- Create a baseline for the requirements (the minimal set)
- From the baseline, identify changes to requirements
- Assess the impact of changes
- Create and maintain the requirements repository
- Implement requirements
- Perform gap analysis between the product specifications and requirements

This is all very generic. How would this translate to gathering and managing requirements for the cloud? The key is that at this stage, the technical requirements are not important yet. Architects shouldn't bother about whether they want Azure Functions or Kubernetes in an AWS EKS cluster. That's technology. It's the business requirements that come first.

Business requirements for the cloud can be categorized into the following quality attributes:

- Interoperability
- Configurability
- Performance
- Discoverability
- Robustness
- Portability
- Usability

The number one business priority must be security. Businesses will host vital data in public clouds, so the cloud provider must provide security measures and tools to secure that data. All cloud providers do so, but be aware that cloud providers are only responsible for the cloud itself, not for what's in the cloud. This shared responsibility model is crucial to understanding how public clouds work. Over the course of this book, you will learn much more about security.

Then, in general, there are three main goals that businesses want to achieve by using cloud technology:

- Scalability: XaaS and subscriptions are likely the future of any business. This comes with peaks in the demand for services and this means that platforms must be scalable, both upward and downward. This will allow businesses to have the right number of resources available at any time. Plus: since the cloud is very much based on OpEx, meaning that the business doesn't have to invest upfront, you will pay for the usage only.
- Reliability: Unlike a proprietary, privately owned datacenter, public clouds are globally available, allowing businesses to have resources copied to multiple zones even in different parts of the world. If a datacenter or a service fails in one zone, it can be switched to another zone or region. Again: the provider will offer the possibilities, but it's up to the business to make use of it. The business case will be important: is an application or data vital to a business and hence must be available at all times? In that case, business continuity and disaster recovery scenarios are valuable to work out in the cloud.
- Ease of use: Ease of administration might be a better word. Operators have access to all resources through comprehensive dashboards and powerful, largely automated tools to manage the resources in the cloud.

In multi-cloud, these requirements can be matched to services of various providers, resulting in solutions that provide the optimal mix for companies. With multi-cloud, other aspects will be evaluated, such as preventing technology lock-in and achieving business agility, where companies can respond fast to changing market circumstances or customer demands. The VOC will be an important driver to eventually choosing the right cloud and the right technology.

Listening to the Voice of the Customer

The problem with TOGAF is that it might look like the customer is a bit far away. TOGAF talks about enterprises, but enterprises exist thanks to customers. Customers literally drive the business, hence it's crucial to capture the needs and requirements of the customers: the VOC. It's part of an architectural methodology called QFD, which is discussed in more detail in the next section.

The challenge enterprises face in digital transformation is that it's largely invisible to customers. Customers want a specific service delivered to them, and in modern society, it's likely that they want it in a subscription format. Subscriptions are flexible by default: customers can subscribe, change subscriptions, suspend, stop, and reactivate them. This requires platforms and systems that can cope with this flexibility. Cloud technology is perfect for this.

However, the customer doesn't see the cloud platform, but just the service they are subscribed to. Presumably, the customer also doesn't care in what cloud the service is running. That's up to the enterprise, and with the adoption of multi-cloud, the enterprise has a wide variety of choices to pick the right solution for a specific customer requirement. Customer requirements must then be mapped to the capabilities of multi-cloud, fulfilling the business requirements such as scalability, reliability, and ease of use.

This insight leads to the following question: how can enterprises expect customers to know what they want and how they want it? Moreover: are customers prepared to pay for new technology that is required to deliver a new service or product?

The VOC can be a great help here. It's part of the **House of Quality (HOQ)** that is discussed in the next section. In essence, the VOC captures the "what"—the needs and wishes of the customer—and also prioritizes these. What is most important to the customer, that is, the must haves? Next, what are the "nice to haves"? These must be categorized into functional and non-functional requirements. Functional is everything that an app must be able to do, for instance, ordering a product. Non-functional is everything that enables an application to operate.

Think of the quality attributes: portability, reliability, performance, and indeed security. These are non-functional requirements, but they are just as important as the one button that allows customers to pay with a single click, integrating payment into an application with the credit card service. The latter is a functional requirement. It must all be prioritized in development since it's rare that everything can be developed at once.

The Open Group has published a new methodology for enterprise architecture that focuses more on the needs of digital enterprises, embracing agile frameworks and DevOps. This method is called **Open Agile Architecture (OAA)**. References can be found at `https://www.opengroup.org/agilearchitecture`.

In the next section, QFD and the HOQ are explained in more detail.

Defining architecture using QFD and the HOQ

The VOC has captured and prioritized the wants and needs of the customer: this is the input for the HOQ. The QFD process consists of four stages:

- **Product definition**: For this, the VOC is used.
- **Product development**: In the cloud, this refers to the development of the application and merging it with the cloud platform, including the development of infrastructure in the cloud.
- **Process development**: Although QFD was not designed specifically for cloud development, this can be easily applied to how development and deployment are done. Think of agile processes and DevOps, including pipelines.
- **Process quality control**: Again, QFD was not designed with the cloud in mind, but this can be applied to several test cycles that are required in cloud development. Think of quality assurance with compliance checks, security analysis, and various tests.

The HOQ shows the relations between the customer requirements and the product characteristics, defining how the customer needs can be translated into the product. The HOQ uses a relationship matrix to do this, listing how product characteristics map to the customer requirements. Every requirement must have a corresponding item in the product mapping. This is a design activity and must be performed on all levels of an application: on the application itself, the data, and every component that is needed to run the application. This includes the pipelines that organizations use to develop, test, and deploy applications in DevOps processes.

A good reference to get a better understanding of QFD is the website of Quality-One: `https://quality-one.com/qfd/`.

As with TOGAF, this book is not about QFD. QFD helps in supporting and improving design processes and with that, improving the overall architecture of business environments. Although ease of use is one of the business requirements of the cloud, it's not a given that using cloud technology simplifies the architecture of that business. Cloud, and especially multi-cloud, can make things complex. A well-designed architecture process is a must, before we dive into the digital transformation itself.

Understanding the business challenges of multi-cloud

This might be a bit of a bold statement, but almost every modern enterprise has adopted multi-cloud already. They may offer Office 365 to their workers, have their customer base in the SaaS proposition of Salesforce, book trips through SAP Concur, manage their workforce in Workday, and run meetings through Zoom. It doesn't get more multi-cloud than that. However, that's not what enterprises mean by going multi-cloud.

It starts with a strategy, something that will be discussed in the next chapter. But before a company sets the strategy, it must be aware of the challenges it will encounter. These challenges are not technical in the first place, but more organizational: a company that plans for digital transformation must create a comprehensive roadmap that includes the organization of the transformation, the assessment of the present mode of operations, and the IT landscape, and next do a mapping of the IT services that are eligible for cloud transformation.

All of this must be included in a business case. In short: what is the best solution for a specific service? Can we move an application to the cloud as is, or is it a monolith that needs redesigning into microservices? What are the benefits of moving applications to the cloud? What about compliance, privacy, and security? A crucial point to consider is: how much effort will the transformation take and how do we organize it? By evaluating the services in various clouds and various solutions, businesses will be able to obtain an optimized solution that eventually fulfills the business requirements. In *Chapter 2, Business Acceleration Using the Multi-Cloud Strategy*, we will learn more about the business drivers for adopting multi-cloud.

In the next section, we will briefly address the organizational challenge before the start of the transformation. *Chapter 2* talks about the business strategy, and *Chapter 3, Starting the Multi-Cloud Journey*, discusses the first steps of the actual transformation.

Setting the scene for cloud transformation

In this section, the first basic steps to start the cloud transformation are discussed. They may seem quite obvious, but a lot of companies try to skip some of these steps in order to gain speed—but all too often, skipping steps lead to failures, slowing down the transformation.

The first step is creating an accurate overview of the IT portfolio of the company: a catalog of all the applications and technology used, including the operational procedures. Simply listing that the company applies IT service management is not sufficient. What matters is how service management is applied. Think of incident, problem, change, and configuration management. Also, list the service levels to which services are delivered to the company, and thus to its business.

The following step is to map this catalog to a cloud catalog. The most important question is: what would be the benefit for the customers if services were transitioned to the cloud? To answer that question, the enterprise architect must address topics such as cloud affinity, various cloud service models, and the agreed **Key Performance Indicators (KPIs)** in terms of the business, security, costs, and technical debt of the company. It all adds up in a business case.

The company can now select the first candidates for cloud migration and start planning the actual migration. From that moment on, the cloud portfolio must be managed.

In *Chapter 3, Starting the Multi-Cloud Journey*, you will learn more about the transition and transformation of environments to the cloud, addressing topics such as microservices and cloud-native.

Addressing organizational challenges

Digital and, as part of that, cloud transformation start with people. The Open Group released a digital operating model that shows how roles and skills are changing in a digital enterprise, using cloud technology. The model is continuously updated, reflecting the continuous change in transformation.

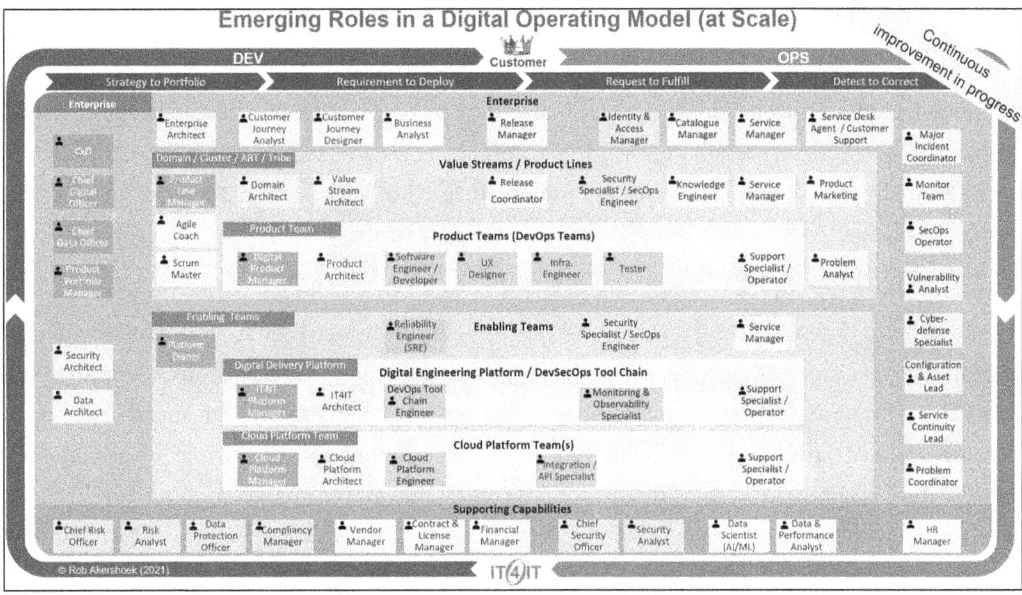

Figure 1.4: New roles in digital transformation (courtesy of Rob Akershoek)

The model shows a demarcation between enabling and product teams. The model assumes that there are more or less centralized teams that take care of the foundation, the landing zone, enabling the product teams to autonomously develop and deploy products.

Important new roles in the model are the customer journey analyst and the customer journey designer. The model places the VOC at the heart of everything.

Another important aspect of this model is the role of architecture. We can clearly see the hierarchy in architecture, starting with enterprise architecture setting the standards. The domain and value stream architects are key in translating the standards to the customer requirements, defining the architectures for the various products and releases. These architects are more business-focused. The IT and cloud platform architects are in the enabling layer of the model.

Organizing the skills of the architect

The role of an architect is changing, due to digital transformation. What is the current role of an architect do? It depends a bit on where the architect sits in the organization.

Let's start at the top of the tower with the enterprise architect. First, what is enterprise architecture? The enterprise architect collects business requirements that support the overall business strategy and translates these into architecture that enables solution architects to develop solutions to deliver products and services. Requirements are key inputs for this and that's the reason why requirements management sits in the middle of TOGAF, the commonly used framework for enterprise architecture.

In modern companies, enterprise architects also need to adapt to the new reality of agile DevOps and overall disruption through digital transformation. In TOGAF, technology is an enabler, something that simply supports the business architecture. That is changing. Technology might no longer be just an enabler, but actually drive the business strategy. With that, even the role of the enterprise architect is changing.

In fact, due to digital transformation, the role of every architect is changing. The architect is shifting toward more of an engineering leader, a servant leader in agile projects where teams work in a completely different fashion with agile trains, DevSecOps, and scrum, all led by product management. Indeed: you build it, you run it. That defines a different role for architecture. Architecture becomes more of a "floating" architecture, developing as the product development evolves and the builds are executed in an iterative way. There's still architecture, but it's more embedded into development. That comes with different skills.

The architect is shifting toward more of an engineering leader, that is, a servant leader. That's true. The logical next question then would be: do we still need architects or do we need leading engineers in agile-driven projects? The answer is both yes and no and it's only valid for digital architecture.

No, if architecture is seen as the massive upfront designs before a project can be started. Be aware: you will still need an architect to design a scanner. Or a house. A car. Something physical. The difference for an IT architect is that IT is software these days; it's all code. Massive, detailed upfront designs are not needed to get started with software, which doesn't mean that upfront designs are not needed at all. But we can start small, with a **minimal viable product** (**MVP**), and then iterate to the eventual product. There will always be a request for a first architecture or a design.

But what is architecture then? Learn from architect and author Gregor Hohpe: architecture is providing options. According to Hohpe, agile methods and architecture are ways to deal with uncertainty and that leads to the conclusion that working in an agile way allows us to benefit from architecture, since it provides options.

That's what this book is about: options. Options in cloud providers, the services that they provide, and how businesses can use these, starting with the next chapter, about defining the multi-cloud strategy for businesses.

Summary

In this chapter, we learned what a true multi-cloud concept is. It's more than a hybrid platform, comprising different cloud solutions such as IaaS, PaaS, SaaS, containers, and serverless in a platform that we can consider to be a best-of-breed mixed zone. You are able to match a solution to a given business strategy. Here, enterprise architecture comes into play: business requirements lead at all times and are enabled by the use of data, applications, and, lastly, technology. Enterprise architecture methodologies such as TOGAF are good frameworks for translating a business strategy into an IT strategy, including roadmaps.

We looked at the main players in the field and the emerging cloud providers. Over the course of this book, we will further explore the portfolios of these providers and discuss how we can integrate solutions, really mastering the multi-cloud domain. In the final section, the organization of the transformation was discussed, as well as how the role of the architect is changing.

In the next chapter, we will further explore the enterprise strategy and see how we can accelerate business innovation using multi-cloud concepts.

Questions

1. Although we see a major move to public clouds, companies may have good reasons to keep systems on-premises. Compliance is one of them. Please name another argument for keeping systems on-premises.

2. The market for public clouds is dominated by a couple of major players, with AWS and Azure being recognized as leaders. Name other cloud providers that have been mentioned in this chapter.
3. We discussed TOGAF as an enterprise architecture methodology. To capture business requirements, we studied a different methodology. What is the name of that methodology and what are the key components of this methodology?
4. IaaS, PaaS, and SaaS are very well-known cloud service models. What does CaaS refer to typically?

Further reading

- *Every business is a software business* by João Paulo Carvalho, available at `https://quidgest.com/en/articles/every-business-software-business/`
- *Multi-Cloud for Architects*, by Florian Klaffenbach and Markus Klein, Packt Publishing

Join us on Discord!

Read this book alongside other users, cloud experts, authors, and like-minded professionals. Ask questions, provide solutions to other readers, chat with the authors via. Ask Me Anything sessions and much more.

Scan the QR code or visit the link to join the community now.

`https://packt.link/cloudanddevops`

2

Collecting Business Requirements

This chapter discusses how enterprises can accelerate business results by implementing a multi-cloud strategy. Typically, this is a task for enterprise or business architects, but in digital transformation, enterprise architecture and cloud architecture are tightly connected. We have to collect business requirements as a first step before we can think of the actual cloud strategy.

Every cloud platform/technology has its own benefits and by analyzing business strategies and defining what cloud technology fits best, enterprises can really take advantage of multi-cloud. A strategy should not be "cloud-first" but "cloud-fit." But before we get into the technical strategy and the actual cloud planning, we must explore the business or enterprise strategy and the financial aspects that drive this strategy.

In this chapter, we're going to cover the following main topics:

- Analyzing the enterprise strategy for the cloud
- Defining the cloud strategy from the enterprise architecture
- Fitting cloud technology to business requirements
- Applying the value streams of IT4IT
- Keeping track of cloud developments—focusing on the business strategy
- Creating a comprehensive business roadmap
- Mapping the business roadmap to a cloud-fit strategy

Analyzing the enterprise strategy for the cloud

Before we get into a cloud strategy, we need to understand what an enterprise strategy is and how businesses define such a strategy. As we learned in the previous chapter, every business should have the goal of generating revenue and earning money. That's not really a strategy. The strategy is defined by how it generates money with the products the business makes or the services that it delivers.

A good strategy comprises a well-thought-out balance between timing, access to and use of data, and something that has to do with braveness—daring to make decisions at a certain point in time. That decision has to be based on—you guessed it—proper timing, planning, and the right interpretation of data that you have access to. If a business does this well, it will be able to accelerate growth and, indeed, increase revenue. The overall strategy should be translated into use cases. Use cases can be:

- Delivering products in new business models, such as SaaS
- Achieving more resilience in business, for instance, by implementing disaster recovery using the cloud
- Faster time to market in product development through the quick deployment of development environments
- Analysis of big data using data lakes in the cloud

These use cases must be reflected by the strategy of the enterprise. They will drive the decisions in designing and implementing cloud solutions and even in the choice of cloud platforms.

The success of a business is obviously not only measured in terms of revenue. There are a lot of parameters that define success as a whole and for that matter, these are not limited to just financial indicators. Nowadays, companies rightfully also have social indicators to report on. Think of sustainability and social return. However, a company that does not earn money, one way or the other, will likely not last long.

What are the drivers for business strategy? Typically, these are categorized into four areas:

- Financial objectives
- Customer objectives
- Product objectives
- Internal objectives

In the first chapter, the customer requirements were discussed alongside the methodologies to capture these, the use of **Quality Function Deployment** (**QFD**), **the House of Quality** (**HOQ**), **and the Voice of the Customer** (**VOC**). By understanding the customer needs, the enterprise is able to design, develop, and deploy products that customers are willing to buy—if the price is right. So, there's another challenge for the enterprise: it needs to deliver products at a price that is acceptable to customers. The price needs to cover the costs and preferably with enough margin to make some profit. However, there's one more aspect that is crucial to enterprise strategy: timing.

Time is one of the most important factors to consider when planning for business acceleration. Having said that, it's also one of the most difficult things to grasp. No one plans for a virus outbreak and yet it happened in 2020, leading to a worldwide pandemic. It was a reason for businesses not to push for growth at that time. The strategy for a lot of companies probably changed from pushing for growth to staying in business by trying to drive costs down. It proved that modern businesses have to be extremely agile.

Business agility has become one of the most important strategic drivers to go to the cloud, but what is business agility? It's the capability of a business to respond quickly to events in rapidly changing markets. This comes with a different enterprise architecture. The enterprise has to become adaptive in every layer. That means that the organization has to change: smaller, agile working teams, interacting closely with customers and focusing on smaller tasks that can be executed in sprints of a couple of weeks.

But it also means that systems must become agile. Changing customer demands must be met quickly, resulting in new features in systems. A lot of enterprises still have a massive technical debt, with big, monolithic systems that are hard to update and upgrade without the need to change the entire system. Modern architecture is about microservices: applications are compiled as a collection of services that are loosely coupled. Teams will work on specific services, independently from other teams. The services will interact with each other through protocols such as HTTP, TCP, and AMQP.

Shifting to a subscription-based economy

In the past decade, the markets for most businesses have changed dramatically. Markets have shifted from privately owned to platform economies and eventually a subscription-based economy. Products and services are used as a subscription, "as a service." Consumers can have services on any device, at any time, at any place they desire. They simply pay a fee to get access to a service; they don't want to own the product or service.

The challenge for businesses is that services are now consumed in a completely different manner. It's not a one-off sale anymore, and customers can actually do something with the subscription: it can be paused, changed, suspended, restated, or stopped completely. Subscriptions are very fluid. Hence, enterprises must have architectures that are capable of addressing this flexibility, and systems must be agile and scalable.

Another aspect crucial to business strategy and agility is access to and the use of data. It looks like the use of data in business is something completely new, but of course, nothing could be further from the truth. Every business in every era can only exist through the use of data. We might not always consider something to be data since it isn't always easy to identify, especially when it's not stored in a central place.

Data is the key. It's not only about raw data that a business (can) have access to, but also about analyzing that data. Where are my clients? What are their demands? Under what circumstances are these demands valid and what makes these demands change? How can I respond to these changes? How much time would I have to fulfill the initial demands and apply these changes if required? This all comes from data. Nowadays, we have a lot of data sources. The big trick is how we can make these sources available to a business—in a secure way, with respect to confidentiality, privacy, and other compliance regulations.

An example will make this clearer. The changes in the global healthcare market make an excellent example of business challenges, due to changing market circumstances such as a lack of skilled staff and a globally aging population, requiring more cure and care. The sustainability of the global health system is under high pressure. Hence, there's a general understanding that there should be more focus on prevention, rather than on treatments. Treatment costs society a lot more than preventing people from getting sick in the first place. Governments and companies are therefore now trying to make sure that people start improving their lifestyles. Data is absolutely essential to start this improvement and then to develop services that will help people develop and maintain a better lifestyle.

Companies are investing in collecting health data. And no surprise, it's big tech that's heading the game, since the place to collect data is the cloud. It's collected from medical devices and also from devices such as smartwatches and equipment in gyms connected to the internet. This data is analyzed so that trends can be made visible. Based on that data, individual health plans can be organized, but the data can also be utilized for commercial goals: selling running shoes, diet products, or health services, preferably on a subscription basis so that services can easily be adapted when customer demands change.

Challenges in this healthcare use case are primarily confidentiality and protection of data, and compliance with international privacy regulations. These are bigger challenges than the technology itself. The technology to collect data from various sources and make it available through data mining and analytics is generally available.

In the following sections, we will address business challenges such as business agility, security, data protection, and time to market and understand why cloud technology can help enterprises in dealing with these challenges.

Considering cloud adoption from enterprise architecture

In the previous section, the changes in the enterprise markets were discussed. The modern economy is changing from ownership to subscription-based models, leading to the need for flexibility, adaptability, agility, and scalability in all layers of the enterprise, including the organization and the systems. Subscriptions come with new payment models such as **pay-as-you-go** (**PAYG**) and freemium concepts, where a basic service is delivered at no charge, but customers can enhance service with paid options. Services must be interoperable, but also capable of interacting with other systems such as payment services.

The overarching enterprise architecture as a result of this digital transformation will inevitably change. The architecture will enable faster development, targeting microservices developed by smaller teams and using cloud-native technology. In the next sections, you will learn what an architect should take into account in defining this new digital, cloud-native strategy. This is typically the work of an enterprise architect but, as we already mentioned in the introduction to this chapter, with digital transformation, enterprise and strategic cloud architecture are getting closely related to each other. The enterprise architect must understand the cloud, and the cloud architect should have knowledge about enterprise architecture since cloud adoption is not purely a technological subject.

Long-term planning

When a business is clear on its position and its core competencies, it needs to set out a plan. Where does the company want to be in 5 years from now? This is the most difficult part. Based on data and data analytics, it has to determine how the market will develop and how the company can anticipate change. Again, data is absolutely key, but so is the swiftness with which companies can change course since market demands do change extremely rapidly.

Financial structure

A business needs a clear financial structure. How is the company financed and how are costs related to different business domains, company divisions, and its assets? As we will find out as part of financial operations, the cloud can be of great help in creating fine-grained insight into financial flows throughout a business. With correct and consistent naming and tagging, you can precisely pinpoint how much cost a business generates in terms of IT consumption. The best part of cloud models is that the foundation of cloud computing is *paying for what you use*. Cloud systems can *breathe* at the same frequency as the business itself. When business increases, IT consumption can increase. When business drops, cloud systems can be scaled down, and with that, they generate lower costs, whereas traditional IT is way more static.

So, nothing is holding us back from getting our business into the cloud. However, it does take quite some preparation in terms of (enterprise) architecture. In the following section, we will explore this further.

Fitting cloud technology to business requirements

We are moving business into the cloud because of the required agility and, of course, to control our costs.

The next two questions will be: with what and how? Before we explore the how and basically the roadmap, we will discuss the first question: with what? A cloud adoption plan starts with business planning, which covers business processes, operations, finance, and, lastly, the technical requirements. We will have to evaluate the business demands and the IT fulfillment of these requirements.

When outsourcing contracts, the company that takes over the services performs so-called due diligence. As per its definition, due diligence is *"a comprehensive appraisal of a business undertaken by a prospective buyer, especially to establish its assets and liabilities and evaluate its commercial potential"* (source: `https://www.lexico.com/en/definition/due_diligence`). This may sound way too heavy of a process to get a cloud migration asset started, yet it is strongly recommended as a business planning methodology. Just replace the words *prospective buyer* with the words *cloud provider* and you'll immediately get the idea behind this.

Business planning

One really important step in the discovery phase that's crucial to creating a good mapping to cloud services is evaluating the service levels and performance indicators of applications and IT systems. Service levels and **key performance indicators** (**KPIs**) will be applicable to applications and the underlying IT infrastructure, based on specific business requirements.

Think of indicators like metrics of availability, durability, and levels of backup, including RTO/RPO specifications, requirements for **business continuity** (**BC**), and **disaster recovery** (**DR**). Are systems monitored 24/7 and what are the support windows? These all need to be considered. As we will find out, service levels and derived service-level agreements might be completely different in cloud deployments, especially when looking at PaaS and SaaS, where the responsibility of platform (PaaS) and even application (SaaS) management is largely transferred to the solution provider.

If you ask a CFO what the core system is, they will probably answer that financial reporting is absolutely critical. It's the role of the enterprise architect to challenge that. If the company is a meat factory, then the financial reporting system is not the most critical system. The company doesn't come to a halt when financial reporting can't be executed. The company does, however, come to a halt when the meat processing systems stop; that immediately impacts the business. What would that mean in planning the migration to cloud systems? Can these processing applications be hosted from cloud systems? And if so, how? Or maybe, when?

In business architecture, the architect defines the purpose, vital functions, critical processes, and interactions between various business components. It describes:

- Business processes.
- Products and services and their respective taxonomies.
- Business capabilities.
- Business architecture defines how the enterprise functions in order to deliver products and services. It also defines what data it needs to be able to function and, lastly, what systems are required.

Business planning involves the following items:

- Discovery of the entire IT landscape, including applications, servers, network connectivity, storage, APIs, and services from third-party providers.
- Mapping of IT landscape components to business-critical or business-important services.
- Identification of commodity services and shared services and components in the IT landscape.
- Evaluation of IT support processes with regard to commodity services, and critical and important business services. This includes the levels of automation in the delivery of these services.

These are the first topics to discuss when initiating a digital transformation that typically involves cloud migration or cloud development plans; from the enterprise strategy, it should be clear what the core competence and, therefore, the core business process is, supported by the company's core systems.

Financial planning

After the business planning phase, we also need to perform financial analyses. After all, one of the main rationales for moving to cloud platforms is cost control. Be aware: moving to the cloud is not always a matter of lowering costs. It's about making your costs *responsive* to actual business activity. Setting up a business case to decide whether cloud solutions are an option from a financial perspective is, therefore, not an easy task. Public cloud platforms offer **Total Cost of Ownership** (**TCO**) calculators.

TCO is indeed the total cost of owning a platform and it should include all direct and indirect costs. What do we mean by that? When calculating the TCO, we have to include the costs that are directly related to systems that we run: costs for storage, network components, compute, licenses for software, and so on. But we also need to consider the costs of the labor that is involved in managing systems for engineers, service managers, or even the accountant that evaluates the costs related to the systems. These are all costs; however, these indirect costs are often not taken into account in the full scope. Especially in the cloud, these should be taken into account. Think of this: what costs can be avoided by, for example, automating service management and financial reporting?

So, there's a lot to cover when evaluating costs and the financial planning driving architecture. Think of the following:

- All direct costs related to IT infrastructure and applications. This also includes hosting and housing costs—for example, (the rental of) floor space and power.
- Costs associated with all staff working on IT infrastructure and applications. This includes contractors and staff from third-party vendors working on these systems.
- All licenses and costs associated with a vendor or third-party support for systems.
- Ideally, these costs can be allocated to a specific business process, division, or even user group so that it's evident where IT operations costs come from.

Why is this all important in drafting architecture? A key financial driver to start a cloud journey is the shift from CapEx to OpEx. In essence, **CapEx—capital expenditure**—concerns upfront investments—for example, buying physical machines or software licenses. These are often one-off investments, of which the value is depreciated over an economic life cycle. **OpEx—operational expenditure**—is all about costs related to day-to-day operations and for that reason is much more granular. Usually, OpEx is divided into smaller budgets, which teams need to have to perform their daily tasks. In most cloud deployments, the client really only pays for what they're using. If resources sit idle, they can be shut down and costs will stop. A single developer could—if mandated for this—decide to spin up an extra resource if required.

That's true for a **PAYG** deployment, but we will discover that a lot of enterprises have environments for which it's not feasible to run in full PAYG mode. You simply don't shut down instances of large, critical ERP systems. So, for these systems, businesses will probably use more stateful resources, such as reserved instances that are fixed for a longer period. For cloud providers, this means a steady source of income for a longer time, and therefore, they offer reserved instances against lower tariffs or to apply discounts. The downside is that companies can be obliged to pay for these reserved resources upfront. Indeed, that's CapEx. To cut a long story short, the cloud is not OpEx by default.

Understanding the cost of delay

We have our foundation or reference architecture set out, but now, our business gets confronted with new technologies that have been evaluated: what will they bring to the business? As we mentioned previously, there's no point in adopting every single new piece of tech that is released. The magic words here are *business case*.

A business case determines whether the consumption of resources supports a specific business need. A simple example is as follows: a business consumes a certain bandwidth on the internet. It can upgrade the bandwidth, but that will take an investment. That is an out-of-pocket cost, meaning that the company will have to pay for that extra bandwidth. However, it may help workers to get their job done much faster. If workers can pick up more tasks just by the mere fact that the company invests in a faster internet connection, the business case will, in the end, be positive, despite the investment.

If market demands occur and businesses do not adapt to this fast enough, a company might lose part of the market share and thereby lose revenue. Adopting new technology or speeding up the development of applications to cater to changing demands will lead to costs. These costs correspond to missing specific timing and getting services or products to the market in time. This is what we call the cost of delay.

Cost of delay, as a piece of terminology, was introduced by Donald Reinertsen in his book *The Principles of Product Development Flow*, published by *Celeritas Publishing, 2009*:

> *We need cost of delay to evaluate the cost of queues, the value of excess capacity, the benefit of smaller batch sizes, and the value of variability reduction. Cost of Delay is the golden key that unlocks many doors. It has an astonishing power to totally transform the mindset of a development organization.*

Although it's mainly used as a financial parameter, it's clear that the cost of delay can be a good driver to evaluate the business case for adopting cloud technology. Using and adopting consumption of cloud resources that are more or less agile by default can mitigate the financial risk of cost of delay.

Moving to the benefit of opportunity

If there's something that we could call the cost of delay, there should also be something that we could call the benefit of opportunity. Where the cost of delay is the risk of missing momentum because changes have not been adopted timely enough, the benefit of opportunity is really about accelerating the business by exploring future developments and related technology. It can be very broad. As an example, let's say a retailer is moving into banking by offering banking services using the same app that customers use to order goods. Alternatively, think of a car manufacturer, such as Tesla, moving into the insurance business.

The accessibility and even the ease of using cloud services enable these shifts. In marketing terms, this is often referred to as blurring. In the traditional world, that same retailer would really have had much more trouble offering banking services to its customers, but with the launch of SaaS in financial apps, from a technological point of view, it's not that hard to integrate this into other apps. Of course, this is not considering things such as the requirement of a banking license from a central financial governing institution and having to adhere to different financial compliance frameworks. The simple message is that with cloud technologies, it has become easier for businesses to explore other domains and enable a fast entrance into them, from a purely technological perspective.

The best example? AWS. Don't forget that Amazon was originally an online bookstore. Because of the robustness of its ordering and delivery platform, Amazon figured out that it also could offer storage systems "for rent" to other parties. After all, they had the infrastructure, so why not fully capitalize on that? Hence, S3 storage was launched as the first AWS cloud service and by all means, it got AWS to become a leading cloud provider, next to the core business of retailing. That was truly a benefit of opportunity.

Technical planning

Finally, we've reached technical planning, which starts with foundation architecture. You can plan to build an extra room on top of your house, but you'll need to build the house first—assuming that it doesn't exist. The house needs a firm foundation that can hold it in the first place but also *carry* the extra room in the future. The room needs to be integrated with the house since it would really be completely useless to have it *stand alone* from the rest of the house. It all takes good planning. Good planning requires information—data, if you like.

In the world of multi-cloud, you don't have to figure it out all by yourself. The major cloud providers all have reference architectures, best practices, and use cases that will help you plan and build the foundation, making sure that you can fit in new components and solutions almost all the time. We will discuss this in detail in *Chapter 3, Starting the Multi-Cloud Journey*.

That's exactly what we are going to do in this book: plan, design, and manage a future-proof multi-cloud environment. In the next section, we will take our first look at the foundation architecture.

Applying the value streams of IT4IT

The problem that many organizations face is controlling the architecture of businesses at large. IT4IT is a framework that helps organizations with that. It's complementary to TOGAF and, for this reason, is also issued as a standard by The Open Group. IT4IT is also complementary to **IT Infrastructure Library** (**ITIL**), where ITIL provides best practices for IT service management. IT4IT provides the foundation to enable IT service management processes with ITIL. It is meant to align and manage a digital enterprise. It deals with the challenges that these enterprises have, such as the ever-speeding push for embracing and adopting new technology. The base concept of IT4IT consists of four value streams:

- **Strategy to portfolio**: The portfolio contains technology standards, plans, and policies. It deals with the IT demands of the business and maps these demands to IT delivery. An important aspect of the portfolio is project management to align business and IT.
- **Requirements to deploy**: This stream focuses on creating and implementing new services or adapting existing services, in order to reach a higher standard of quality or to obtain a lower cost level. According to the documentation of The Open Group, this is complementary to methods such as Agile Scrum and DevOps.
- **Request to fulfill**: To put it very simply, this value stream is all about making life easy for the end customers of the deployed services. As companies and their IT departments adopt structures such as IaaS, PaaS, and SaaS, this stream enables service brokering by offering and managing a catalog and, with that, speeds up the fulfillment of new requests made by end users.
- **Detect to correct**: Services will change. This stream enables monitoring, management, remediation, and other operational aspects that drive these changes.

The following diagram shows the four streams of IT4IT:

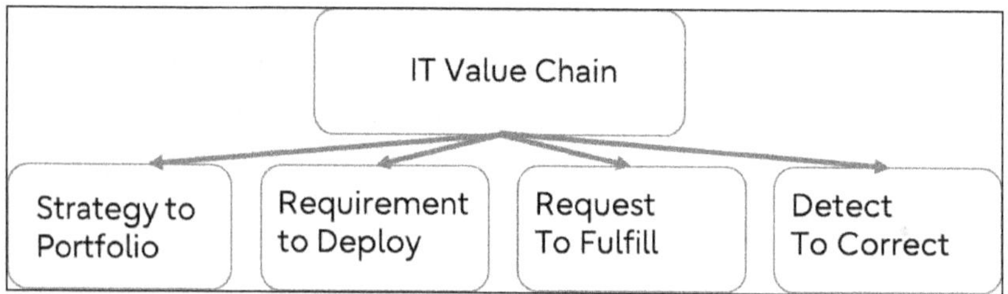

Figure 2.1: IT4IT value streams (The Open Group)

Frameworks such as IT4IT are very valuable in successfully executing this mapping. In the next section, we will focus on how cloud technology fits into the business strategy.

With that, we have not only defined a strategy; we have also learned how to apply the value streams of IT4IT, and how to realistically keep track of the ever-increasing cloud developments. We know where we're coming from, and we know where we're heading to. Now, let's bring everything together and make it more tangible by creating the business roadmap and finally mapping that roadmap to our cloud strategy, thereby evaluating the different deployment models and cloud development stages.

Keeping track of cloud developments—focusing on the business strategy

Any cloud architect or engineer will tell you that it's hard to keep up with developments. Just for reference, AWS and Azure issued over 2,000 features in their respective cloud platforms in just over 1 year. These can be big releases or just some minor tweaks.

The major clouds Azure, AWS, Google Cloud Platform, and Alibaba release thousands of new features every year (refer to `https://www.gartner.com/doc/reprints?id=1-2AOZQAQL&ct=220728&st=sb`). These can be simple additions to existing services such as virtual machines or storage, but also new services with pre-packaged code that enable the fast deployment of web services.

But there's much more. Think of the possibilities of **Virtual Reality (VR)**, **Augmented Reality (AR)**, and **Artificial Intelligence (AI)** from cloud platforms or building digital twins in the cloud. Cloud providers already also offer quantum simulation and it's expected that this will grow significantly in the near future. However, there are innovations that even go beyond this. Think of blockchain, Web 3.0, and the Metaverse.

Web 3.0 is based on blockchain technology and promises to hand back control over data to the user. The other promise is the Metaverse: a 3D version of the internet, something that is still hard to really understand since definitions of the Metaverse differ quite a lot. In most definitions, the Metaverse is presented as a sort of parallel universe, where we have digital twins moving in that digital world. Predictions by Bloomberg are that by 2024, the Metaverse will hold over 80 billion USD worth of business (refer to `https://www.bloomberg.com/professional/blog/metaverses-80-billion-etf-assets-by-2024-virtually-a-reality/`). Google, Microsoft, Meta, and Apple are heavily investing in the technology.

It is a constant stream of innovations: big, medium, and small. We have to keep one thing in mind: whether it's AR, VR, blockchain, or the Metaverse, somewhere there's technology involved at a low level. What do we mean by that? Someone has to develop, program, test, and deploy the technology. Anyway, trying to keep up with all these innovations, releases, and new features is hard, if not impossible.

And it gets worse. It's not only the target cloud platforms but also a lot of tools that we need to execute, in order to migrate or develop applications and prepare our businesses for the next big thing. Just have a look at the Periodic Table of DevOps Tools, which is currently managed by `Digital.ai` and is continuously refreshed with new technology and tools. It's just a matter of time before blockchain technology, Web 3.0, and Metaverse toolkits are added to this table.

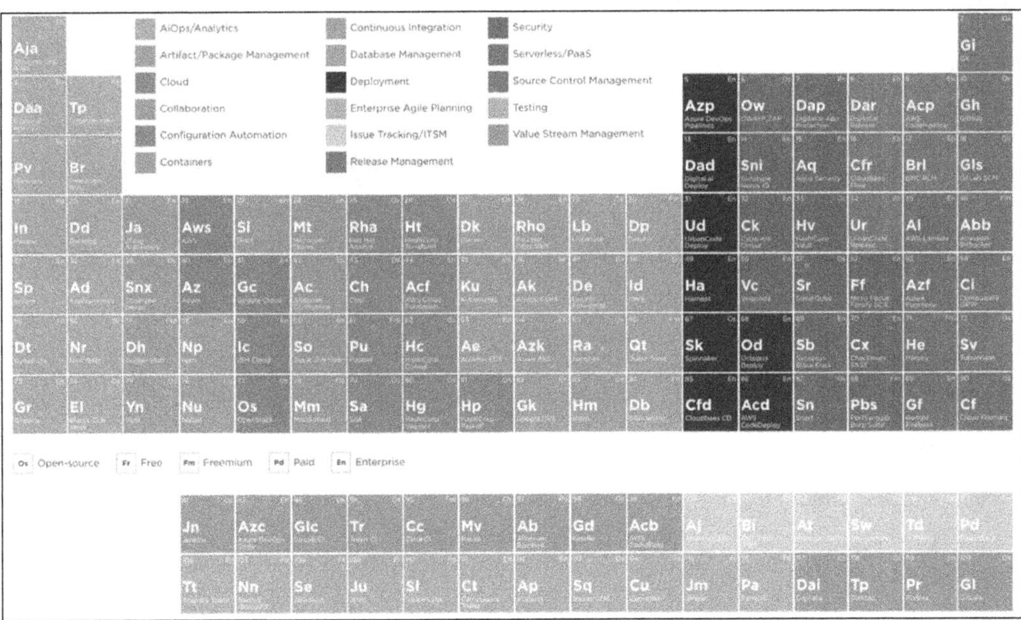

Figure 2.2: Periodic Table of DevOps Tools by Digital.ai

An interactive version of the Periodic Table of DevOps Tools can be found at `https://digital.ai/devops-tools-periodic-table`. For a full overview of the cloud-native landscape, refer to the following web page, which contains the *Cloud Native Trail Map of the Cloud Native Computing Foundation*: `https://landscape.cncf.io/`.

There's no way to keep up with all the developments. A business needs to have focus, and that should come from the strategy that we discussed in the previous sections. In other words, don't get carried away with all the new technology that is being launched.

There are three important aspects that have to be considered when managing cloud architecture: the foundation architecture, the cost of delay, and the benefit of opportunity.

Creating a comprehensive business roadmap

There are stores filled with books on how to create business strategies and roadmaps. This book absolutely doesn't have any pretensions of condensing this all into just one paragraph. However, for an enterprise architect, it is important to understand how the business roadmap is evaluated:

- The mission and vision of the business, including the strategic planning of how the business will target the market and deliver its goods or services.
- Objectives, goals, and direction. Again, this includes planning, in which the business sets out when and how specific goals are met and what it will take to meet the objectives in terms of resources.
- **Strengths, weaknesses, opportunities, and threats** (**SWOT**). The SWOT analysis shows whether the business is doing the right things at the right time or that a change in terms of the strategy is required.
- Operational excellence. Every business has to review how it is performing on a regular basis. This is done through KPI measurements: is the delivery on time? Are the customers happy (**customer satisfaction—CSAT**)?

Drivers for a business roadmap can be very diverse, but the most common ones are as follows:

- Revenue
- Gross margin
- Sales volume

- Number of leads
- Time to market
- Customer satisfaction
- Brand recognition
- Return on investment

These are shared goals, meaning that every division or department should adhere to these goals and have their planning aligned with the business objectives. These goals end up on the business roadmap. These can be complex in the case of big enterprises but also rather straightforward, as shown in the following screenshot:

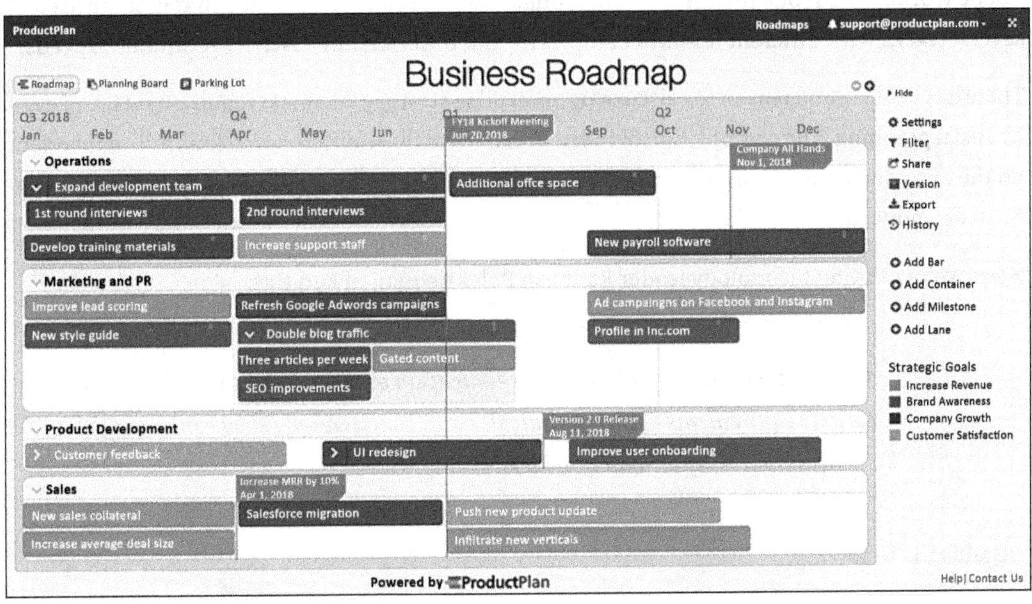

Figure 2.3: Template for a business roadmap (by ProductPlan)

IT is the engine that drives everything: development, resource planning, **Customer Relationship Management (CRM)** systems, websites for marketing, and customer apps. And these days, the demands get more challenging; the life cycle is getting shorter and the speed is getting faster. Where IT was the bottleneck for ages, it now has all the technology available to facilitate the business in every aspect. IT is no longer considered a cost center but a business enabler.

Mapping the business roadmap to a cloud-fit strategy

Most businesses start their cloud migrations from traditional IT environments, although a growing number of enterprises are already quite far into cloud-native development too. We don't have to exclude either one; we can plan to migrate our traditional IT to the cloud, while already developing cloud-native applications in the cloud itself. Businesses can have separate cloud tracks, running at different speeds. It makes sense to execute the development of new applications in cloud environments using cloud-native tools. Next, the company can also plan to migrate its traditional systems to a cloud platform. There are a number of ways to do that. We will be exploring these, but also look at drivers that start these migrations. The key message is that it's likely that we will not be working with one roadmap. Well, it might be one roadmap, but one that is comprised of several tracks with different levels of complexity and different approaches, at different speeds.

There has been a good reason for discussing enterprise strategy, business requirements, and even financial planning. The composition of the roadmap with these different tracks is fully dependent on the outcome of our assessments and planning. And that is architecture too—let there be no mistake about that.

Here, we're quoting technology leader Radhesh Balakrishnan of Red Hat:

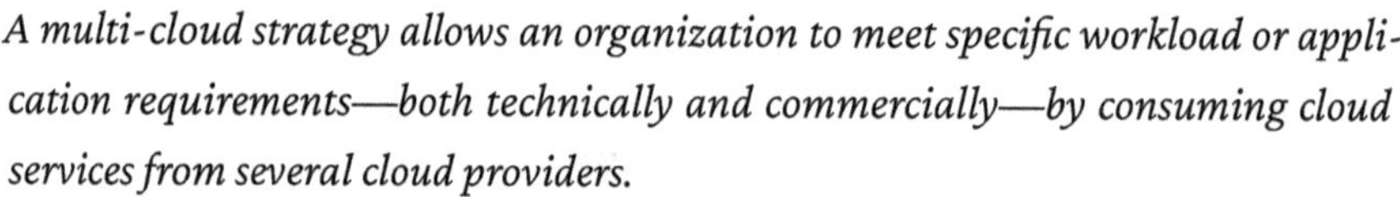

> *A multi-cloud strategy allows an organization to meet specific workload or application requirements—both technically and commercially—by consuming cloud services from several cloud providers.*

He adds the following:

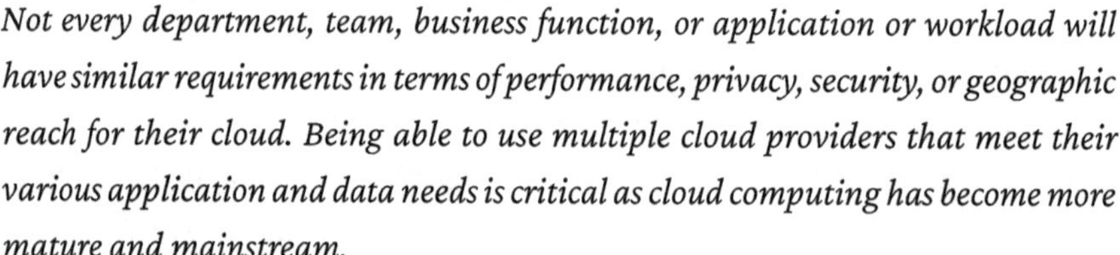

> *Not every department, team, business function, or application or workload will have similar requirements in terms of performance, privacy, security, or geographic reach for their cloud. Being able to use multiple cloud providers that meet their various application and data needs is critical as cloud computing has become more mature and mainstream.*

These business requirements will drive the cloud migration approach. We recognize the following technological strategies:

- **Rehost**: The application, data, and server are migrated as is to the target cloud platform. This is also referred to as **lift and shift**. The benefits are often quite low. This way, we're not taking any advantage of cloud-native services.
- **Replatform**: The application and data are migrated to a different target technology platform, but the application architecture remains as is. For example, let's say an application with a SQL database is moved to PaaS in Azure with Azure SQL Server. The architecture of the application itself is not changed.
- **Repurchase**: In this scenario, an existing application is replaced by SaaS functionality. Note that we are not really repurchasing the same application. We are replacing it with a different type of solution.
- **Refactor**: Internal redesign and optimization of the existing application. This can also be a partial refactoring where only parts of an application are modified to operate in an optimized way in the cloud. In the case of full refactoring, the whole application is modified for optimization in terms of performance and at lower costs. Refactoring is, however, a complicated process. Refactoring usually targets PaaS and SaaS.
- **Rearchitect**: This is one step further than refactoring and is where the architecture of an application, as such, is modified. This strategy does comprise an architectural redesign to leverage multi-cloud target environments.
- **Rebuild**: In this strategy, developers build a new cloud-native application from scratch, leveraging the latest tools and frameworks.
- **Retire**: This strategy is valid if an application is not strategically required for the business going forward. When an application is retired, data needs to be cleaned up and often archived, before the application and underlying infrastructure are decommissioned. Of course, an application can be retired as a follow-up strategy when the functionality in an application is refactored, rearchitected, rebuilt, or repurchased.
- **Retain**: Nothing changes. The existing applications remain on their current platform and are managed as-is they are.

It's not easy to predict the outcomes of the business case as many parameters can play a significant role. Benchmarks conducted by institutes such as Gartner point out that total cost savings will vary between 20 and 40 percent in the case of rehosting, replatforming, and repurchasing. Savings may be higher when an application is completely rearchitected and rebuilt. Retiring an application will lead to the highest savings, but then again, assuming that the functionality provided by that application is still important for a business, it will need to fulfill that functionality in another way—generating costs for purchasing, implementation, adoption, and/or development.

This all has to be taken into account when drafting the business case and choosing the right strategy.

On top of that, multi-cloud will bring its own set of challenges. There might be good reasons to have a data lake in one cloud and have office applications hosted in another cloud. Rationales for this might be specific capabilities that cloud providers have developed in various domains. Think of the Microsoft 365 suite and the capabilities in data and AI that Google has developed. There's a major business benefit in a strategy that is initiated by the best of breed: choosing the best solution and best provider for specific business use cases.

Another reason might be that developers have a preference to work with certain tools in a specific cloud, or that applications must be hosted in different clouds for compliance reasons. If countries or regions have restrictions on using certain cloud providers or a provider isn't available in a region, then companies are forced to use multiple providers. Think of companies that can use AWS and Azure in Europe but have to use Alibaba Cloud or Tencent in the Chinese region. Companies would still like to offer the same type of services, regardless of the cloud provider, but it will bring extra challenges in portability and the integration of services.

At a high level, we can plot these strategies into three stages:

- **Traditional**: Although organizations will allow developers to work with cloud-native tools directly on cloud platforms, most of them will still have traditional IT. When migrating to the cloud, they can opt for three scenarios:
    - Lift and shift systems "as is" to the cloud and start the application's modernization in the cloud.
    - Modernize the applications before migrating them to the target cloud platform.
    - A third scenario would be a mix between the first two.
- **Rationalized**: This stage is all about modernizing the applications and optimizing them for usage in the target cloud platform. This is the stage where PaaS and SaaS are included to benefit from cloud-native technology.
- **Dynamic**: This is the final stage where applications are fully cloud-native and therefore completely *dynamic*; they're managed through agile workstreams using **continuous improvement** and **continuous development** (**CI/CD**), fully scalable using containers and serverless solutions, and fully automated using the principle of everything as code, making IT as agile as the business requires.

It all comes together in the following model. This model suggests that the three stages are sequential, but as we have already explained, this doesn't have to be the case:

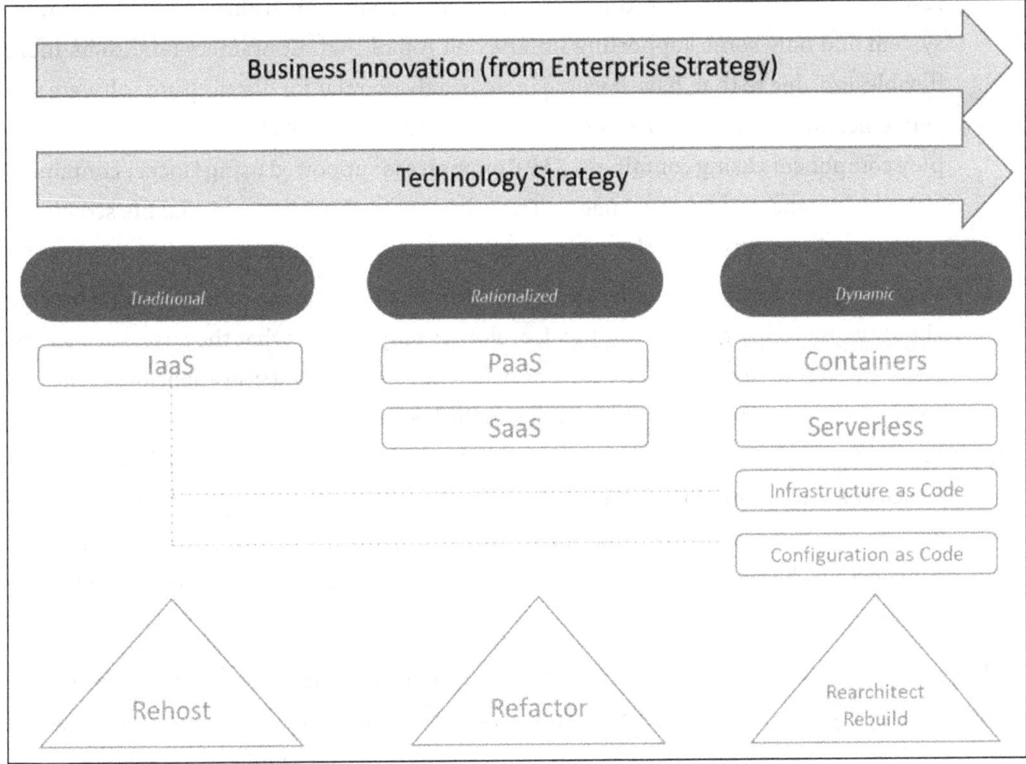

Figure 2.4: Technology strategy following business innovation

This model shows three trends that will dominate the cloud strategy in the forthcoming years:

- **Software to services:** Businesses do not have to invest in software anymore, nor in infrastructure to host that software. Instead, they use software that is fully managed by external providers. The idea is that businesses can now focus on fulfilling business requirements using this software, instead of having to worry about the implementation details and hosting and managing the software itself. It's based on the economic theory of endogenous growth, which states that economic growth can be achieved through developing new technologies and improvements in production efficiency. Using PaaS and SaaS, this efficiency can be sped up significantly.

- **VM to container**: Virtualization brought a lot of efficiency to data centers. Yet, containers are even more efficient in utilizing compute power and storage. VMs still use a lot of system resources with a guest operating system, whereas containers utilize the host operating system and only some supporting libraries on top of that. Containers tend to be more flexible and, due to that, have become increasingly popular for distributing software in a very efficient way. Even large software providers such as SAP already distribute and deploy components using containers. SAP Commerce is supported using Docker containers, running instances of Docker images. These images are built from special file structures mirroring the structures of the components of the SAP Commerce setup.
- **Serverless computing**: Serverless is about writing and deploying code without worrying about the underlying infrastructure. Developers only pay for what they use, such as processing power or storage. It usually works with triggers and events: an application registers an event (a request) and triggers an action in the backend of that application—for instance, retrieving a certain file. Public cloud platforms offer different serverless solutions: Azure Functions, AWS Lambda, and Google Knative. Serverless offers the maximum scalability with the greatest cost control. One remark has to be made at this point: although serverless concepts will become more and more important, it will not be technically possible to fit everything into serverless propositions.

In the next chapter, we will get into the more technical details, learning about the transformation from monolithic architectures into microservices, and the benefits of containers and serverless concepts, while keeping the infrastructure across systems consistent and easy to manage.

Summary

In this chapter, we explored the methodologies that are used to analyze enterprise or business strategies and mapped these to a cloud technology roadmap. We also learned that it is close to impossible to keep track of all the new releases and features that are launched by cloud and technology providers. We need to determine what our business goals and objectives are and define a clear architecture that is as future-proof as possible, yet agile enough to adopt new features if the business demands this.

Business agility must be the focus of a modern, digital enterprise. Business drivers are derived from financial, customer, product, and internal objectives, but these are rapidly changing in the current markets. One of the major trends is the upcoming subscription-based economy that forces businesses to create agile organizations, and systems that are able to respond to these changes quickly.

Enterprise architectures, using frameworks such as IT4IT, help us to design and manage a multi-cloud architecture that is robust but also scalable in every aspect. We have also seen how IT will shift, along with the business demands coming from traditional to rationalized and dynamic environments using SaaS, containers, and serverless concepts.

In the next chapter, we will learn how to translate the business requirements to cloud requirements, starting our multi-cloud journey.

Questions

1. How would you define business agility?
2. What would be the first thing to define if we created a business roadmap?
3. In this chapter, we discussed cloud transformation strategies. Rehost and replatform are two of them. Name two more.
4. In this chapter, we identified major developments in the cloud market. What is recognized as being the main change in business models that can be facilitated by using cloud-native technology?

Further reading

- *How Competitive Forces Shape Strategy*, by Michael E. Porter (Harvard Business Review)

3

Starting the Multi-Cloud Journey

HashiCorp published a survey in 2022 that showed that almost 75% of all businesses were executing a multi-cloud strategy. That number is expected to grow to over 86% by 2025. Many more companies claim that multi-cloud is critical for the success of the company. The challenge is how to start a multi-cloud strategy while keeping operations lean and IT infrastructure consistent, but creating development platforms that provide agility to business. It only works with a solid strategy, and that's what this chapter is about.

In this chapter, we're going to cover the following main topics:

- Understanding cloud vocabulary
- Planning assessments
- Planning transition and transformation
- Exploring options for transformation
- Executing technology mapping and governance

Understanding cloud vocabulary

In IT we have a common language, meaning that the industry has words for common technologies that are used. Whatever technology a company uses, most brands have a commonly accepted name for it. Take networks, for instance. **Wide Area Networks (WANs)**, **Local Area Networks (LANs)**, and software-defined LAN: irrespective of the brand that is used to set up a network, the terminology is the same. It's crucial to have a common understanding of what we mean and what we're talking about when it comes to the cloud.

That common vocabulary is not a given in the cloud. It already starts with creating a space in a public cloud. In Azure, we're getting a subscription with a tenant. In AWS, we're creating accounts, while in Google Cloud, we talk about projects. On the other hand, don't be distracted by the commercial names that providers give to their services.

Although each provider uses its own terms to refer to similar concepts during implementation, there are certain terms that are often used to talk about the cloud in general. Let's look at some:

- **CMP**: A **Cloud Management Platform** provides a platform that enables management across various clouds, public and private. It allows management with one single interface, instead of having to go through different cloud portals. It can help in provisioning services, metering and billing, and configuration management of resources that are deployed in various cloud platforms.
- **Cloud Native**: This indicates that applications and software were particularly written for cloud usage.
- **Host**: This is the physical machine that sits in the data center of the cloud provider that hosts the guest environments of customers. Remember the cliché: the cloud is just someone else's computer. You're using it.
- **Virtual Machine**: A virtual machine runs on the physical host, but acts as a complete server.
- **Hypervisor**: This allows a host machine to be able to run multiple virtual machines. The technology of hypervisors was essentially the start of the cloud since it allows servers to act as multiple servers. The best-known hypervisor and market leader is VMWare's vSphere, followed by Microsoft's Hyper-V, Citrix Xen, and Red Hat Enterprise Virtualization. But there are many more.
- **Hybrid**: Typically, this is to indicate that a cloud platform uses both public clouds, such as AWS or Azure and private clouds that run in the data center of the company itself or a third-party data center.
- **Load Balancing**: If we run multiple resources such as servers, databases, and storage spaces, we need a mechanism that divides the load over these resources. In the cloud, we have many resources, some dedicated to a specific customer, but also a lot of shared resources. Load balancing is essential for spreading traffic and data across the resources.
- **Multi-tenancy**: This is probably the most important term to remember in the cloud, especially in terms of privacy, and keeping our data safe and secure in our own environment, meaning that only authorized users can view the data inside that tenant. Imagine the cloud as a big house with multiple tenants. These tenants will have a room for themselves, but also share facilities. Multi-tenancy indicates that software is running on shared components, where resources are dynamically assigned to software.

- **Resource:** This is the term that is used for components that are used in the cloud to run applications, save data, and execute services. A server is a resource. So is a database or a firewall.
- **Scalability:** Because of virtualization, shared resources, load balancing, and multi-tenancy, the cloud is scalable. Companies use what they need at a certain time. They can scale up when demand is high and scale down in more quiet times. This is the big advantage over environments in privately owned data centers, where equipment (such as servers and storage cabinets) is simply there but sits idle when there's no demand for it. The inverse is true as well in on-premises: when we need more capacity, such as servers or storage, we will likely need to order the equipment and wait for delivery, whereas in the cloud, we can spin up instances immediately.
- **UI:** The user interface. At the end of the day, operators and developers must be able to work with the resources in the cloud. This is done through the UI in most cases, a portal with all the services that can be deployed in a cloud platform. Having said that, most developers and operators use other mechanisms to manage workloads in the cloud, since a portal with a graphical UI is too slow for large-scale operations. Then scripting comes into play, for instance, using command-line interfaces with PowerShell that can be used in Azure and AWS, and GCP commands in GCP. To summarize, public clouds provide a UI for users and APIs for large-scale operations. In the latter case, we will need to script in, for instance, Powershell to interact with the APIs.

Speaking the same language helps in understanding what specific clouds we are talking about and how we can use them in the most optimal way. It's like going on a real journey: it helps when we can speak at least some basic lines in the language of the country we are traveling to so that we are sure that we reach our planned destination. We have to plan our journey. Hence, the next step is planning and executing an assessment to validate if we are ready to migrate to the cloud.

Planning assessments

You have to know where you're coming from in order to define where you're going. There's absolutely no use in building a cloud landing zone without knowing what will land on the platform. The next question is how applications should be migrated to the platform—if the business case is positive and the decision has been made to start the migration. Note that migration is one of the cloud adoption paths. We can also develop cloud-native solutions, but in that case, companies will likely have to redesign and rebuild applications. Migration is typically about moving existing applications to the cloud, typically referred to as rehosting or replatforming. Part of the migration can be to start using PaaS services, for instance, managed database services from the cloud. However, the application itself usually remains as it is. We will talk about this in more detail later in this chapter.

In this section, we will assess the readiness of the organization to adopt the cloud, just like the Cloud Adoption Framework advises. Relevant questions include "Who are all the stakeholders that need to be involved in the activities?" Most importantly, "Will the migration impact the customer, thus the business, and in what way? What are the drivers to start the transition and transformation to a cloud platform?" These are just a few questions that must be answered before we start planning the actual activities. They will show if the organization is ready to start adopting cloud technology. The assessment must reveal the business advantages, the effort the business needs to invest, and if the right people with the right skills are available.

Is an assessment necessary? The short answer is: yes—without exceptions. Assessments will also reveal the level of maturity of an organization. This level of maturity helps an organization to understand whether they are ready to move to the cloud at all.

Perhaps this is a good spot to briefly talk about maturity. For starters, a mature IT organization doesn't necessarily mean a mature organization working in the cloud. Most maturity models work according to five levels. *Figure 3.1* shows an example, in this case, the **Capability Maturity Model**, or **CMM**.

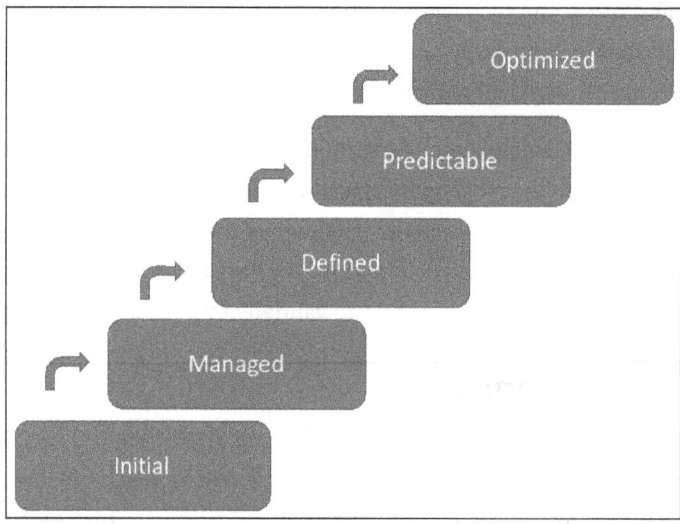

Figure 3.1: The Capability Maturity Model (CMM)

Level 1—initial is the lowest level, where processes are poorly controlled and outcomes are highly unpredictable. In the cloud, this would mean that organizations would simply start building environments in a cloud without proper processes or defining what the result should be. Level 5 is the highest level, where organizations can focus on improvements since projects and management are well-defined, processes are under control, and outcomes are predictable and measurable.

Most organizations are somewhere between levels 3 and 4. But are these organizations not working on improvements? They are, but not as part of the standing organization. Typically, improvements and innovations are done separately from production and need handover procedures to implement into production.

One more time: enterprises that have well-defined and controlled IT organizations aren't necessarily also well-equipped for starting a transition into the cloud. To put it in different words: the cloud is a different beast. Where IT organizations run the entire platform, including the hosting infrastructure, they are now running on "someone else's computer." Here's where the segregation of duties plays a very important role. Cloud providers are responsible for the cloud (the data center), and customers are responsible for what's in the cloud (the virtual servers in the data center and the data). Note that this is the case for IaaS. For PaaS and SaaS, this changes. For example, in the case of SaaS, the provider is responsible for the entire solution.

Is an organization ready to start working in such a model? Notice that this is not a technical question: we're not starting with mapping the existing technology to the cloud. We're starting with organizational readiness since the operating model in the cloud is fundamentally different from privately owned data centers. Readiness is the first topic in any assessment.

Let's explore the common steps in an assessment:

1. **Set the scope**. This should not be executed from the technology but from the business perspective. What are the business functions that would benefit from a transformation and the usage of cloud technology?
2. **Identify all infrastructure components**. This results in a complete list of all infrastructure, such as network equipment, servers, storage, and security appliances. Can we map these components to solutions in the cloud? What solutions do we need to implement infrastructure in the cloud?
3. **Identify all applications and application components**. This results in an application mapping showing what business functionality is fulfilled by what application and specific application components. Are applications cloud-ready, can they be migrated to cloud infrastructure, or do we need to refactor, replatform, or even redesign applications before we can host them in the cloud? What would be the effort to get applications cloud ready?
4. **Identify security requirements, including existing policies and security guardrails**. What security policies are in place and how do these translate in the cloud? Are existing policies including compliance statements, and if so, do cloud solutions adhere to these compliance rules?

5. **Identify all stakeholders that must be involved in migrating to the cloud, including business and application owners.**

The information that is discovered through these steps could determine if an organization is ready to move to the cloud or if they need more time. This will drive the business case.

One of the challenges that many enterprises face is the availability of resources. It's recommended to assess the availability of resources: does the company have the right people with the required skills to execute the transformation? Competition is fierce in the war on talent; this must be addressed in the earliest stage. This will also impact the business case; timelines and budgets will be heavily put under pressure if resource planning is not considered.

Executing technology mapping and governance

Companies have an ambition and a strategy to fulfill that ambition. In the modern, digital company, technology will undoubtedly play an important role in fulfilling that ambition. But, as with every other aspect within the governance of a company, the deployment and management of technology needs proper planning. First, technology must be assessed against the goals a company wants to achieve. In other words: technology must add value to the business goals. Second, a company must be ready to adopt new technology. This means that it must have trained, skilled staff that is able to work with it. With that, we have identified the two main blockers to the successful implementation of technology, including the cloud:

- It might not add value to the business
- There are no human resources available who can work with it

Technology mapping can help here. It typically starts with defining the use cases: what will the technology be used for? Next, it's assessed against the existing portfolio to determine how the technology must be implemented and integrated within the processes and technology base of the company. The final step is actual planning, wherein the adoption of change and training are key elements.

It starts with the ambition and the strategy. In a world where markets and customer demands change continuously, business agility is a must. Companies must be able to respond to changes fast and act accordingly. Innovations can't be neglected, but it's important to pick the right innovations.

Keeping track of innovation

Cloud technology evolves fast. Remember that AWS started in 2006 with its **Simple Queue Service (SQS)**, a messaging service that could be publicly used. Today, AWS offers well over two hundred services from data centers all over the world. New services are launched every year. How do companies keep track of all the innovations?

First of all, not every new service might be beneficial for your company. Services need to fit the business ambition, strategy, and goals—as with everything. Having said that, most innovations can be categorized as features that either make working in the cloud easier (automation), more efficient (cost and efficient use of multiple data sources), or open up new possibilities for businesses. An example of the latter is services that enable the use of artificial intelligence or even quantum simulation.

It's important to have a clear vision of where a business is going. In the digital era, that is already very hard. Customer demand changes constantly and at an ever-increasing speed. Innovations can surely help businesses to fulfill these demands. The challenge is how to assess and judge if innovations are really need-to-haves or nice-to-haves.

The major cloud providers organize yearly conferences where they announce major new features and releases of new technology. There's something to be aware of with these events: they are targeting tech-savvy audiences.

Microsoft Ignite: This is the event for technology professionals and developers working with Microsoft technologies, including Azure. It focuses on new technology that Microsoft has developed and released in Azure.

Microsoft Inspire: This is the business event for partners offering services using Microsoft technologies, such as Azure. The conference focuses on how to use Azure services to grow and manage the business and is less technical than Ignite.

AWS Re.invent: The technology conference of AWS, which annually shows new products and releases.

AWS business events: This platform is for business leaders, providing opportunities to discuss how AWS can help expand the business.

Google I/O: The big technology conference of Google covers more than just Google Cloud. It addresses every technology that Google provides, including Android and Google services.

Google Cloud Next: The main event for Google Cloud.

Oracle Cloud World: The main event to learn everything about new releases and features in Oracle Cloud Infrastructure.

These are the conferences where the latest information about products and platforms is shared. Next to this, there are user communities around the world that are worthwhile attending. Experts Live, focusing on Microsoft technologies, is a good example of such a user event. Lastly, social media such as Twitter, Mastodon, LinkedIn, and Reddit are great sources to keep track of innovations.

Adopting innovations

Adopting innovations is a matter of culture in the first place, but the most important thing is that innovations must be triggered by the business. Innovations must lead to added value. The challenge for every innovation is to question what benefit it will bring to the customers of a company, and what added value is involved. This can be a direct benefit leading to a better customer experience, but also as a result of optimized processes within the company. In the end, that will also lead to improved customer experience.

Let's have a look at the common ambitions of modern companies. These are commonly grouped into customer experience, sustainability, and financial efficiency. If these are the greater ambitions of a company, then innovations must be challenged against these ambitions.

Is it improving the customer experience—will it make customers happier?

Is it improving the sustainability of the enterprise and/or society at large?

Given the overall business ambition, is it worth the investment, and how will it impact financial performance?

Let's use an example. Serverless functions were a great innovation in cloud technology. When used in event-driven use cases, it's likely that they improve the customer experience since actions are immediately started at a customer trigger and delivery of services commences without manual interventions. They also improve sustainability since there's no use for heavy VMs that need to be deployed and managed. Fewer resources are invoked, resulting in less impact. Are they worth the investment? The answer is probably yes, given the fact that they will add value to the business.

But, it's also likely a huge change in the architecture of systems. That must be taken into account: it will require skilled resources and effort to adopt the innovation and make the required changes in the landscape. That is part of the total investment. Then we must agree on the importance of customer experience and reaching sustainability goals.

One important aspect that we must consider in introducing innovations is that not everyone will be willing to adopt these. Operators have a task to keep systems as stable as possible. Innovations causing changes to systems might lead to failures and even outages of systems, resulting in a lot of extra work for them. This is the reason to always bring developers and operators together in introducing and implementing innovations, indeed in a DevOps manner. They all need to be on the same page.

That is culture: reaching common agreements on the business goals and a common understanding of the impact of change by adopting innovations. There will be a lot more on this topic during the course of this book.

To summarize, innovations must be significant to the business. They must lead to added value and add to the overall business ambition of a company. With that, we are at defining that ambition in a North Star and how multi-cloud can fit into that.

Defining roadmaps and business alignment

There's a good chance that you have heard about the North Star architecture. It's commonly used to define the future ambition of an enterprise, especially in digital transformation. Almost every enterprise has defined a North Star. It sets the ambition and lets teams iterate toward the realization of that ambition: it's the fundamental idea of agile working. The architecture evolves in every iterative step, adapted, and refactored whenever necessary to move toward the North Star.

However, teams need guidance and guardrails. The risk of iterative working and only having an ambition defined in the North Star is that teams develop code and applications without guidance on interoperability and integration throughout systems. Or, teams spend a lot of time developing something that another team has already done. They might develop a feature that is readily available as a service in a cloud platform.

Roadmaps will provide guidance, but only if the roadmap itself is "business integrated." What do we mean by that? Business ambition and goals must be aligned with the capabilities of the enterprise and next, the technology stack of the enterprise. Combined, these should converge into an actionable strategy telling teams what to use to build specific business functionality.

So, in building roadmaps, we start with the business model and the business functionality. The business model provides insights into how the enterprise works, translated into business functionality. This maps to application functionality and the usage of data inside the functionality. What business function does an application serve and what data does the function need, when, and why? Lastly, the technology mapping is done.

What sort of databases would serve data processing best? Which technology does the enterprise need to deliver application, and thus business, functionality? In a roadmap, this can then be matched with announced propositions from cloud providers and other suppliers. But, it must be clear from the roadmap how a new feature, release, product, or service will enhance the functionality.

In short, a technology roadmap defines strategic planning to achieve specific business goals and ambitions using technology. Technology roadmaps include not only the technology itself but map these to business functions.

Architects are encouraged to use templates to build roadmaps. An example is provided in the following diagram. Tools such as Miro are very useful in building roadmaps.

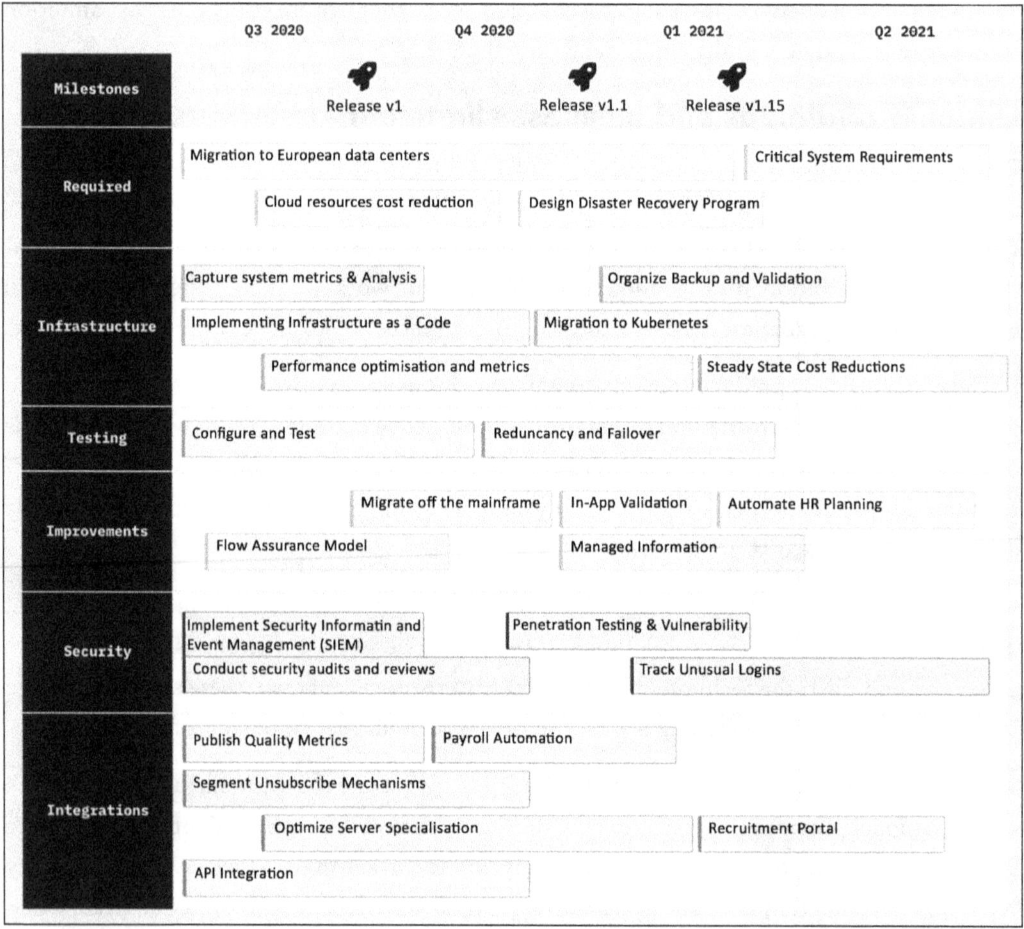

Figure 3.2: Example of a Miro board showing a technology roadmap (`https://miro.com/templates/technology-roadmap/`)

Roadmaps will help architects to define strategies for migrating or updating software systems, and planning the development of new technology or other projects. But architects have to keep in mind that there's one starting point, and that's the business. In the next chapter, we will discuss service designs in the cloud and how business **key performance indicators** (**KPIs**) and **service-level agreements** (**SLAs**) will help companies get the best out of the cloud. Cloud providers offer excellent tools for that with the **Cloud Adoption Framework** (**CAF**) and **Well-Architected Framework** (**WAF**). We will learn how to use these in architecture.

Planning transition and transformation

It must be clear that the planning of the transition and transformation to the cloud starts with the business case. Architects are advised to follow the business-data-application-technology rule to design the roadmap for applications that are designated to be moved to cloud platforms. But before companies can really start migrations, there's a lot of work that must be done.

Assuming that, from the assessment, we have a clear overview of the **current mode of operations**, the **CMO**. From the business case, the company also has set the ambition, goals, and strategy to transform the business to the **future mode of operations** (**FMO**)—and the rationale as to why it's necessary or desirable to transform. As said, this goes beyond technology. The transition and transformation plans should include at least the following topics on the various tiers as shown in the diagram:

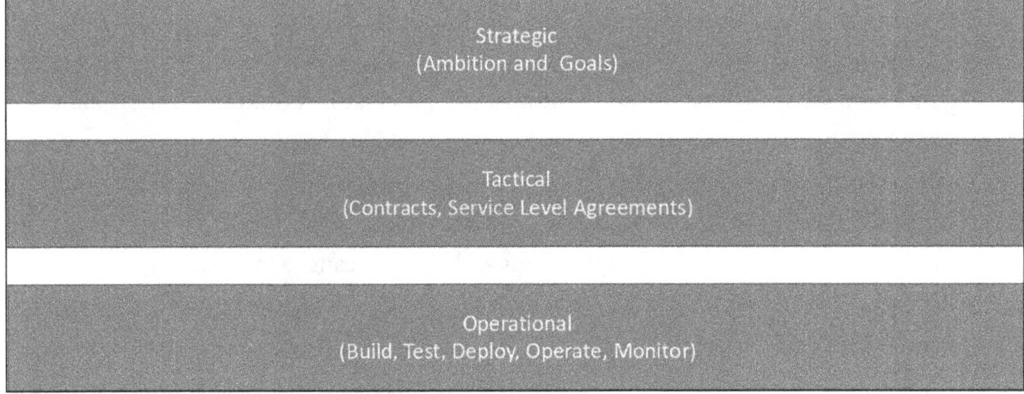

Figure 3.3: Simplified enterprise governance model

The highest tier is strategic. This is the tier where the actual business strategy is defined, the topic of the previous chapter. This is also the level where all the governance processes are defined to help govern and control all activities in migration and, eventually, operations.

The next tier is tactical. An enterprise that is moving applications—or starting to develop applications—in public cloud platforms will have to think about service levels and KPIs. There will be contracts and agreements with the cloud providers that must match the business objectives. All major cloud providers offer SLAs. These contracts define things such as uptime guarantees and credit policies. These are typically legal documents; hence it's advised to involve a legal counselor to study these documents.

The SLAs for Microsoft Azure can be retrieved from `https://www.microsoft.com/licensing/docs/view/Service-Level-Agreements-SLA-for-Online-Services?lang=1`; for AWS, refer to `https://aws.amazon.com/legal/service-level-agreements/`, and for GCP, to `https://cloud.google.com/pubsub/sla`.

The final tier is the operational tier. This is the tier where all operational processes are defined: build, test, deploy, operate, and monitor. In every operational process, it must be clear how cloud services can and should be used. This includes the management of infrastructure, data, and applications.

This means that in the planning of transition and transformation, the architect must start with translating the business ambition and goals to tactical parameters, and lastly, to technical solutions.

A transition plan will consist of the following items to cover it all:

- **Discovery of assets**: We need to have a complete and comprehensive overview of the environment and how it's serving the business processes. The latter is important: there will be easy, standalone applications that can be migrated to the cloud without a severe impact on critical business processes. There will be, without a doubt, applications and data that are crucial to the business. When the current environments suffer from outages, the lack of access to these crucial applications and data can cause a lot of damage, even when outages are planned. From the discovery, it must be clear from the business functionality and criticality what environments are crucial. This will influence the migration strategy.
- One other important topic in discovery is that we need to establish all interactions in the environment. What workload is connected to other workloads? This involves routing, database usage, (shared) storage, backups, and the security settings in an environment.
- Last but not least, we need to identify how the assets are operated and by whom.

- **Assessing the assets against the business plan**: The architect needs to define a target architecture and a target operating model. Both must comply with some principles before the actual build can start. A multi-cloud strategy and architecture must ensure consistency across the targeted platforms. This is not only a matter of avoiding the lock-in in a specific cloud but more so a matter of reducing complexity in managing workloads. We will be distributing workloads across platforms, even in various service models such as IaaS, PaaS, and SaaS. Then, it's important to take quality attributes into account:
    - Reliability
    - Maintainability
    - Usability
    - Portability
    - Correctness
    - Efficiency
    - Integrity or security
    - Testability
    - Flexibility
    - Reusability
- If the architect works against these quality attributes, the company will get the benefits of multi-cloud:
    - **Best of breed solutions**: Since this is reason number one to start a multi-cloud journey. We want to use the best of the various worlds and have the best solution for our enterprise proposition, delivering value to the customers.
    - **Improved resiliency**: Workloads are distributed across various platforms. This will decrease the risk of failures and outages with a severe impact on the business.
    - **Business agility**: Multi-cloud leverages agility, by taking advantage of the opportunity to choose the right tools and solutions for business demand.
- There are two more things that architects already must consider in this stage:
    - **Security**: Here's the statement: already, public clouds are extremely well secured. Cloud providers don't have a choice in securing their platforms, with thousands of customers that work with their services and run (critical) business processes. But the provider is responsible for the cloud, and the customer for what's in the cloud. AWS, Azure, and GCP offer a wide variety of security tools, but it's up to the customer to use them wisely.

- **Cost control**: Companies must be aware of the costs that they are generating by using cloud services. The cloud can be complex in terms of costs, metering, and billing. Services constantly change, and new features are continuously released, making it hard to always have a proper understanding of where costs are coming from. This is the domain of FinOps, which we will discuss in depth in this book.

- **Planning**: This will involve all activities needed to actually perform the migration and the transformation of the workloads. Typically, the migration is planned in waves, where Wave 0 is used to build and test the landing zone and the migration procedures, as defined in the target architecture and target operating model. It's advised to have simple applications in Wave 1 and more complex environments in later waves.

We now have a plan and we can start the build. It's the topic of the next section.

Starting the build

One of the key decisions a company makes when defining cloud strategy is which cloud providers to choose. Then, they should build landing zones in each cloud of choice. This should include the basics, such as communications, policies, security, monitoring, business continuity and disaster recovery, processes, and training. From there, the business can start working on planning and building specialized environments per workload. Remember: there's no point in building and configuring a landing zone without knowing what will be landing on it.

What is a landing zone? Typically, a cloud landing zone is defined as the pre-configured basic environment that is provisioned in a public cloud so that it's ready to start hosting applications. Landing zone is a common term, but—remember the first section in this chapter about the common language—all cloud providers have different names and services to create landing zones.

There's an important step before we start configuring the landing zone itself: making sure that we can reach the landing zone. We need connectivity.

Setting up simple connectivity

You can make a subscription to any cloud provider and start working in it, but a company would want to have an enrolment to which its workers can securely connect from the company's domain to a specific cloud service. Basically, there are three options to enable that connection: a **virtual private network** (**VPN**), direct connections, and using a fully managed broker service from a telecom company or connectivity partner. In the next sections, we are going to have an in-depth look at each of these options.

One of the most used technologies is the VPN. In essence, a VPN is a tunnel using the internet as a carrier. It connects from a certain IP address or IP range to the IP address of a gateway server in the public cloud.

Before we get into this, you have to be aware of what a public cloud is. If you as a business deploy services in Azure, AWS, **GCP**, or any other public cloud (remember, there are more public clouds, such as Oracle, OpenStack, IBM Cloud, and Alibaba, and the basic concepts are all more or less the same), you are extending your data center to that cloud. It, therefore, needs a connection between your data center and that extension in the public cloud.

The easiest and probably also the most cost-efficient and secure way to get that connection fast is through a VPN. The internet is already there, and all you would have to do in theory is assign IP addresses or the IP range that is allowed to communicate to that extension, creating a tunnel. That tunnel can be between an office location (site) or from just one user connecting to the cloud. The latter is something we refer to as a point-to-site or site-to-site VPN.

In the public cloud itself, that connection needs to terminate somewhere, unless you want all resources to be opened up for connectivity from the outside. That is rarely the case, and it's certainly not advised. Typically, a business would want to protect workloads from direct and uncontrolled external connections. When we're setting up VPNs, we need to configure a zone in the public cloud with a gateway where the VPN terminates. From the gateway, the traffic can be routed to other resources in the cloud, using routing rules and tables in the cloud. It works the same way as in a traditional data center, where we would have a specific connectivity zone or even a **demilitarized zone** (**DMZ**) before users actually get to the systems. The following architecture shows the basic principle of a VPN connection to a public cloud:

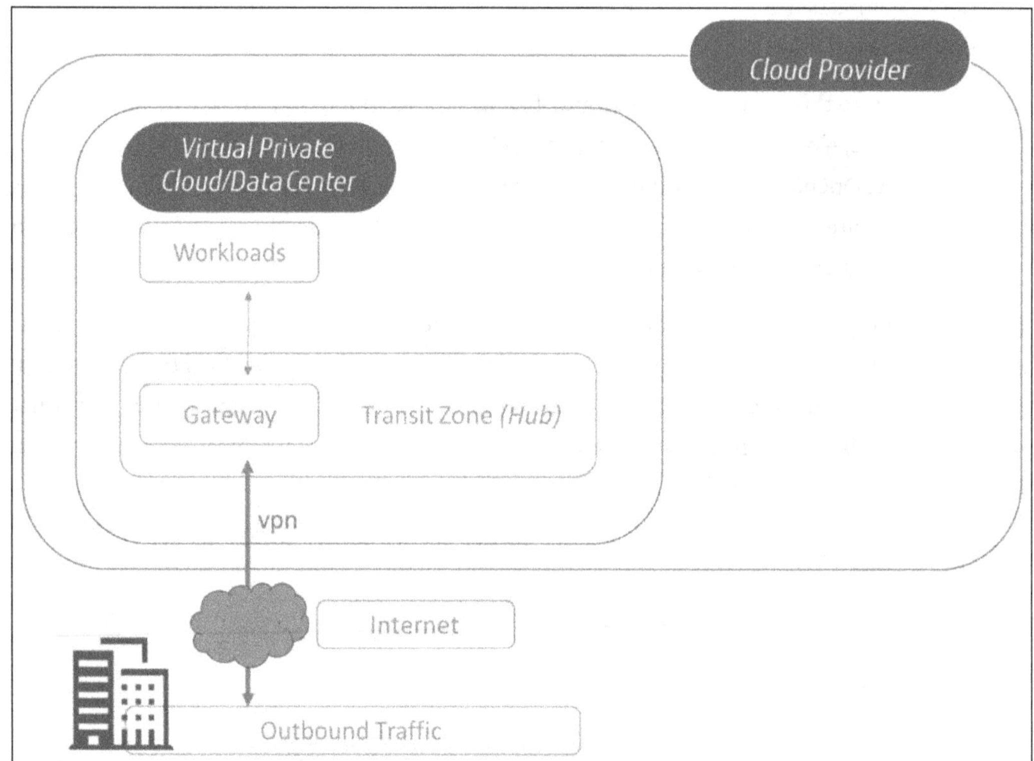

Figure 3.4: The basic architecture of VPN connectivity

One remark: there are other ways to connect to public clouds, including direct, dedicated connectivity with, for instance, Azure ExpressRoute, AWS Direct, and Google Direct Connect. These are typically considered by enterprises in case of large, mission-critical deployments in public clouds and substantial transfer of data that requires high performance. This type of connection is also recommended—sometimes even required—to adhere to compliance statements or to ensure reliability and consistent latency.

The next step is to launch the landing zone, our workspace in the cloud.

Setting up landing zones

In Azure, the landing zone starts with a subscription. AWS deploys accounts and GCP provisions projects. But they all do sort of the same thing, and that is configuring a basic environment where you can start working with the respective services that these providers offer. Looking at frameworks such as Well-Architected and Cloud Adoption, these landing zones offer:

- Security controls
- **Identity and Access Management** (IAM)
- Tenancy ("workspace" allocated to a specific customer)
- Network configurations
- Resources related to operations such as monitoring and backup

With landing zones, customers get a standardized environment. Yet, it's good practice to think about setting up landing zones so that they can be managed from day one. The phases for setting up landing zones are:

- **Design**: Defining the basic parameters for security controls, IAM, and networking.
- **Deploy**: In Azure, we can use Azure Blueprints to deploy a landing zone. It offers standardized templates to roll out a landing zone, with the possibility to integrate design principles. In AWS, the service is simply called AWS Landing Zone, which sets up all the basic services in AWS using Cloud Formation templates and services such as CloudTrail for monitoring. Google offers the Google Deployment Manager, which allows developers to deploy templates in GCP using YAML, Python, or Jinja2.
- **Operate**: As soon as a customer starts building things, the landing zone will expand. It needs to be managed, just as any other infrastructure. If companies run multiple environments in the mentioned clouds, it's wise to set up tools that make operations across these multiple environments easier. We will discuss this in the section about keeping infrastructure consistent.
- **Update**: Updates from operations must be looped back into the design so that architectures are continuously improved and stay consistent, documented, and transferable.

Following the standardized approaches of the cloud providers, the landing zone will enable organizations to consistently manage:

- Resources in the cloud platform
- Access

- Monitoring
- Costs
- Governance and security

This chapter focused more on the generic principles for setting up landing zones. In *Chapter 6, Controlling the Foundation Using Well-Architected Frameworks*, we will discuss the various landing zone propositions in AWS, Azure, GCP, Alibaba, and OCI in more detail.

With a landing zone in place in our preferred cloud or multiple clouds, we can start thinking of migrating or deploying workloads.

Exploring options for transformation

In developing and migrating workloads to the cloud, there are a number of options that architects must consider from the beginning. In this section, we will elaborate on these choices.

From monolith to microservices

A lot of companies will have technical debt, including monolithic applications. These are applications where services are tightly coupled and deployed as one environment. It's extremely hard to update or upgrade these applications; updating a service means that the whole application must be updated. Monolithic applications are not very scalable and agile. Microservices might be a solution, wherein services are loosely coupled.

Transforming a monolithic application to microservices is a very cumbersome process. First of all, the question that must be answered is: is it worthwhile? Does the effort and thus costs weigh up to the benefits of transformation? It might be better to leave the application as-is, maybe lift-and-shift it to the cloud, and parallel design, build, and deploy a replacement application using a microservices architecture.

With that, we have identified the possible cloud transformation strategies that we discussed in *Chapter 2, Collecting Business Requirements:*

- Rehost
- Replatform
- Repurchase
- Refactor
- Rearchitect
- Rebuild

- Retire
- Retain

Microservices typically involve a rearchitect and refactor. The functionality of the application and the underlying workloads is adapted to run in microservices. That includes the rearchitecture of the original monolithic application and refactoring the code. This might sometimes be a better option, with the development of cloud-native applications replacing the old application, especially when the original application is likely to consume a lot of heavy—costly—resources in the cloud or prevent future updates and upgrades.

Enterprises can make the decision to retain the old environment for backup or disaster recovery reasons. This will definitely lead to extra costs: costs of having to manage the old environment and investing in the development of a new environment. Part of the strategy can also be "make or buy" with either in-house development or buying software "off the shelf." It all needs to be considered in drawing the plans.

More technology has been emerging to create cloud-native architectures, moving applications away from the classical VMs to containers and serverless. We'll review this in the next sections.

From machines to serverless

Public clouds started as a copy of existing data centers, with servers, storage, and network connectivity like companies had in their own data centers. The only difference was that this data center was operated by a different company that "rented" equipment to customers. To enable a lot of customers to make use of the data center, most of the equipment was virtualized by implementing software that divided a server into multiple, software-defined servers. One server hosted multiple customers in a multi-tenant mode. The same principle was applied to storage and network equipment, enabling very efficient usage of all available resources.

Virtual machines were the original model that was used by customers in the public cloud. A service was hosted on a machine with a fixed number of CPUs, memory, and attached disks. The issue was that, in some cases, services didn't need the full machine all of the time, but nonetheless, the whole machine was charged to the customer for that time. There were even services that only were short-lived: rapidly scaled up and down again, as soon as the service was not running anymore. Especially in microservices, this is becoming common practice. Event-based architectures are where a service is triggered by an action of a customer and stopped as soon as the action has been executed. Full-blown virtual machines are too heavy for these services.

Serverless options in the public cloud are a solution for this. In the traditional sense, you use your own server on which your own software runs. With a serverless architecture, the software is run in the cloud only when necessary. There's no need to reserve servers, saving costs. Don't be fooled by the term "serverless" as there are still servers involved, but this time, the service only uses a particular part of the server and only for a short amount of time. Serverless is a good solution when an organization doesn't want to bother about managing the infrastructure.

But the biggest advantage of using serverless options is the fact that it can help organizations in becoming event-driven. With microservices, there are software components that focus on one specific task, such as payments or orders. These are transactions that follow a process. Serverless functions each perform their own step in the process, only consuming the resources the function needs to execute that specific task in the process. Then, serverless is a good option to include in the architecture.

Major public clouds have these solutions: Azure Functions, AWS Lambda, and Google Cloud Functions.

Containers and multi-cloud container orchestration

Serverless is a great solution to run specific functions in the cloud, but they are not suitable to host applications or application components. Still, companies want to get rid of heavy virtual machines. VMs are generally an expensive solution in the cloud. The machine has a fixed configuration and runs a guest operating system for itself. So, the hosting machine is virtualized, allowing multiple workloads to run on one machine. But these workloads that run in a VMs, still require their own operating system. Each VM runs its own binaries, libraries, and applications, causing the VM to become quite big.

Containers are a way to use infrastructure in a more efficient way to host workloads. Containers work with a container engine and only that engine requires the operating system. Containers share the host operating system kernel but also the binaries and libraries. This makes the containers themselves quite light and much faster than VMs. When an architecture is built around microservices, containers are a good solution.

Containers are a natural solution to run microservices, but there are other scenarios for containers. A lot of applications can be migrated to containers easily and with that, moved to the cloud quickly—making containers a good migration tactic.

Each container might run a specific service or application component. In case of upgrades or updates, only a few containers might be "impacted."

The following diagram explains the difference between a VM and a container:

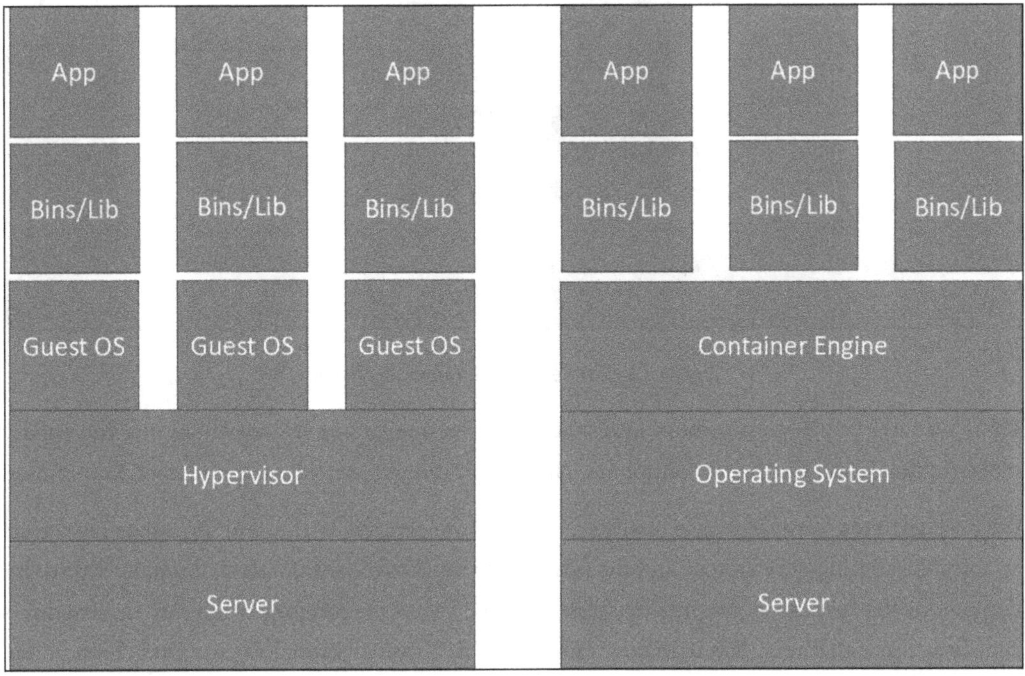

Figure 3.5: Virtual machine (left) versus containers (right)

This is a good point to explain something else in working with a container architecture: sidecars. Sidecar containers run along with the main container holding a specific functionality of the application. If we only want to change that functionality but nothing else, we can use sidecar containers. The sidecar container holds the functionality that shouldn't be changed, while the functionality in the main container is updated. The following diagram shows a simple example of a sidecar architecture:

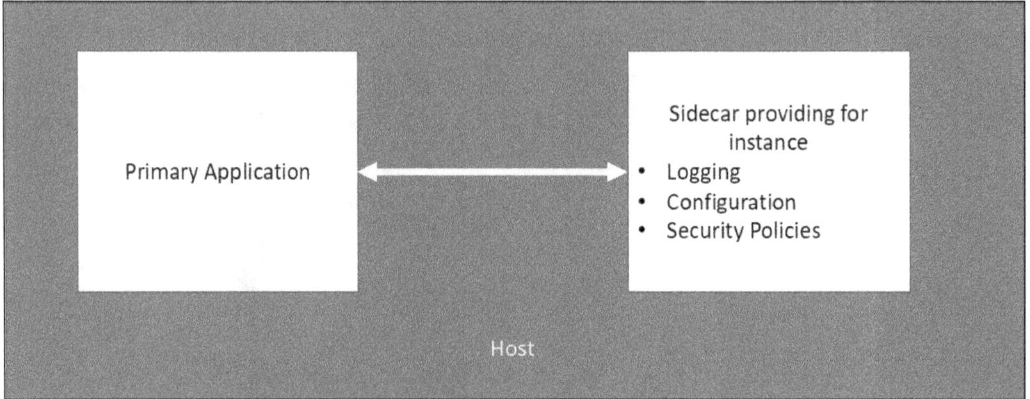

Figure 3.6: Simple architecture for a sidecar

There's a catch to using containers, and that's the aforementioned container engine. You need a platform that is able to run the containers. The default industry standard has become Kubernetes.

With Kubernetes, containers are operated on compute clusters with a management layer that enables the sharing of resources and the scheduling of tasks to workloads that reside within the containers. Resources are compute clusters, a group of servers –commonly referred to as nodes—that host the containers. The management or orchestration layer makes sure that these nodes work as one unit to run containers and execute processes—the tasks—that are built inside the containers.

The cluster management tracks the usage of resources in the cluster such as memory, processing power, and storage, and then assigns containers to these resources so that the cluster nodes are utilized in an optimized way and applications run well.

In other words, scaling containers is not so much about the containers themselves but more about scaling the underlying infrastructure. Kubernetes uses Pods, enabling the sharing of data and application code among different containers, acting as one environment. Pods work with the share fate principle, meaning that if one container dies in the Pod, all containers go with it.

All major cloud providers offer solutions to run containers using Kubernetes:

- **Azure Kubernetes Services (AKS)**: The managed service to deploy and run Kubernetes clusters in Azure. Azure also offers Azure Container Apps and Azure Container Instances as serverless options.
- **Elastic Kubernetes Services (EKS)**: The AWS-managed service for Kubernetes platforms in AWS. EKS Anywhere uses EKS Distro, the open-source distribution of Kubernetes. Since it's a managed service, AWS takes care of testing and tracking Kubernetes updates, dependencies, and patches. To be clear, the same applies to AKS. AWS also offers a serverless solution to run container environments: Fargate. This removes the need to provision and manage servers, and simply allocates the right amount of compute, eliminating the need to choose instances and scale cluster capacity.
- **Google Kubernetes Engine (GKE)**: The managed service of Google Cloud to deploy and run Kubernetes clusters in GCP.
- **Alibaba Cloud Container Service for Kubernetes (ACK)**: The managed service of Alibaba Cloud to deploy and run Kubernetes clusters on Alibaba Cloud.
- **Oracle Container Engine for Kubernetes (OKE)**: The managed container service that we can use in OCI. This service also includes serverless options with OKE Virtual Nodes.

All the mentioned providers also offer unmanaged services to run containers and container orchestrations, but the advantage of managed services is that the functionality of, for instance, scaling and load balancing across clusters is completely automated and taken care of by the provider.

The Kubernetes clusters with Pods and nodes must be configured on that specific platform, using one of the services mentioned above. There are technologies that provide tools to manage Kubernetes clusters across various cloud platforms, such as VMWare Tanzu, NetApp Astra, and Azure Arc:

- **VMWare Tanzu:** This is the suite of products that VMware launched to manage Kubernetes workloads and containers across various cloud platforms. The offering was launched with Tanzu Mission Control but has evolved over the past years with offerings that allow for application transformation (Tanzu Application Platform) and cloud-native developments (Tanzu Labs).
- **NetApp Astra:** NetApp started as a company that specialized in storage solutions, specifically in **network attached storage (NAS)**, but over the years, NetApp evolved to a cloud management company with a suite of products, including Astra, that allows management of various Kubernetes environments.

- **Azure Arc**: Azure Arc-enabled Kubernetes allows you to attach and configure Kubernetes clusters running anywhere. You can connect your clusters running on other public cloud providers (such as GCP or AWS) or clusters running on your on-premises data center (such as VMware vSphere or Azure Stack HCI) to Azure Arc.

The interesting part of these products is that they are also able to manage lightweight Kubernetes, known as K3S, in environments that are hosted in on-premises private stacks, allowing for seamless integration of Kubernetes and container management through one console.

Keeping the infrastructure consistent

Microservices, serverless, containers, and legacy environments that run virtual machines in a more traditional way all need to be operated from the cloud. The big challenge is to keep the infrastructure consistent. In this section, we will briefly discuss methodologies and tools to achieve this.

The preferable way of keeping infrastructure consistent is by working through templates. Such a template contains all the configurations with which an infrastructure component should comply. We can take a virtual machine as an example. A VM can be deployed straight from the marketplace of a cloud provider. Typically, companies have specific demands for servers: they must be configured with specific settings that define the configuration of the operating system, level of access, and security parameters of the server. First, we don't want to do this manually every time we enroll a server, and second, if we do it manually, the chances are that there will be deviations. Hence, we use templates to automate the enrollment of the desired state of servers and to keep all servers consistent with policies.

Let's look at an example of how a template for a VM could look. Be aware that this list is not meant to be exhaustive:

- Sizing of the VM
- Operating system
- Configuration parameters of the operating system
- Access policies
- Network settings
- Workgroup or domain settings
- Security parameters
- Boot sequence

This can, of course, be done for every component in our infrastructure: storage, databases, routers, gateways, and firewalls, for example. There are a couple of methods to create templates. The two common ones are:

- Manual configuration and saving the template in a repository
- Cloning a template from an existing resource

There are a number of tools that can help in maintaining templates and keeping the infrastructure consistent:

- **Terraform**: This is an open-source tool by HashiCorp and became the industry standard for **Infrastructure as Code** (**IaC**). Terraform allows you to create, deploy, and manage infrastructure across various cloud platforms. Users define and deliver data center infrastructure using a declarative configuration language known as **HashiCorp Configuration Language** (**HCL**), or optionally, **JavaScript Object Notification** (**JSON**).
- **Bicep**: Bicep files let users define infrastructure in Azure in a declarative way. These files can be used multiple times so that resources are deployed in a consistent way. The advantage of Bicep is that it has an easy syntax compared to JSON templates. Bicep addresses Azure Resources directly through **Azure Resource Manager** (**ARM**), whereas in JSON, these resources must first be defined. Quickstart templates for Bicep are available through `https://github.com/Azure/azure-quickstart-templates/tree/master/quickstarts`.
- **CloudFormation**: What Bicep does for Azure, CloudFormation does for AWS. It provisions IaC to AWS, using CloudFormation templates. CloudFormation templates are available on Github: `https://github.com/awslabs/aws-cloudformation-templates`.

All these technologies evolve at the speed of lightning. The challenge that every company faces is how to keep up with all the developments. We will try to give some guidance in the next section.

Summary

The goal of this chapter was to provide some common understanding of different cloud concepts and how companies could use these to get the best-of-breed solutions to improve the business. Starting a multi-cloud journey requires proper preparation to get the best out of cloud technology, including emerging technologies such as micro-services, containers, and serverless. We noticed that it can become very complex. We must make sure that we keep the platforms consistent, by decreasing complexity and enabling effective management of workloads. Next, we started our journey by creating a plan for transition and transformation, starting with connectivity, and defining landing zones.

Cloud technology evolves at an extremely high pace. It's hard to keep up with all the new developments and the release of new features. In the last section, we learned how to stay in control with technology mapping, using the principles of the North Star architecture and technology roadmaps.

With a common understanding of the cloud and the underlying technology, we are ready to start using the cloud. In the next chapter, we will further elaborate on defining our cloud migration strategy by designing service models with the guidance of the Cloud Adoption Framework and tools that are provided through the Well-Architected Framework. We will learn how to set up a service catalog to get the best out of the cloud.

Questions

1. What is a CMP?
2. This chapter discussed various Kubernetes deployments in public clouds. Name the managed Kubernetes services of Azure, AWS, and GCP.
3. What are Ignite and Re:Invent?
4. True or false: A North Star is a detailed enterprise architecture.

Join us on Discord!

Read this book alongside other users, cloud experts, authors, and like-minded professionals. Ask questions, provide solutions to other readers, chat with the authors via. Ask Me Anything sessions and much more.

Scan the QR code or visit the link to join the community now.

`https://packt.link/cloudanddevops`

4

Service Designs for Multi-Cloud

All cloud providers offer a cloud adoption framework that helps businesses to implement governance and deploy services, while controlling service levels and **key performance indicators (KPIs)** in the cloud. In multi-cloud environments, businesses would have to think about how to implement governance over different cloud components, coming from different providers, while still being able to manage it as a single environment.

This chapter will introduce the base pillars for a unified service design and governance model, starting with identities. Everything in the cloud is an identity; anything that has a role and therefore needs access to resources is an identity. It requires a different way of thinking—users, VM, a piece of code even. We will look at the different pillars of multi-cloud governance and study the various stages in cloud adoption, using the Well-Architected and cloud adoption frameworks of cloud providers.

In this chapter, we will cover the following topics:

- Introducing the scaffold for multi-cloud environments
- Working with Well-Architected Frameworks
- Understanding cloud adoption
- Translating business KPIs into cloud SLAs
- Using cloud adoption frameworks to align between cloud providers
- Understanding identities and roles in the cloud
- Creating the service design and governance model

Introducing the scaffold for multi-cloud environments

How does a business start in the cloud? You would be surprised, but a lot of companies still just start without having a plan. How difficult can it be, after all? You get a subscription and begin deploying resources. That probably works fine with really small environments, but you will soon discover that it grows over your head. Think about it—would you start building a data center just by acquiring a building and obtaining an Ethernet cable and a rack of servers? Of course not. So why would you just start building without a plan in the public cloud? You would be heading for disaster, and that's no joke. As we saw in *Chapter 1, Introduction to Multi-Cloud*, a business will need a clear overview of costs, a demarcation of who does what, when, and why in the cloud, and, most importantly, it all needs to be secure by protecting data and assets, just as a business would do in a traditional data center.

If there's one takeaway from this book, it's this: you are building a data center. You are building it using public clouds, but it's a data center. Treat it as a data center.

Luckily, all major cloud providers feel exactly the same way and have issued cloud adoption frameworks. Succinctly put, these frameworks help a business in creating the plan and, first and foremost, help to stay in control of cloud deployments. These frameworks do differ on certain points, but they also share a lot of common ground.

Now, the title of this section contains the word *scaffold*. The exact meaning of **scaffold** is a structure that supports the construction and maintenance of buildings; the term was adopted by Microsoft to support, build, and manage environments that are deployed in Azure. It's quite an appropriate term, although in the cloud, it would not be a temporary structure. It's the structure that is used as the foundation to build and manage the environments in a cloud landing zone.

Scaffolding comprises a set of pillars, which will be covered in the following sections.

Working with Well-Architected Frameworks

Well-Architected Frameworks were invented to help customers build environments in public clouds by providing them with best practices. This book assumes that you are working in multi-cloud and facing the challenge that every cloud provider has its own tools, workflows, set of commands, service naming, and mapping. The good news is that all major cloud providers provide Well-Architected Frameworks. Even better news: these frameworks all do sort of the same thing.

The frameworks typically contain five pillars, or scaffolds. The frameworks for AWS, Azure, and GCP all provide guidelines for:

- Operational excellence
- Security
- Cost optimization
- Reliability
- Performance efficiency
- Sustainability (at the time of writing, only included by AWS)

Architects can now use these frameworks to design solutions in the cloud. The Well-Architected Frameworks will show how to optimize workloads in the cloud, with control of costs and optimized serviceability. It also contains guardrails to properly secure workloads to keep data in the cloud protected. The frameworks help in implementing **financial controls (FinOps)**, **security controls (SecOps)**, and **management of workloads (BaseOps)**. All of these will be extensively discussed during the course of this book.

The Well-Architected Frameworks offer best practices to validate cloud architectures. These frameworks contain the following principles, guidelines, and best practices.

First, Well-Architected Frameworks offer guidelines to run workloads in the most efficient way, using the best solution in a specific cloud. Solutions can be VMs, but also containers and serverless options. **Operational excellence** will guide us in setting up efficient operations and management of cloud environments, for instance, by using automation and concepts such as **Infrastructure as Code (IaC)**.

We must keep our environments secure and compliant. Well-Architected Frameworks provide best practices to manage the security posture of workloads that are hosted in the cloud. Remember that the cloud provider is responsible for the cloud, but the customer is responsible for what is in the cloud. AWS, Azure, GCP, and other providers offer a variety of tools to secure workloads and data, but it's up to the customer to use these tools. The most important question to be answered is: *who is allowed to do what, when, and why?* This is addressed in implementing **Identity and Access Management (IAM)**.

Managing security also includes compliancy. Companies often need to be compliant with industry standards and regulations. Well-Architected Frameworks will guide you in setting up compliance with the most common-known compliance frameworks, such as PCI-DSS, NIST, and ISO.

Cost management: Companies must know what they spend in the cloud and if it's adding value to the business. Cost management, or **FinOps**, monitors cloud spend, but also offers tools to control costs with, for instance, budgeting, alerting, and rightsizing workloads. Rightsizing may involve the choice of serverless solutions instead of VMs, or using reserved instances when workloads are needed for a longer period. With reserved instances, companies can save up to 70 percent of the VM cost compared to pay as you go.

Establish reliability: The cloud provides solutions that ensure high availability by spreading workloads over various zones in the cloud data centers and even (global) regions. But reliability is also about automation, for instance, through automatic failovers, recovery, and scaling solutions to cover peak loads.

As mentioned, sustainability is a relatively new pillar in frameworks. This pillar focuses on using solutions that require less energy and making efficient use of resources in the cloud in order to reduce energy consumption by optimizing the utilization of power and cooling. This topic is covered in *Chapter 19, Introducing AIOps and GreenOps*, where we will introduce and discuss GreenOps in more detail.

In the next sections, we will elaborate a bit more on topics that must be addressed regarding the Well-Architected Frameworks.

Identity and Access Management (IAM)

Who may do what, when, and why? It is key to understand this in the public cloud, so we will devote more words to identity and access later on in this chapter, in the section *Understanding identities and roles in the cloud*. The most important thing to bear in mind is that virtually everything in the cloud is an identity. We are not only talking about people here, but also about resources, functions, APIs, machines, workloads, and databases that are allowed to perform certain actions. All these resources need to be uniquely identified in order to authenticate them in your environment.

Next, specific access rules need to be set for these identities; they need to be authorized to execute tasks. Obviously, identity directory systems such as Active Directory or OpenLDAP are important as identity providers or identity stores. The question is whether you want this store in your public cloud environment or whether you want an authentication and authorization mechanism in your environment, communicating with the identity store. As everything is transforming into an "as a service" model, architects could also choose identity as a service. An example of this is Okta, which manages identity and access across multiple clouds.

As stated previously, we will look into this in more detail later on in this chapter under the heading *Understanding identities and roles in the cloud*, since this is the key starting point in multi-cloud architecture and governance.

Security

Here's a bold statement: platforms such as Azure, AWS, and GCP are likely the most secure platforms in the world. They have to be, since thousands of companies host their systems on these platforms.

Yet, security remains the responsibility of the business itself. Cloud platforms will provide you with the tools to secure your environment. Whether you want to use these tools and to what extent is entirely down to the business. Security starts with policies. Typically, businesses will have to adhere to certain frameworks that come with recommendations or even obligations to secure systems. As well as these industry standards, there are horizontal security baselines, such as the **Center for Internet Security** (**CIS**). The CIS baseline is extensive and covers a lot of ground in terms of hardening resources in the cloud. The baseline has scored and non-scored items, where the scored items are obviously the most important. Auditors will raise an issue on these items when the baseline is not met.

One framework deserves a bit more attention: MITRE ATT&CK. Having a baseline implemented is fine, but how do you retain control over security without being aware of actual attack methods and exploited vulnerabilities? MITRE ATT&CK is a knowledge base that is constantly evaluated with real-world observations in terms of security attacks and known breaches. The keyword here is real-world. They keep track of real attacks, and mitigation of these attacks, for the major public clouds—Azure, AWS, and GCP—but also for platforms that are enrolled on top of these platforms, such as Kubernetes for container orchestration.

The MITRE ATT&CK matrices can be found at `https://attack.mitre.org/`.

There's a lot to say about securing your cloud platform. *Chapter 16, Defining Security Policies for Data*, provides best practices in relation to cloud security.

Cost management

You've likely heard this before from companies that enter the cloud: "the public cloud is not as cheap as the traditional stack that I run in my own data center." That could be true. If a business decides to perform a lift and shift from traditional workloads to the public cloud, without changing any parameters, they will realize that hosting that workload 24/7/365 in the public cloud is probably more expensive than having it in an on-premises system. There are two explanations for that:

- Businesses have a tendency not to fully calculate all related costs to workloads that are hosted on on-premises systems. Very often, things such as power, cooling, but also labor (especially involved in changes) are not taken into consideration.
- Businesses use the public cloud like they use on-premises systems, but without the functionality that clouds offer in terms of flexibility. Not all workloads need to be operational 24/7/365. Cloud systems offer a true pay-as-you-go model, where businesses only pay for the actual resources they consume—even all the way down to the level of CPUs, memory, storage, and network bandwidth.

Cost management is of the utmost importance. You want to be in full control of whatever your business is consuming in the cloud. But that goes hand in hand with IAM. If a worker has full access to platforms and, for instance, enrolls a heavy storage environment for huge data platforms such as data lakes, the business will definitely get a bill for that. There are two possible problems here:

- First of all, was that worker authorized to perform that action?
- Second, has the business allocated a budget for that environment?

Let's assume that the worker was authorized, and that budget is available. Then we need to be able to track and trace the actual consumption and, if required, be able to put a chargeback in place for a certain division that uses the environment. Hence, we need to be able to identify the resources. That's where naming and tagging in resources come into play. We have to make sure that all resources can uniquely be identified by name. Tags are used to provide information about a specific resource, for example, who owns the resource and hence will have to pay to use that resource.

Chapter 12, Cost and Value Modeling in Multi-Cloud Propositions, provides best practices on how to implement FinOps.

Monitoring

We need visibility on and in our platform. What happens and for what reason? Is our environment still healthy in terms of performance and security? Cloud providers offer native monitoring platforms: Azure Monitor, AWS CloudWatch, Google Cloud Monitoring (previously known as Stackdriver), and CloudMonitor in Alibaba Cloud. The basic functionality is comparable all in all: monitoring agents collect data (logs) and metrics, and then store these in a log analytics space where they can be explored using alert policies and visualization with dashboards. All these monitoring suites can operate within the cloud itself, but also have APIs to major **IT Service Management** (**ITSM**) systems such as ServiceNow, where the latter will then provide a single pane of glass over different platforms in the IT environment.

There are quite a number of alternatives to the native toolsets and suites. Good examples of such tools are Splunk and Datadog. Some of these suites operate in the domain of **AIOps**, a rather new phenomenon. AIOps does a lot more than just execute monitoring: it implies intelligent analyses conducted on metrics and logs. In *Chapter 19, Introducing AIOps and GreenOps*, we will cover AIOps, since it is expected that this will become increasingly popular in multi-cloud environments.

Automation

We have already talked about costs. By far the most expensive factor in IT is costs related to labor. For a lot of companies, this was the driver to start offshoring labor when they outsourced their IT. But we have already noticed that IT has become a core business again for these companies. For example, a bank is an IT company these days. They almost completely outsourced their IT in the late nineties and at the beginning of the new millennium, but we now see that their IT functions are back in-house. However, there is still the driver to keep costs down as much as possible. With concepts such as Infrastructure as Code and Configuration as Code, repositories to store that code, and having the deployment of code fully automated using DevOps pipelines, these companies try to automate to the max. By doing this, the amount of labor is kept to a minimum.

Costs are, however, not the only reason to automate, although it must be said that this is a bit of an ambiguous argument. Automation tends to be less fault tolerant than human labor (*robots do it better*). Most large outages are caused by human error that could have been avoided by automation. However, automation only works when it's tested thoroughly. And you still need good developers to write the initial code and the automation scripts. True, the cloud does offer a lot of good automation tools to make the lives of businesses easier, but we will come to realize that automation is not only about tools and technology.

It's also about processes; methodologies such as **Site Reliability Engineering** (**SRE**)—invented by Google—are gaining a lot of ground.

These pillars are described in almost all frameworks. In the following sections, we will study these pillars as we go along the various stages involved in cloud adoption.

Understanding cloud adoption

You may have come across one other term: the **cloud landing zone**. The landing zone is the foundation environment where workloads eventually will be hosted. Picture it all like a house. We have a foundation with a number of pillars. On top of that, we have a house with a front door (be aware that this house should not have a back door), a hallway, and a number of rooms. These rooms will be empty: no decoration, no furniture. That all has yet to be designed and implemented, where we have some huge shops (portals) from where we can choose all sorts of solutions to get our rooms exactly how we want them. The last thing to do is to actually move the residents into the house. And indeed, these residents will likely move from room to room. Remember: without scaffolding, it's hard to build a house in the first place.

This is what cloud adoption frameworks are all about: how to adopt cloud technology.

The respective cloud adoption frameworks can be found as follows:

- The Azure Cloud Adoption Framework: `https://azure.microsoft.com/en-us/cloud-adoption-framework/#cloud-adoption-journey`
- AWS: `https://aws.amazon.com/professional-services/CAF/`
- GCP: `https://cloud.google.com/adoption-framework/`

Often, this is referred to as a journey. Adoption is defined by a number of stages, regardless of the target cloud platform. Governing and managing are cross-processes supporting those stages.

The following diagram shows the subsequent stages of cloud adoption:

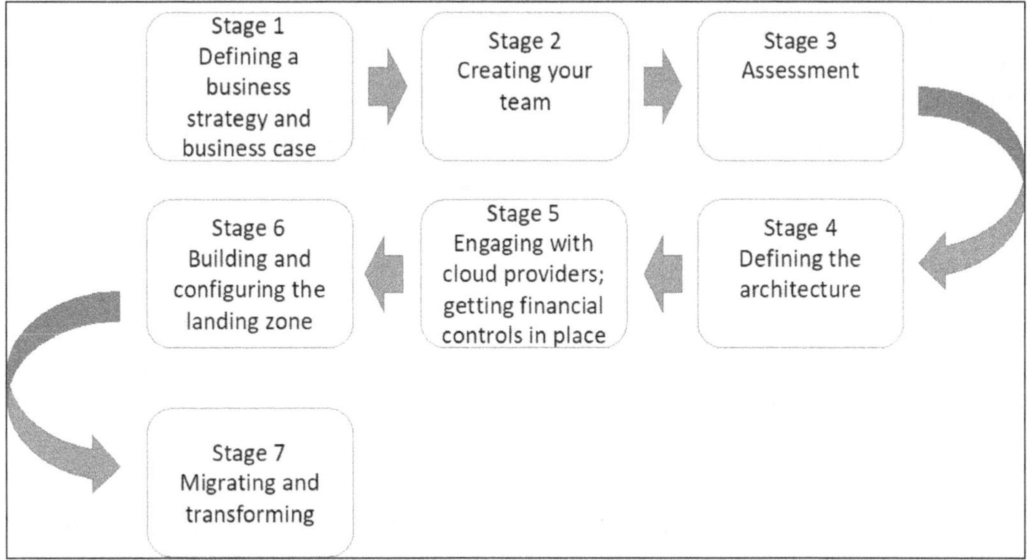

Figure 4.1: The seven steps in cloud adoption

Let's discuss each stage in detail in the following sections.

Stage 1—Defining a business strategy and business case

In *Chapter 2, Business Acceleration Using a Multi-Cloud Strategy*, we looked extensively at the business strategy and how to create a business case. A business needs to have clear goals and know where cloud offerings could add value. We have also looked at technical strategies such as rehost, replatform, and rebuild. Rebuild might not always be the cheapest solution when it comes to **total cost of ownership** (**TCO**). That's because businesses tend to forget that it takes quite some effort to rearchitect and rebuild applications to cloud-native environments. This architecture and build costs should be taken into consideration. However, the native environment will undoubtedly bring business benefits in terms of flexibility and agility. In short, this first stage is crucial in the entire adoption cycle.

Stage 2—Creating your team

Let's break this one immediately: there's no such thing as a T-shaped professional, someone who can do everything in the cloud, from coding applications to configuring infrastructure. The wider the T, the less deep the stroke under the T gets. In other words, if you have professionals who are generic and basically have a knowledge of all cloud platforms, they will likely not have a deep understanding of one specific cloud technology. But also, a software developer is not a network engineer, and vice versa.

Of course, someone who's trained in architecting and building cloud environments in Azure or AWS will probably know how to deploy workloads and a database and have a basic understanding of connectivity. But what about firewall rules? Or specific routing? Put another way: a firewall specialist might not be very skilled at coding in Python. If you have these types of people on your team, congrats. Pay them well, so they won't leave. But you will probably have a team with mixed skills. You will need developers and staff that are proficiently trained in designing and configuring infrastructure, even in the cloud.

Some adoption frameworks refer to this as the cloud center of excellence or—as AWS calls it—the Cloud Adoption Office, which is where the team with all the required skills is brought together. Forming this center of excellence is an important step in adopting cloud technology.

Stage 3—Assessment

The assessment phase is a vital step in ensuring proper migration to a target cloud environment. Before we start migrating or rebuilding applications in our landing zone, we need to know what we have. This is also important for the landing zone that we are planning to build. Look at it this way: you need to know what the house will look like before you start laying the foundation of that house. You need a blueprint.

First and foremost, we need to assess our business strategy. Where do we want to go with our business and what does it take to get there? The next question would be: is our current IT state ready to deliver that strategy? Which applications do we have and what business function do they serve? Are applications and the underlying infrastructure up to date or are they (near) the end of service, or even the end of life? What support contracts do we have and do we need to extend these during our transformation to the cloud, or can we retire these contracts?

You get the point: a proper assessment takes some time, but don't skip it. Our end goal should be clear; at the end of the day, all businesses want to become digital companies steered by data-driven decisions. We need an environment in which we can disclose data in a secure way and make the environment as flexible and scalable as possible so it can breathe at business speed. The problem is our traditional IT environment, which has been built out over many years. If we're lucky, everything will be well documented, but the hard reality is that documentation is rarely up to date. If we start our transformation without a proper assessment, we are bound to start pulling bricks out of an old wall without knowing how stable that wall really is. Again, here too, we require scaffolds.

Stage 4—Defining the architecture

This is the stage where you will define the **landing zone**—the foundation platform where the workloads will be hosted, based on the outcomes of the assessment. It starts with connectivity so that administrators and developers can reach the workloads in the cloud. Typically, connections will terminate in a transit zone or hub, the central place where inbound and outbound traffic is regulated and filtered by means of firewalls, proxies, and gateways.

Most administrators will access systems through APIs or consoles, but in some cases, it might be recommended to use a jump server, a stepping stone, or a bastion server, the server that forms the entrance to the environment before they can access any other servers. Typically, third parties such as system integrators use this type of server. In short, this transit zone or hub is crucial to the architecture.

The next thing is to define the architecture according to your business strategy. That defines how your cloud environment is set up. Does your business have divisions or product lines? It might be worthwhile having the cloud environment corresponding to the business layout, for instance, by using different subscriptions or the **Virtual Private Cloud** (**VPC**) per division or product line.

There will be applications that are more generically used throughout the entire business. Office applications are usually a good example. Will these applications be hosted in one separate subscription? And what about access for administrators? Does each of the divisions have its own admins controlling the workloads? Has the business adopted an agile way of working, or is there one centralized IT department that handles all of the infrastructure? Who's in charge of the security policies? These policies might differ by division or even workload groups. These security policies might also not be cloud friendly or applicable in a cloud-native world. They might need to be updated based on feedback from your cloud **Subject Matter Experts** (**SMEs**), the specialists.

One major distinction that we can already make is the difference between systems of record and systems of engagement, both terms first used by Microsoft. Systems of record are typically backend systems, holding the data. Systems of engagement are the frontend systems used to access data, work with the data, and communicate said data. We often find this setup reflected in the tiering of environments, where tier 1 is the access layer, tier 2 is the worker (middleware), and tier 3 is the database layer. A common rule in architecture is that the database should be close to the application accessing the database. In the cloud, this might work out differently, since we will probably work with **Platform as a Service** (**PaaS**) as a database engine.

These are the types of questions that are addressed in the cloud adoption frameworks. They are all very relevant questions that require answers before we start. And it's all architecture. It's about mapping the business to a comprehensive cloud blueprint. *Chapter 5, Managing the Enterprise Cloud Architecture*, of this book is all about architecture.

Stage 5—Engaging with cloud providers; getting financial controls in place

In this stage, we will decide on which cloud platform we will build the environment and what type of solutions we will be using: **Infrastructure as a Service** (**IaaS**), PaaS, **Software as a Service** (**SaaS**), containers, or serverless. This solution should be derived from the architecture that we have defined in stage 3. We will have to make some make-or-buy decisions: can we use native solutions, off the shelf, or do we need to develop something customized to the business requirements?

During this stage, we will also have to define business cases that automatically come with make-or-buy analyses. For instance, if we plan to deploy VMs on IaaS, we will have to think of the life cycle of that VM. In the case of a VM that is foreseen to live longer than, let's say for the sake of argument, 1 year, it will be far more cost efficient to host it on reserved instances as opposed to using the pay-as-you-go deployment model. Cloud providers offer quite some discounts on reserved instances, for a good reason: reserved instances mean a long-term commitment and, hence, a guaranteed income. But be aware: it's a long-term commitment. Breaking that commitment comes at a cost. Do involve your financial controller to have it worked out properly.

Development environments will generally only exist for a shorter time. Still, cloud providers do want businesses to develop as much as possible on their cloud platforms and offer special licenses for developers that can be really interesting. At the end of the day, cloud providers are only interested in one thing: the consumption of their platforms. There are a lot of programs that offer all sorts of guidance, tools, and migration methods to get workloads to these platforms.

Stage 6—Building and configuring the landing zone

There are a number of options to actually start building the landing zone. Just to ensure we understand: the landing zone is the foundation platform, typically the transit zone or hub and where the basic configuration of VNets, VPCs, or projects is carried out. We aim to have it automated as much as we can, right from the start. Hence, we will work according to the Infrastructure as Code and Configuration as Code, since we can only automate when components are code-based.

However, there are, of course, other ways to start your build, for instance, using the portals of the respective cloud providers.

If you are building a small, rather simple environment with just a few workloads, then the portal is the perfect way to go. But assuming that we are working with an enterprise, the portal is not a good idea to build your cloud environment.

It's absolutely fine to start exploring the cloud platform, but as the enterprise moves along and the environments grow, we need a more flexible way of managing and automating workloads. As has already been said, we want to automate as much as we can. How do we do that? By coding our foundation infrastructure and defining that as our master code. That master code is stored in a repository. Now, from that repository, we can fork the code if we need to deploy infrastructure components. It is very likely that every now and then, we have to change the code due to certain business requirements. That's fine, as long as we merge the changed, approved code back into the master repository. By working in this way, we have deployed an infrastructure pipeline, shown as follows:

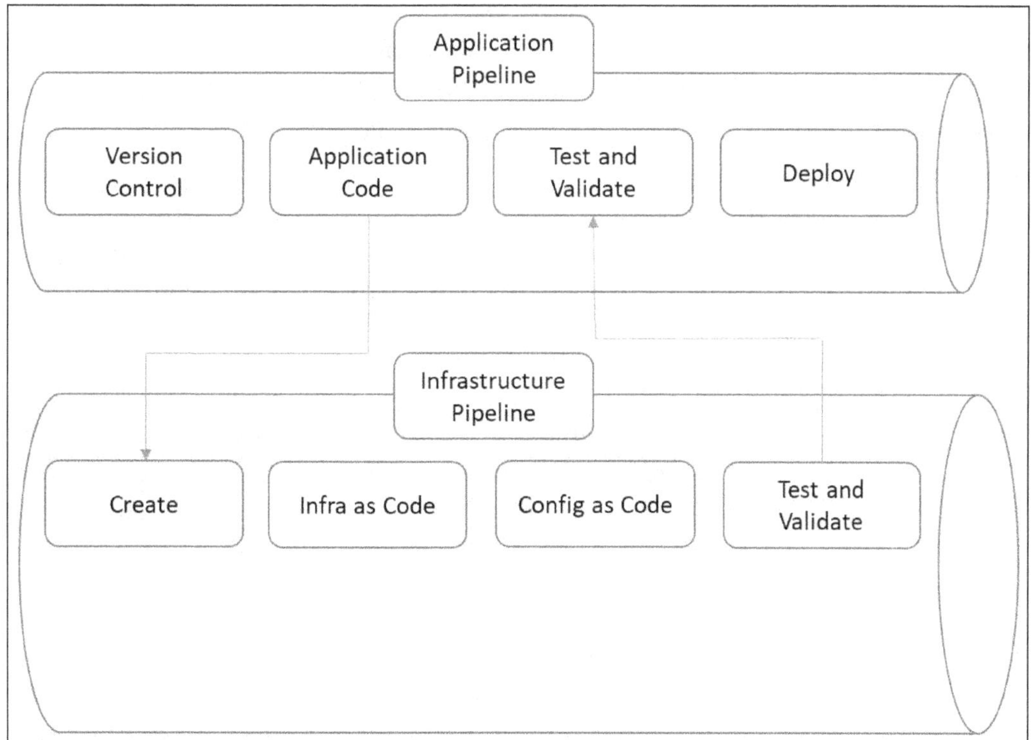

Figure 4.2: Basic pipeline structure

In a multi-cloud environment, the biggest challenge would be having a consistent repository and pipeline that could expand over multiple cloud platforms. After all, although the base concepts of AWS, Azure, and GCP are more or less the same, they do differ in terms of applied technology.

In this section, we are mainly talking about IaaS components. A popular tool in this area is Terraform, an open-source software by HashiCorp that is specifically designed to deploy data center infrastructure. Terraform abstracts the code of these clouds into cloud-agnostic code, based on HCL or JSON. **HashiCorp Configuration Language (HCL)** is the native language of Terraform, while **JavaScript Object Notation (JSON)** is more commonly known as a markup language.

If you search for alternatives, you may find tools such as Ansible, Chef, Puppet, or SaltStack. However, these are configuration tools and work quite differently from a provisioning tool such as Terraform.

Stage 7—Migrating and transforming

Now we have our landing zone and we're ready to start deploying services and workloads into our cloud environments. This is what we call **cloud transformation**. This is the stage where we will implement the technical strategies that we discussed in *Chapter 2, Business Acceleration Using a Multi-Cloud Strategy*, such as rehost, replatform, rebuild, retain, and retire. Following the assessment, we have defined for each application or application group what the business focus is and what the technical strategy is. In this final stage, we will shift our applications to the new platform, apply new cloud services, or rebuild applications native to the cloud. Obviously, we will do that in a DevOps way of working, where we do all the building and configuration from code, as we discussed in stage 5.

DevOps typically comprises the agile concepts of epics and features. Don't overcomplicate things. The epic could be *implementing new cloud architecture* and the transformation of an application or application group can be the feature. With the knowledge that we acquired from the assessment phase, combined with the technical strategy, the DevOps team should be able to do a good refinement, breaking down the feature into tasks that can be executed and completed in a limited number of sprints.

There are two really important items in a transformation stage: going live and the exit strategy. Before going live, testing is required. The need for good testing is really underestimated: it must be part of the DevOps CI/CD pipeline. It is very strongly advised to run the full test cycle: unit, integration, and end user test. From a process point of view, no company should allow anything to go live before all test findings have proven to be addressed. The same applies to the exit strategy: no company should allow anything to go live without a clearly defined exit strategy, that is, how to roll back or move environments out of the cloud again, back to the original state.

It's one more reason to consider rebuilding environments, so that there's always the old environment as a fallback if things turn out not to be working as designed. Of course, testing should prevent things from going wrong, but we have to be realistic too: something will break.

Translating business KPIs into cloud SLAs

Frankly, infrastructure in the cloud should be a black box for a business. Infrastructure is like turning on a water tap. Some IT companies referred to operating cloud infrastructure as liquid or fluid IT for that reason: it was simply there, all the time. As a consequence, the focus on SLAs shifted to the business itself. Also, that is part of cloud adoption. As enterprises are moving ahead in the adoption process, a lot of businesses are also adopting a different way of working. If we can have flexible, agile infrastructure in the cloud, we can also speed up the development of environments and applications. Still, also in the cloud, we have to carefully consider service-level objectives and KPIs.

Let's have a look at the cloud SLA. What are topics that would have to be covered in an **SLA**? The *A* stands for *agreement* and, from a legal perspective, it would be a contract. Therefore, an SLA typically has the format and the contents that belong to a contract. There will be definitions for contract duration, start date, legal entities of the parties entering into the agreement, and service hours. More important are the agreements on the KPIs. What are we expecting from the cloud services that we use and are paying for? And who do we contact if something's not right or we need support for a certain service? Lastly, what is the exact scope of the service?

These are all quite normal topics in any IT service contract. However, it doesn't work the same way when we're using public cloud services. The problem with the SLA is that it's usually right-sized per business. The business contracts IT services, where the provider tailors the services to the needs of that business. Of course, a lot of IT providers standardize and automate as much as they can to obtain maximum efficiency in their services, but nonetheless, there's room for tweaking and tuning. In the public cloud, that room is absolutely limited. A company will have a lot of services to choose from to tailor to its requirements, but the individual services are as they are. Typically, cloud providers offer an SLA per service. Negotiating the SLA per service in a multi-cloud environment is virtually impossible.

It's all part of the service design: the business will have to decide what components they need to cater to for their needs and assess whether these components are fit for purpose. The components—the services themselves—can't be changed. In IaaS, there will be some freedom, but when we're purchasing PaaS and SaaS solutions, the services will come out of the box.

The business will need to make sure that an SaaS solution really has the required functionality and delivers at the required service levels.

Common KPIs in IT are availability, durability, **Recovery Time Objective** (**RTO**), and **Recovery Point Objective** (**RPO**), just to name a few important ones. How do these KPIs work out in the major public clouds?

Defining availability

Availability is defined as the time when a certain system can actually be used.

Availability should be measured from end to end. What do we mean by this? Well, a VM with an operating system can be up and functioning alright, but if there's a failure in one software component running on top of that VM and the operating system, the application will be unusable, and hence unavailable to the user. The VM and operating system are up (and available), but the application is down. This means that the whole system is unavailable, and this has some serious consequences. It also means that if we want to obtain an overall availability of 99.9 percent of our application, this means that the platform can't have an availability below that 99.9 percent. And then nothing should go wrong.

In traditional data centers, we needed to implement specific solutions to guarantee availability. We would have to make sure that the network, compute layers, and storage systems can provide for the required availability. Azure, AWS, and GCP largely take care of that. It's not that you get an overall availability guarantee on these platforms, but these hyperscalers—Azure, AWS, and Google Cloud—do offer service levels on each component in their offerings.

By way of an example, a single-instance VM in Azure has a guaranteed connectivity of 99.9 percent. Mind the wording here: the connectivity to the VM is guaranteed. Besides, you will have to use premium storage for all disks attached to the VM. You can increase the availability of systems by adding availability, zones, and regions in Azure. Zones are separate data centers in an Azure region. At the time of writing, Azure has 60+ regions worldwide. In summary, you can make sure that a system is always online somewhere around the globe in Azure, but it takes some effort to implement such a solution using load balancing and Traffic Manager over the Azure backbone. That's all a matter of the following:

- Business requirements (is a system critical and does it need high availability?)
- The derived technical design
- The business case, since the high-availability solution will cost more money than a single-ended VM

Comparing service levels between providers

It's not an easy task to compare the service levels between providers. As with a lot of services, the basic concepts are all more or less the same, but there are differences. Just look at the way Azure and AWS detail their service levels on compute. In Azure, these are VMs. In AWS, the service is called **EC2—Elastic Compute Cloud** in full. Both providers work with service credits if the monthly guaranteed uptime of instances is not met and, just to be clear, system infrastructure (the machine itself!) is not available. If uptime drops below 99.99 percent over a month, then the customer receives a service credit of 10 percent over the monthly billing cycle. Google calculates credits if the availability drops below 99.5 percent.

Again, requirements for service levels should come from the business. For each business function and corresponding system, the requirements should be crystal clear. That ultimately drives the architecture and system designs. The cloud platforms offer a wide variety of services to compose that design, making sure that requirements are met. To put it in slightly stronger terms, the hyperscalers offer the possibility to have systems that are ultimately resilient. While in traditional data centers, **disaster recovery** (**DR**) and business continuity meant that a company had to have a second data center as a minimum, cloud platforms offer this as a service.

Azure, AWS, and GCP are globally available platforms, meaning that you can actually have systems available across the globe without having to make enormous investments. The data centers are there, ready to use. Cloud providers have native solutions to execute backups and store these in vaults in different regions, or they offer third-party solutions from their portals so that you can still use your preferred product.

However, it should be stressed once again that the business should define what their critical services and systems are, defining the terms for recovery time and recovery points. The business should define the DR metrics and, even more importantly, the processes when a DR plan needs to be executed. Next, it's up to IT to fulfill these requirements with technical solutions. Will we be using a warm standby system, fully mirrored from the production systems in a primary region? Are we using a secondary region and what region should that be then? Here, compliancy and public regulations such as the **General Data Protection Regulation** (**GDPR**) or data protection frameworks in other parts of the world also play an important role. Or will we have a selection of systems in a second region?

One option might be to deploy acceptance systems in another region and leverage these to production in case of a failover in DR. That implies that acceptance systems are similar to production systems.

How often do we have to back up the systems? A full backup once a week? Will we be using incremental backups and, if so, how often? What should be the retention time of backup data? What about archiving? It's all relatively easy to deploy in cloud platforms, but there's one reason to be extremely careful in implementing every available solution. It's not about the upfront investment, as with traditional data centers (the real cash out for investment in capital expenditure, or **CAPEX**), but you will be charged for these services (the operational expenditure, or **OPEX**) every month.

In short, the cloud needs a plan. That's what we will explore in the next sections, eventually in creating a service design.

Using cloud adoption frameworks to align between cloud providers

The magic word in multi-cloud is a single pane of glass. What do we mean by that? Imagine that you have a multi-cloud environment that comprises a private cloud running VMware and a public cloud platform in AWS, and you're also using SaaS solutions from other providers. How would you keep track of everything that happens in all these components? Cloud providers might take care of a lot of things, so you need not worry about, for example, patches and upgrades. In SaaS solutions, the provider really takes care of the full stack, from the physical host all the way up to the operating systems and the software itself. However, there will always be things that you, as a company, will remain responsible for. Think of matters such as IAM and security policies. Who has access to what and when?

This is the new reality of complexity: multi-cloud environments consisting of various solutions and platforms. How can we manage that? Administrators would have to log in to all these different environments. Likely, they will have different monitoring solutions and tools to manage the components. That takes a lot of effort and, first and foremost, a lot of different skills. It surely isn't very efficient. We want to have a single pane of glass: one ring to rule them all.

Let's look at the definition of a single pane of glass first. According to TechTarget (`https://searchconvergedinfrastructure.techtarget.com/definition/single-pane-of-glass`), it's

> *a management console that presents data from multiple sources in a unified display. The glass, in this case, is a computer monitor or mobile device screen.*

The problem with that definition is that it's purely a definition from a technological perspective: just one system that has a view of all the different technology components in our IT landscape. However, a single pane of glass goes way beyond the single monitoring system. It's also about unified processes and even unified automation. Why is the latter so important? If we don't have a unified way of automation, we'll still be faced with a lot of work in automating the deployment and management of resources over different components in that IT landscape. So, a single pane of glass is more an approach that can be visualized in a triangle:

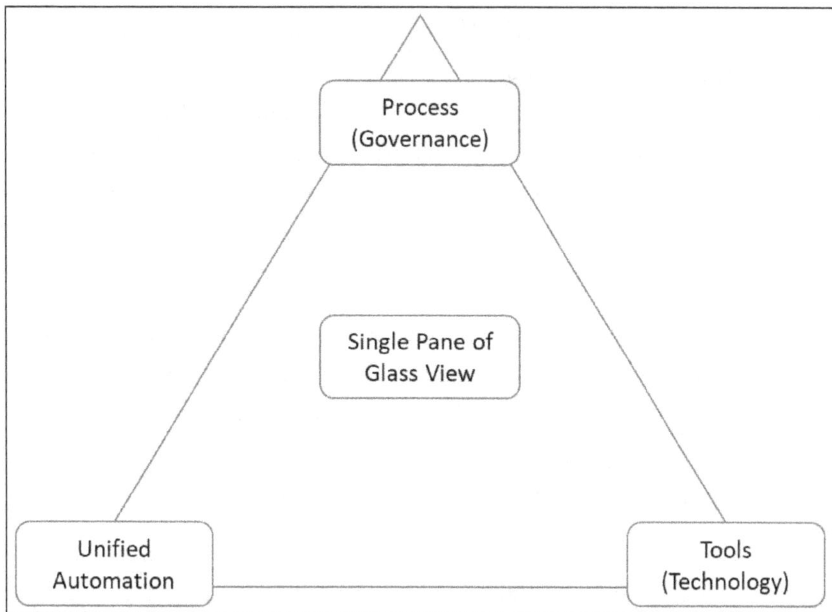

Figure 4.3: Graphic representation of a single pane of glass

Do such systems exist that embrace that approach? Yes, there are full IT service management systems such as BMC Helix Multi-Cloud Management and ServiceNow with the Now platform. There are certainly more alternatives, but these are considered to be the market leaders, according to Gartner's Magical Quadrant for ITSM systems.

ITSM, that's what we are talking about. The tool—the technology—is one thing, but the processes are just as important. IT service management processes include, as a minimum, the following processes:

- **Incident management**: Tracking and resolving incidents in the platform itself or resources hosted on the platform
- **Problem management**: Tracking and resolving recurring incidents

- **Change management**: Tracking and controlled implementation of changes to the platform or resources hosted on the platform
- **Configuration management**: Tracking the state of the platform and the resources hosted on the platform

The cornerstone of ITSM is knowledge: you need an indisputable insight into what your platform looks like, how it is configured, what types of assets live on your platform, and which assets and resources are deployed on the platform. Assets and resources are often referred to as configuration items and they are all collected and tracked in a **Configuration Management Database** (**CMDB**) or master repository (**Master Data Records**, **MDRs**). Here, the challenge in the cloud really starts. With scalable, flexible resources and even resources that might only live for a very brief period of time in our estate, such as containers or serverless functions, we are faced with the risk that our CMDB or asset repository will never be as accurate as we would like it to be, although the leading ITSM systems have native APIs to the monitoring tools in the different clouds that are truly responsive and collect asset data in (near) real time.

The agility of the cloud makes change management probably the most important process in multi-cloud environments. Pipelines with infrastructure and configuration as code help. If we have a clear process of how the code is forked, tested, validated, and merged back into the master branch, then changes are fully retrievable. If a developer skips one step in the process, we are heading for failure and worse: without knowing what went wrong.

All cloud adoption frameworks stress the importance of governance and processes on top of the technology that clouds provide. All frameworks approach governance from the business risk profiles. That makes sense: if we're not in agreement about how we do things in IT, at the end of the day, the business is at risk. Basically, the ultimate goal of service management is to reduce business risks that emanate from a lack of IT governance. IT governance and ITSM are a common language among technology providers, for a very good reason.

Back to our *one ring to rule them all*. We have unified processes, defined in ITSM. There are different frameworks for ITSM (ITIL or Cobit, for example), but they all share the same principles. Now, can we have one single dashboard to leverage ITSM, controlling the life cycle of all our assets and resources? We already mentioned BMC Helix and ServiceNow as technology tools. But can we also have our automation through these platforms? Put another way, can we have automation that is fully cross-platform? This is something that Gartner calls **hyperautomation**.

Today, automation is often executed per component or, in the best cases, per platform. By doing that, we're not reaching the final goal of automation, which is to reduce manual tasks that have to be executed over and over again. We're not reducing the human labor and, for that matter, we're not reducing the risk of human error. On the contrary, we are introducing more work and the risk of failure by having automation divided into different toolsets on top of different platforms, all with separate workflows, schedules, and scripts. Hyperautomation deals with that. It automates all business processes and integrates these in a single automated life cycle, managed from one automation platform.

In summary, cloud adoption frameworks from Azure, AWS, and GCP all support the same principles. That's because they share the common IT service management language. That helps us to align processes across platforms. The one challenge that we have is the single dashboard to control the various platforms and have one source of automation across the cloud platforms—hyperautomation. With the speed of innovation in the cloud, that is becoming increasingly complex, but we will see more tools and automation engines coming to market over the next years, including the rise and adoption of AIOps.

Understanding identities and roles in the cloud

Everything in the cloud has an identity. There are two things that we need to do with identities: authenticate and authorize. For authentication, we need an identity store. Most enterprises will use **Active Directory** (**AD**) for that, where AD becomes the central place to store the identities of persons and computers. We won't be drilling down into the technology, but there are a few things you should understand when working with AD. First of all, an AD works with domains. You can deploy resources—VMs or other virtual devices—in a cloud platform, but if that cloud platform is not part of your business domain, it won't be very useful. So, one of the key things is to get resources in your cloud platform domain-joined. For that, you will have to deploy domain services with domain controllers in your cloud platform or allow cloud resources access to the existing domain services. By doing that, we are extending the business to the cloud platform.

That sounds easier than it is in practice. Azure, AWS, and GCP are public clouds. Microsoft, Amazon, and Google are basically offering big chunks of their platforms to third parties: businesses that host workloads on a specific chunk. But they will still be on a platform that is owned and controlled by the respective cloud providers. The primary domain of the platform will be `onmicrosoft.com` or `aws.amazon.com`: this makes sense if you think of all the (public) services they offer on their platforms.

If we want our own domain on these platforms, we will need to ring-fence a specific chunk by attaching a registered domain name to the platform. Let's, for example, have a company with the name `myfavdogbiscuit.com`. On Azure, we can specify a domain with `myfavdogbiscuit.onmicrosoft.com`. Now we have our own domain on the Azure platform. The same applies obviously to AWS and GCP. Resources deployed in the cloud domains can now be domain-joined, if the domain on the cloud platform is connected to the business domain. That connection is provided by domain controllers. The following diagram shows the high-level concept of AD Federation:

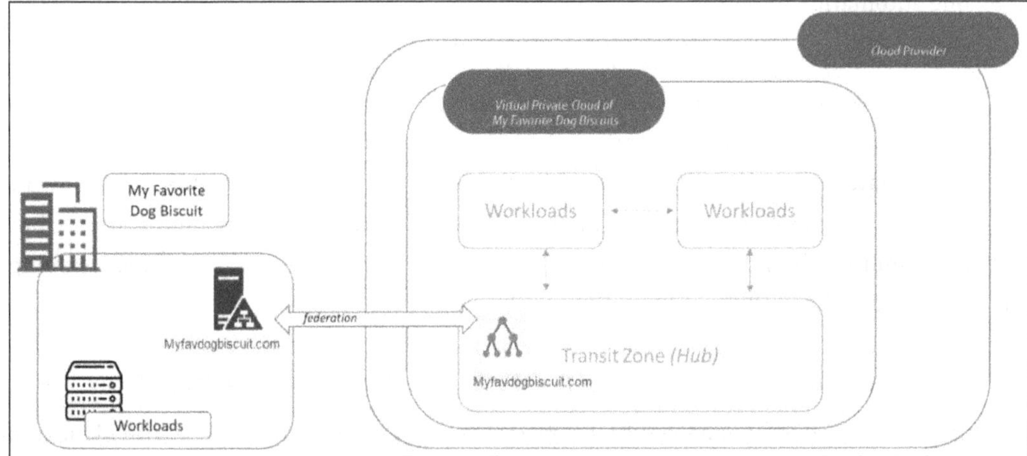

Figure 4.4: Active Directory Federation

In AD, we have all our resources and persons that are allowed inside our domain. Authentication is done through acknowledgment: an identity is recognized in the domain or rejected. This AD uses Kerberos to verify an identity. It's important to know that all cloud providers support AD, the underlying **Lightweight Directory Access Protocol (LDAP)** standard, and Kerberos.

If a resource or person can't be identified in the directory, it simply won't get access to that domain, unless we explicitly grant them access. That can be the case when a person doesn't have an account in our domain, but needs to have access to resources that are on our platform. We can grant this person access using a business-to-business connection. In Azure, that is conveniently called B2B where Azure AD is used for external identities. In AWS it's called Cognito, and in GCP, Cloud Identity.

We have identified a person or a resource using the directory, but now we have to make sure that this resource can only do what we want or allow it to do and nothing more. This is what we call authorization: we specify what a resource is allowed to do, when certain criteria are met. First, we really want to make sure that the resource is whatever it claims it is.

For people logging in, it is advised to use multi-factor authentication. For compute resources, we will have to work with another mechanism and typically that will be based on keys: a complex, unique hash that identifies the resource.

One more time: we have defined identities in our environment, either human personnel or system identities. How can we define what an identity is allowed in our environment? For that, we need **Role-Based Access Control** (**RBAC**). RBAC in Azure, IAM in AWS, and Cloud Identity in GCP let you manage access to (parts of) the environment and what identities can do in that environment. We can also group identities to which a specific RBAC policy applies.

We have already concluded that all cloud platforms support AD and the underlying protocols. So, we can federate our AD with domains in these different clouds. Within Azure, the obvious route to do so would be through connecting AD to Azure AD. Although it might seem like these are similar solutions, they are totally different things. AD is a real directory, whereas Azure AD is a native identity provider within Azure, able to authenticate cloud-native identities. It does not have the authentication mechanisms that AD has, such as Kerberos. Azure AD will authenticate using AD. And with that, it's quite similar to the way the other platforms federate with AD. In both AWS and GCP, you will need identities that can be federated against AD. In other words, your AD will always remain the single source of truth for identity management, the one and only identity store.

Creating the service design and governance model

The final thing to do is to combine all the previous sections into a service design and governance model for multi-cloud environments. So, what should the contents be of a service design? Just look at everything we have discussed so far. We need a design that covers all the topics: requirements, identities and access management, governance, costs, and security. Let's discuss these in detail.

Requirements

This includes the service target that will comprise a number of components. Assuming that we are deploying environments in the public cloud, we should include the public cloud platform as such as a service target. The SLA for Microsoft Online Services describes the SLAs and KPIs committed to by Microsoft for the services delivered on Azure. These are published on `https://azure.microsoft.com/en-us/support/legal/sla/`. For AWS, the SLA documentation can be found at `https://aws.amazon.com/legal/service-level-agreements/`. Google published the SLAs for all cloud services on GCP at `https://cloud.google.com/terms/sla/`. These SLAs will cover the services that are provided by the respective cloud platforms; they do not cater to services that a business builds on top of these cloud-native services.

By way of an example, if a business builds a tiered application with frontends, worker roles, and databases, and defines that as a service to a group of end users, this service needs to be documented separately as a service target.

Next, we will list the relevant requirements that have to be addressed by the service target:

- **Continuity requirements**: This will certainly be relevant for business-critical services. Typically, these are addressed in a separate section that describes RTO/RPO, backup strategies, business continuity plans, and DR measures.
- **Compliance requirements**: You will need to list the compliance frameworks that the company is constrained by. These can be frameworks related to privacy, such as the EU GDPR, but also security standards such as ISO 27001. Keep in mind that Microsoft, AWS, and Google are global companies but are based in the US. In some industry sectors outside the US (this applies typically to EU countries), working with US-based providers is allowed only under strict controls. The same applies to agreements with Chinese providers such as Alibaba, one of the up-and-coming public clouds. Always consult a legal advisor before your company starts deploying services in public clouds or purchasing cloud services.
- **Architectural and interface requirements:** Enterprises will likely have an enterprise architecture, describing how the company produces goods or delivers services. The business architecture is, of course, a very important input for cloud deployment. It will also contain a listing of various interfaces that the business has, for example, with suppliers of components or third-party services. This will include interfaces within the entire production or delivery chain of a company—suppliers, HR, logistics, and financial reporting.
- **Operational requirements**: This section has to cover life cycle policies and maintenance windows. An important requirement that is set by the business is so-called **blackout periods**, wherein all changes to IT environments are halted. That may be the case, for example, at year-end closing or in the case of expected peaks in production cycles. The life cycle includes all policies related to upgrades, updates, patches, and fixes to components in the IT environment.
- **As with the continuity requirements, this is all strongly dependent on the underlying cloud platform**: Cloud providers offer a variety of tools, settings, and policies that can be applied to the infrastructure to prevent the downtime of components. Load balancing, backbone services, and planning components over different stacks, zones (data centers), and even regions are all possible countermeasures to prevent environments from going offline for any reason, be it planned or unplanned. Of course, all these services do cost money, so a business has to define which environments are business-critical so as to set the right level of component protection and related operational requirements.

- **Security and access requirements:** As stated previously, cloud platforms all offer highly sophisticated security tools to protect resources that are deployed on their platforms, yet security policies and related requirements should really be defined by the business using the cloud environments. That all starts with who may access what, when, and why. A suitable RBAC model must be implemented for admin accounts.

Next, we will look at the **Risks, Assumptions, Issues, and Dependencies** (**RAID**) and the service decomposition.

RAID

A service design and governance model should contain a RAID log. This RAID log should be maintained so that it always represents the accurate status, providing input to adjust and adapt principles, policies, and the applied business and technical architecture.

Service decomposition

The next part is service decomposition, in other words, the product breakdown of the services. What will we be using in our cloud environment?

- **Data components:** What data is stored, where, and in what format, using which cloud technology? Think of SQL and NoSQL databases, data lakes, files, and queues, but also in terms of the respective cloud storage solutions, such as Blob Storage in Azure, S3, Glacier, or Google Cloud Storage.
- **Application components:** Which applications will be supported in the environment, and how are these built and configured? This defines which services we need to onboard and makes sure there's a clear definition between at least business-critical systems and systems that are not critical. A good method is to have systems categorized, for example, into gold, silver, and bronze, with gold denoting business-critical systems, silver denoting other important production and test systems, and bronze for development systems.
    - However, be careful in categorizing systems. A development system can be critical in terms of finances. Just imagine having a hundred engineers working on a specific system under time pressure to deliver and the development systems become unavailable. This will surely lead to delays and a hundred engineers sitting idle, thereby costing a lot of money. We cannot stress enough how important business cases are.

- **Infrastructure components:** VMs, load balancers, network appliances, firewalls, databases, storage, and so on. Remember that these will all configure items in a CMDB or MDR.
- **Cloud-native components**: Think of PaaS services, containers, and serverless functions. Also, list how cloud components are managed: through portals, CLIs, or code interfaces such as PowerShell or Terraform.
- **Security components:** Security should be intrinsic on all layers. Data needs to be protected, both in transit and at rest, applications need to be protected from unauthorized access, infrastructure should be hardened, and monitoring and alerting should be enabled on all resources.

Typically, backup and restoration are also part of the security components. Backup and restore are elements related to protecting the IT estate, but ultimately protecting the business by preventing the loss of data and systems. Subsequently, for business-critical functions, applications, and systems, a **business continuity and disaster recovery** (**BCDR**) plan should be identified. From the BCDR plan, requirements in terms of RPO/RTO and retention times for backups are derived as input for the architecture and design for these critical systems.

As you can tell, we follow the **The Open Group Architecture Framework** (**TOGAF**) **Architecture Development Method** (**ADM**) cycle all the way: business requirements, data, applications, and the most recent technology. Security is part of all layers. Security policies are described in a separate section.

Roles and responsibilities

This section in the service design defines who's doing what in the different service components, defining roles and what tasks identities can perform when they have specific roles. Two models are detailed in the following sections.

Governance model

This is the model that defines the entities that are involved in defining the requirements for the service, designing the service, delivering the service, and controlling the service. It also includes lines of report and escalation. Typically, the first line in the model is the business setting the requirements, the next line is the layer comprising the enterprise architecture, responsible for bridging business and IT, and the third line is the IT delivery organization. All these lines are controlled by means of an audit as input for risk and change control:

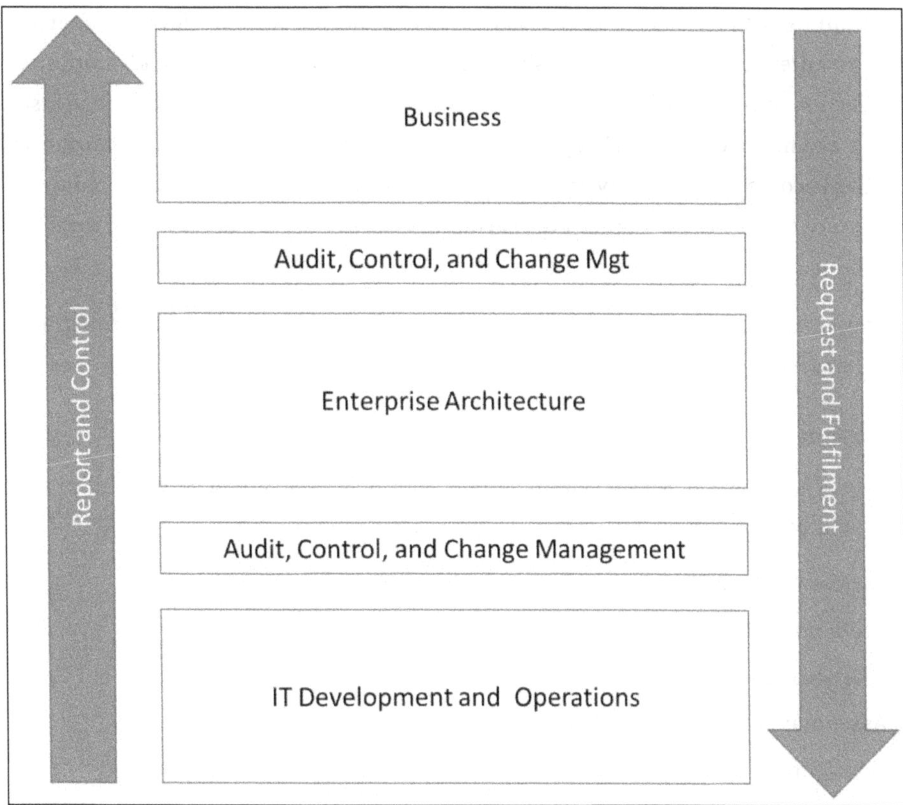

Figure 4.5: Governance model (high-level example)

Now, this is a very simple, straightforward presentation of a governance model. It can have many variations. For one, IT delivery is often done in Agile teams, working according to DevOps principles. The main principle stays intact, even with Agile teams. Even Agile teams need to be controlled and work under definitions of change management. Just one tip: don't overcomplicate models; keep it as simple as possible.

Support model

The support model describes who's delivering the support on the various components. Is that solely an internal IT operations division, or do we need to have support contracts from third parties? For that matter, an enterprise will very likely require support from the cloud provider. The support model also defines to what degree that support needs to be provided. In some cases, having the option to have an occasional talk with a subject matter expert from a provider might be sufficient, but when a company has outsourced its cloud operations, full support support is a must.

An important topic that needs to be covered in the support model is the level of expertise that a company requires in order to run its business in cloud platforms. As we already noticed in *Chapter 2, Business Acceleration Using a Multi-Cloud Strategy*, there's a strong tendency to insource IT operations again, since the way IT is perceived has dramatically changed. For almost every company, IT has become a core activity and not just a facilitator. Since more and more companies are deploying services in the cloud, they are also building their own cloud centers of excellence, but with the subject matter expertise and support of cloud providers.

If we follow all the stages so far as a full cycle, then we will end up with a service catalog, describing exactly what services are in scope, how they are delivered, and who's supporting these services.

Processes

This section defines what processes are in place and how these are managed. All processes involved should be listed and described:

- Incident management
- Problem management
- Change management
- Asset management and CMDB (or MDR)
- Configuration management
- Reporting
- Knowledge base

These are the standard processes as used in ITSM. For cloud environments, it's strongly advised to include the automation process as a separate topic. In the DevOps world that we live in, we should store all code and scripts in a central repository. The knowledge base can be stored in Wikipages in that repository. The DevOps pipeline with the controlled code base and knowledge wikis must be integrated into the ITSM process and the ITSM tooling. Remember: we want to have a single-pane-of-glass view of our entire environment.

There's one important process that we haven't touched on yet and that's the request and request fulfillment. The request starts with the business demand. The next step in the process is to analyze the demand and perform a business risk assessment, where the criticality of the business function is determined and mapped to the required data. Next, the request is mapped to the business policies and the security baseline. This sets the parameters on how IT components should be developed and deployed, taking into account the fact that RBAC is implemented, and resources are deployed according to the services as described in the service decomposition. That process can be defined as request fulfillment.

If different services are needed for specific resources, then that's a change to the service catalog.

We also need to look at cost management and consistent security postures across our cloud environments. These topics will be extensively discussed in separate chapters.

Summary

In this chapter, we've explored the main pillars of cloud adoption frameworks, and we learned that the different frameworks overlap quite a bit. We've identified the seven stages of cloud adoption up until the point where we can really start migrating and transforming applications to our cloud platforms. In multi-cloud environments, control and management are challenging. It calls for a single-pane-of-glass approach, but, as we have also learned, there are just a few tools—*the one ring to rule them all*—that would cater to this single pane of glass.

One of the most important things to understand is that you first have to look at identities in your environment: who, or what, if we talk about other resources on our platform, is allowed to do what, when, and why? That is key in setting out the governance model. The governance model is the foundation of the service design.

In the last section of this chapter, we've looked at the different sections of such a service design. Of course, it all starts with the architecture and that's what we'll be studying in the next chapter: creating and managing our architecture.

Questions

1. You are planning a migration of a business environment to the public cloud. Would an assessment be a crucial step in designing the target environment in that public cloud?
2. You are planning a cloud adoption program for your business. Would you consider cost management as part of the cloud adoption framework?
3. IAM plays an important role in moving to a cloud platform. What is the environment most commonly used as an identity directory in enterprise environments?

Further reading

Alongside the links that we mentioned in this chapter, check out the following books for more information on the topics that we have covered:

- *Mastering Identity and Access Management with Microsoft Azure*, by Jochen Nickel, published by Packt Publishing
- *Enterprise Cloud Security and Governance*, by Zeal Vora, published by Packt Publishing

5

Managing the Enterprise Cloud Architecture

In the previous chapters, we learned about different cloud technology strategies and started drafting a service model, including governance principles. Where do we go from here? From this point onward, you will be—as a business—managing your IT environments in multi-cloud. Successfully managing this new estate means that you will have to be very strict in maintaining the enterprise architecture. Hence, this chapter is all about maintaining and securing the multi-cloud architecture.

This chapter will introduce the methodology to create an enterprise architecture for multi-cloud using **The Open Group Architecture Framework** (**TOGAF**) and other methodologies, including **Continuous Architecture** and the recently published **Open Agile Architecture**. We will study how to define architecture principles for various domains such as security, data, and applications using quality attributes. We will also learn how we can plan and create the architecture in different stages. Lastly, we will discuss the need to validate the architecture and how we can arrange it.

In this chapter, we will cover the following topics:

- Defining architecture principles for multi-cloud
- Using quality attributes in architecture
- Creating the architecture artifacts
- Change management and validation as the cornerstone
- Validating the architecture

Defining architecture principles for multi-cloud

We'll start this chapter from the perspective of enterprise architecture using the **Architecture Development Method** (**ADM**) cycle in TOGAF as a guiding and broadly accepted framework for enterprise architecture. In *Chapter 2, Business Acceleration Using a Multi-Cloud Strategy*, we learned that the production cycle for architecture starts with the business, yet there are two steps before we actually get to defining the business architecture: we have a preliminary phase where we set out the framework and the architecture principles. These feed into the very first step in the actual cycle, known as architecture vision, as shown in the following diagram:

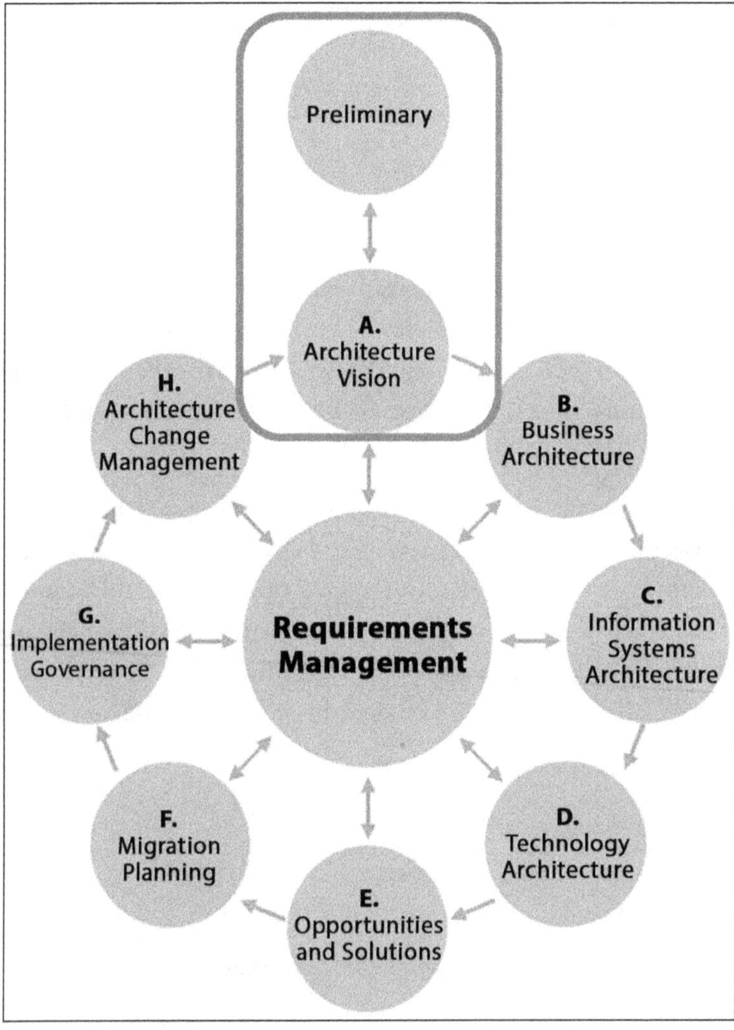

Figure 5.1: The preliminary phase and architecture vision in TOGAF's ADM cycle

The key to any preliminary phase is the architecture principles; that is, your guidelines for fulfilling the architecture. There can be many principles, so the first thing that we have to do is create principle groups that align with our business. The most important thing to remember is that principles should enable your business to achieve its goals. Just to be very clear on this aspect: going to the cloud is not a business goal, just like cloud-first is not a business strategy. These are technology statements at best, nothing more. But principles should do more: they have to support the architectural decisions being made and for that reason, they need to be durable and consistent.

When businesses decide that the cloud might be a good platform to host business functions and applications, the most used principles are flexibility, agility, and cost-efficiency. The latter is already highly ambiguous: what does cost-efficient mean? Typically, it means that the business expects that moving workloads to cloud platforms is cheaper than keeping them on-premises. This could be the case, but the opposite can also be true if incorrect decisions in the cloud have been made by following bad architectural decisions based on ambiguous principles. In short, every principle should be challenged:

- Does the principle support the business goals?
- Is the principle clear so that it can't be subject to multiple interpretations?
- Is the principle leading toward a clearly defined solution?

Some suggested groups for defining principles are as follows:

- Business
- Security and compliance
- Data principles
- Application principles
- Infrastructure and technology principles
- Usability
- Processes

Let's talk about each category in detail, but before we do so, we must learn by what values we are defining the architecture. These values are captured in quality attributes, to be discussed in the next section.

Using quality attributes in architecture

Quality attributes is a term that sprouts from a different framework, called **Continuous Architecture**. Cloud architectures tend to be fluid, meaning that they are very dynamic since they have to respond to changing customer demands fast and continuously. Cloud services are used to enable the business agility that we discussed in the previous chapter. Cloud services, therefore, need to be scalable, but still reliable. Lastly, enterprises do not want to be locked in on any platform, so environments must be able to communicate with other systems or even move to other platforms.

Architectures of systems might change constantly, due to fast developments and continuous new releases enabled by DevOps. Architects have a tedious task to accomplish this. Continuous architecture might be a good guide.

This framework defines quality attributes to which architecture must comply:

- **Operability**: This part of the architecture covers automation in the first place, but also monitoring and logging. In essence, operability is everything that is needed to keep systems operational in a secure state. This requires monitoring:
    - A key decision in monitoring is not what monitoring tool we will use, but what we have to monitor and to what extent. In multi-cloud, monitoring must be cross-platform since we will have to see what's going on in the full chain of components that we have deployed in our multi-cloud environment. This is often referred to as end-to-end monitoring: looking at systems from an end user perspective. This is not only related to the health status of systems, but also whether the systems do what they should do and are free from bugs, crashes, or unexpected stops.
    - Monitoring systems collect logs. We will also need to design where these logs will have to go and how long these will have to be stored. The latter is important if systems are under an audit regime. Auditors can request logs.
    - Monitoring is also maybe even more related to the performance of these systems. From an end user perspective, there's nothing more frustrating than systems that respond slowly.
- **Performance**: Where an architect can decide that a system that responds within 1 second is fast, the end user might have a completely different definition of fast. Even if they agree that the performance and responsiveness of the system is slow, the next question is how to determine what the cause of degrading performance is. Monitoring the environment from the end user's perspective, all the way down to the infrastructure, is often referred to as end-to-end.

There are monitoring environments that really do end-to-end, typically by sending transactions through the whole chain and measuring health (heartbeat, to check if a system is still alive and responding) and performance by determining how fast transactions are processed. Keep in mind that this type of monitoring usually deploys agents on various components in your environment. In that case, we will have to take into consideration how much overhead these agents create on systems, thus taking up extra resources such as CPU and memory. The footprint of agents should be as small as possible, also given the fact that there will undoubtedly be more agents or packages running on our systems. Just think of endpoint protection, such as virus scanning, as an example.

- **Configurability**: Companies use cloud technology to be able to respond quickly to changes, and to become agile. As a result, systems are not static in the cloud. Systems in the cloud must be easy to configure. This is what configurability means: a set of parameters that define how systems must behave. Configurations set the desired state of a resource. As an example, we can use a virtual machine: developers can pick a VM from a portal of a cloud provider. This will be a default VM with default configurations. These configurations might not match the desired state for which the company has defined guidelines and guardrails. In that case, the specific desired configuration must be added separately. Configurations are defined in settings and configuration files. They might include:
  - CPU and memory allocation
  - Storage settings and disk usage
  - Boot sequence
  - Security settings such as hardening the system
  - Access management
  - Operating system configurations
- The configuration tells the resource how it should operate, and how it should "act." Our example is a VM, but it applies to every resource that we use in the cloud. Developers or administrators need to specify how a resource must be deployed, including all components that are part of the resource. Obviously, this is not something that developers and administrators want to do every time a specific resource is deployed. Configurations will have standards, and if there are standards, they can be automated. The moment a new resource is deployed, a process is automatically triggered to apply the desired settings to that resource. In that case, only the master configuration files have to be managed. Cloud providers offer tools to manage configurations, for instance, with **Desired State Configuration** (**DSC**), which is used in PowerShell.

DSC is declarative scripting that defines and automates the application of settings to Microsoft Windows and Linux systems. AWS offers Systems Manager for the same purpose. GCP has this embedded in Compute Engine.

- **Discoverability**: This is exactly what it says it is. It's about finding resources in cloud platforms. Resources must be visible for other resources in order to be able to communicate with each other and, obviously, these resources must be visible for monitoring. But there's more to discoverability. It also includes **service discovery**: the automatic detection of instances and services that run on these instances within a network, for instance, a tenant on a cloud platform:
    - This is extremely important in a microservices architecture: each microservice needs to know where instances are located to discover the services that are hosted on these instances. Service discovery allows the discovery of services dynamically without the need for IP addresses of the instances. These IP addresses will change dynamically, but the services will still be able to "find" each other. The technology typically involves a central services registry.
- **Security**: Here, we are focusing on data protection. After all, data is the most important asset in your multi-cloud environment. The architecture should have one goal: safeguarding the integrity of data. The best way to start thinking of security architecture is to think from the angle of possible threats. How do you protect the data, how do you prevent the application from being breached, and what are the processes in terms of security policies, monitoring, and following up on alerts? The latter is a subject for **Security Operations** (**SecOps**). Security will be extensively discussed in part 4 of this book: *Controlling Security with DevSecOps.*
- **Scalability**: The killer feature of the public cloud is scalability. Whereas previously developers had to wait for the hardware to be ready in the data center before they could actually start their work, we now have the capacity right at our fingertips. But scalability goes beyond that: it's about full agility and flexibility in the public cloud and being able to scale out, scale up, and scale down whenever business requirements call for that.
- Scaling out is also referred to as horizontal scaling. When we scale out an environment, we usually add systems such as virtual machines to that environment. In scaling up—or vertical scaling—we add resources to a system, typically CPUs, memory, or disks in a storage system. Obviously, when we can scale out and up, we can also go the opposite direction and scale systems down. Since we are paying for what we really use in the public cloud, the costs will come down immediately, which is not the case if we have invested in physical machines sitting in a traditional, on-premises environment.

You will be able to lower the usage of these physical machines, but this will not lower the cost of that machine since it's fully **CAPEX—capital expenditures or investments.** We will discuss financials in part 3 of this book: *Controlling Costs in Multi-Cloud Using FinOps.*

- Typical domains for the scalability architecture are virtual machines (also as hosts for container clusters), databases, and storage, but network and security appliances should also be considered. For example, if the business demand for scaling their environment up or out increases, typically, the throughput also increases. This has an impact on network appliances such as switches and firewalls: they should scale too. However, you should use the native services from the cloud platforms to avoid scaling issues in the first place.
- Some things you should include in the architecture for scalability are as follows:
    - **Definition of scale units**: This concerns scalability patterns. One thing you have to realize is that scaling has an impact. Scaling out virtual machines has an impact on scaling the disks that these machines use, and thus the usage of storage.
    - But there's one more important aspect that an architect must take into account: can an application handle scaling? Or do we have to rearchitect the application so that the underlying resources can scale out, up, or down without impacting the functionality and, especially, the performance of the application? Is your backup solution aware of scaling?
    - Defining scale units is important. Scale units can be virtual machines, including memory and disks, database instances, storage accounts, and storage units, such as blobs in Azure or buckets in AWS. We must architect how these units scale and what the trigger is to start the scaling activity.
    - **Allowing for autoscaling**: One of the ground principles in the cloud is that we automate as much as we can. If we have done a proper job of defining our scale units, then the next step is to decide whether we allow autoscaling on these units or allow an automated process for dynamically adding or revoking resources to your environment. First, the application architecture must be able to support scaling in the first place. Autoscaling adds an extra dimension. The following aspects are important when it comes to autoscaling:
        - The trigger that executes the autoscaling process.
        - The thresholds of autoscaling, meaning to what level resources may be scaled up/out or down. Also, keep in mind that a business must have a very good insight into the related costs.

- **Partitioning**: Part of architecting for scalability is partitioning, especially in larger environments. By separating applications and data into partitions, controlling scaling and managing the scale sets becomes easier and prevents large environments from suffering from contention. Contention is an effect that can occur if application components use the same scaling technology but resources are limited due to set thresholds, which is often done to control costs.

Now, we have to make sure that our systems are not just scalable but are also resilient and robust, ensuring high availability:

- **Robustness**: Platforms such as Azure, AWS, and GCP are just there, ready to use. And since these platforms have global coverage, we can rest assured that the platforms will always be available. Well, these platforms absolutely have a high availability score, but they do suffer from outages. This is rare, but it does happen. The one question that a business must ask itself is whether it can live with that risk—or what the costs of mitigating that risk are, and whether the business is willing to invest in that mitigation. That's really a business decision at the highest level. It's all about business continuity.
- Robustness is about resilience, accessibility, retention, and recovery. In short, it's mostly about availability. When we architect for availability, we have to do so at different layers. The most common are the compute, application, and data layers. But it doesn't make sense to design availability for only one of these layers. If the virtual machines fail, the application and the database won't be accessible, meaning that they won't be available.
- In other words, you need to design availability from the top of the stack, from the application down to the infrastructure. If an application needs to have an availability of 99.9 percent, this means that the underlying infrastructure needs to be at a higher availability rate. The underlying infrastructure comprises the whole technology stack: compute, storage, and network.
- A good availability design counters for failures in each of these components, but also ensures the application—the top of the stack—can operate at the availability that has been agreed upon with the business and its end users. However, failures do occur, so we need to be able to recover systems. Recovery has two parameters:
  - **Recovery Point Objective (RPO)**: RPO is the maximum allowed time that data can be lost. An RPO could be, for instance, 1 hour of data loss. This means that the data that was processed within 1 hour since the start of the failure can't be restored. However, it's considered to be acceptable.
  - **Recovery Time Objective (RTO)**: RTO is the maximum duration of downtime that is accepted by the business.

- RPO and RTO are important when designing the backup, data retention, and recovery solution. If a business requires an RPO of a maximum of 1 hour, this means that we must take backups every hour. A technology that can be used for this is snapshotting or incremental backups. With snapshots, we take an instant copy of the data and save that copy, while incremental backups will take backups of the changes that have occurred since the last backup was made. Taking full backups every hour would create too much load on the system and, above that, implies that a business would need a lot of available backup storage.
- It is crucial that the business determines which environments are critical and need a heavy regime backup solution. Typically, data in such environments also needs to be stored for a longer period of time. Standard offerings in public clouds often have 2 weeks as a standard retention period for storing backup data. This can be extended, but it needs to be configured and you need to be aware that it will raise costs significantly.
- One more point that needs attention is that backing up data only makes sense if you are sure that you can restore it. So, make sure that your backup and restore procedures are tested frequently—even in the public cloud.
- **Portability:** There are multiple definitions of portability. Often, the term is referring to application portability. When the same application has to be installed on multiple platforms, portability can significantly reduce costs. The only requirement for portability is that there is a general abstraction between the application logic and the system interfaces. With cloud portability, we mean that applications can run on different cloud platforms: applications and data can be moved between clouds without (major) disruption. The topic is mostly related to the so-called "vendor" or "cloud" lock-in. Companies want to have the freedom to switch providers and suppliers, including cloud platforms. The term is also frequently mixed with interoperability, indicating that applications are cloud agnostic and can be hosted anywhere. Container technology using, for example, Kubernetes can run on any platform supporting Kubernetes.
- However, in practice, portability and interoperability are much harder than the theory claims. The underlying technology that cloud providers use might differ, causing issues in migrating from one cloud to another, especially with microservices that use specific native technology. One other aspect that plays an important role is data gravity, where the data literally attracts the applications toward the data. This can be a massive amount of data that can't be easily moved to another platform. Data gravity can heavily influence the cloud strategy of a company.

- **Usability**: This is often related to the ease of use of apps, with clear interfaces and transparent app navigation from a user's perspective. However, these topics do imply certain constraints on our architecture. Usability requires that the applications that are hosted in our multi-cloud environment are accessible to users. Consequently, we will have to think of how applications can or must be accessed. This is directly related to connectivity and routing: do users need access over the internet or are certain apps only accessible from the office network? Do we then need to design a **Demilitarized Zone (DMZ)** in our cloud network? And where are jump boxes positioned in multi-cloud?
    - Keep in mind that multi-cloud application components can originate from different platforms. Users should not be bothered by that: the underlying technical setup should be completely transparent for users. This also implies architectural decisions: something we call technology transparency. In essence, as architects, we constantly have to work from the business requirements down to the safe use of data and the secured accessibility of applications to the end users. This drives the architecture all the way through.

Continuous architecture is not the only framework that helps in defining agile architecture. While we mentioned TOGAF in this chapter, The Open Group, as an organization that develops standard methodologies for architecture, has issued a different framework that is more suitable for agile architecture than TOGAF, which in itself is rather static. For all good reasons, this new framework is called **Open Agile Architecture (O-AA)**.

The framework addresses business agility as the main purpose for doing architecture in enterprises, embracing the digital transformation these enterprises are going through. O-AA puts the customer and the customer experience at the heart of the architecture. Since customer demands and experiences change continuously, the architecture must be able to adopt this: it catches the interactions between the customer and all the touchpoints the customer has with the enterprise, for example, through the products of the enterprise, but also the website, social media, customer contact centers and even the **Internet of Things (IoT)** such as mobile devices and sensors that are connected to the internet.

This book will not provide deep dives into TOGAF or O-AA. Please refer to the website of The Open Group for more information and the collateral. O-AA was released in 2020. Van Haren Publishing recently released the methodology in book form, available from TOGAF at `https://publications.opengroup.org/`

Next to the quality attributes, we also must define some principles for our multi-cloud architecture. This is discussed in the next section.

Defining principles from use cases

Before we start defining the architecture, we must define the business use case.

A business use case describes the actions a business must take in a specific, defined order to deliver a product or a service to a customer. That product or service must represent a value to that customer since that value can be represented by a price. That price in turn must match the costs that are related to the delivery of the product or service, topped by a margin so that the business makes a profit. So, the business use case describes the workflow to produce and deliver but also defines how the business adds value for its customers while making a profit.

From the use case, the architect derives the principles, starting with the business principles.

Business principles

Business principles start with business units setting out their goals and strategy. These adhere to the business mission statement and, from there, describe what they want to achieve in the short and long term. This can involve a wide variety of topics:

- Faster response to customers
- Faster deployment of new products (time to market)
- Improve the quality of services or products
- Engage more with employees
- Real digital goals such as releasing a new website or web shop
- Increase revenue and profit, while controlling costs

The last point, controlling costs, is a topic of FinOps, which will be discussed in chapters 10, *Managing Costs with FinOps*, and 11, *Improving Cost Management with the FinOps Maturity Model.*

As with nearly everything, goals should be **SMART**, which is short for **specific, measurable, attainable, relevant, and timely**. For example, a SMART-formulated goal could be "*the release of the web shop for product X in the North America region on June 1.*" It's scoped to a specific product in a defined region and targeted at a fixed date. This is measurable as a SMART goal.

Coming back to TOGAF, this is an activity that is performed in phase B of the ADM cycle; that is, the phase in the architecture development method where we complete the business architecture. Again, this book is not about TOGAF, but we do recommend having one language to execute the enterprise architecture in. TOGAF is generally seen as the standard in this case. Business principles drive the business goals and strategic decisions that a business makes. For that reason, these principles are a prerequisite for any subsequent architectural stage.

Principles for security and compliance

Though security and compliance are major topics in any architecture, the principles in this domain can be fairly simple. Since these principles are of extreme importance in literally every single aspect of the architecture, it's listed as the second most important group of principles, right after business principles.

Nowadays, we talk a lot about zero trust and security by design. These can be principles, but what do they mean? **Zero trust** speaks for itself: organizations that comply with zero trust do not trust anything within their networks and platforms. Every device, application, or user is monitored. Platforms are micro-segmented to avoid devices, applications, and users from being anywhere on the platform or inside the networks: they are strictly contained. The pitfall here is to think that zero trust is about technological measures only. It's not. Zero trust is first and foremost a business principle and looks at security from a different angle: zero trust assumes that an organization has been attacked, with the only question left being what the exact damage was. This is also the angle that frameworks such as MITRE ATT&CK take.

Security by design means that every component in the environment is designed to be secure from the architecture of that component: built-in security. This means that platforms and systems, including network devices, are hardened and that programming code is protected against breaches via encryption or hashing. This also means that the architecture itself is already micro-segmented and that security frameworks have been applied. An example of a commonly used framework is the **Center for Internet Security** (**CIS**) framework, which contains 20 critical security controls that cover various sorts of attacks on different layers in the IT stack. As CIS themselves rightfully state, it's not a one size fits all framework. An organization needs to analyze what controls should be implemented and to what extent.

We'll pick just one as an example: data protection, which is control 13 in the CIS framework. The control advises that data in transit and data at rest are encrypted. Note that CIS doesn't say what type of **Hardware Security Modules** (**HSMs**) an organization should use or even what level of encryption.

It says that an organization should use encryption and secure this with safely kept encryption keys. It's up to the architect to decide on what level and what type of encryption should be used.

In terms of compliance principles, it must be clear what international, national, or even regional laws and industry regulations the business has to adhere to. This includes laws and regulations in terms of privacy, which is directly related to the storage and usage of (personal) data.

An example of a principle is that the architecture must comply with the **General Data Protection Regulation** (**GDPR**). This principle may look simple—comply with GDPR—but it means a lot of work when it comes to securing and protecting environments where data is stored (the systems of record) and how this data is accessed (systems of engagement). Technical measures that will result from this principle will vary from securing databases, encrypting data, and controlling access to that data with authentication and authorization. In multi-cloud, this can be even more challenging than it already was in the traditional data center. By using different clouds and PaaS and SaaS solutions, your data can be placed anywhere in terms of storage and data usage.

Data principles

As we mentioned previously, here's where it really gets exciting and challenging at the same time in multi-cloud environments. The most often used data principles are related to data confidentiality and, from that, protecting data. The five most important data principles are:

- **Accuracy**: Data should be accurate, complete, and reliable. Inaccurate data can lead to flawed insights and decisions and can have serious consequences.
- **Relevance**: Data should be relevant to the problem or question at hand. Unnecessary or irrelevant data can add noise to the analysis and make it harder to extract meaningful insights.
- **Timeliness**: Data should be timely and up to date. Outdated or stale data can lead to incorrect or misleading conclusions.
- **Consistency**: Data should be consistent in terms of format, definitions, and units of measurement. Inconsistent data can create confusion and make it difficult to combine or compare different sources of information.
- **Privacy and security**: Data should be protected from unauthorized access and use. Sensitive data, such as personal or financial information, should be handled with care and stored securely to avoid breaches or leaks.

So, how do you ensure that these principles are adhered to? We briefly touched on two important technology terms that have become quite common in cloud environments earlier in this chapter:

- **Systems of record**: Systems of record are data management or information storage systems; that is, systems that hold data. In the cloud, we have the commonly known database, but due to the scalability of cloud platforms, we can now deploy huge data stores comprising multiple databases that connect thousands of data sources. Public clouds are very suitable to host so-called data lakes.
- **Systems of engagement**: Systems of engagement are systems that are used to collect or access data. This can include a variety of systems: think of email, collaboration platforms, and content management systems, but also mobile apps or even IoT devices that collect data, send it to a central data platform, and retrieve data from that platform.

A high-level overview of the topology for holding systems of record and systems of engagement is shown in the following diagram, with **Enterprise Resource Planning** (**ERP**), **Content Management** (**CMS**), and **Customer Relationship Management** (**CRM**) systems being used as examples of systems of record:

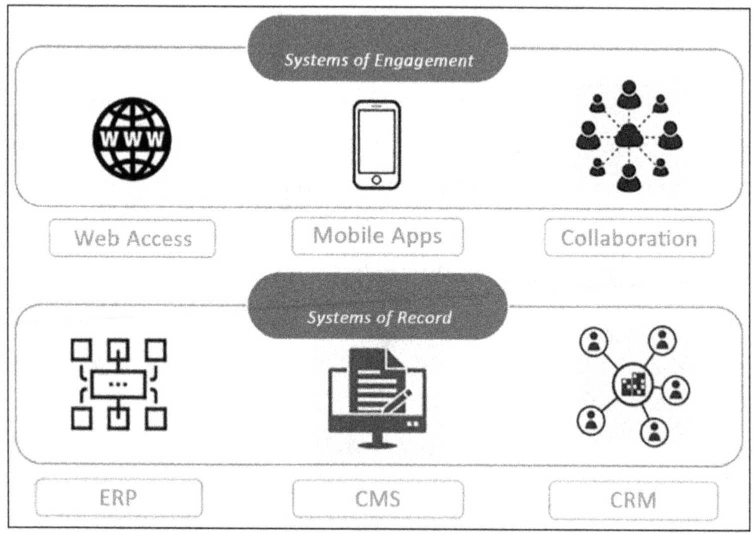

Figure 5.2: Simple representation of systems of engagement and systems of record

The ecosystem of record and engagement is enormous and growing. We've already mentioned data lakes, which are large data stores that mostly hold raw data. In order to work with that data, a data scientist would need to define precise datasets to perform analytics. Azure, AWS, and Google all have offerings to enable this, such as Data Factory and Databricks in Azure, EMR and Athena in AWS, and BigQuery from Google.

Big data and data analytics have become increasingly important for businesses in their journey to become data-driven: any activity or business decision, for that matter, is driven by actual data. Since clouds can hold petabytes of data and systems need to be able to analyze this data fast to trigger these actions, a growing number of architects believe that there will be a new layer in the model. That layer will hold "systems of intelligence" using machine learning and **artificial intelligence (AI)**. Azure, AWS, and Google all offer AI-driven solutions, such as Azure ML in Azure, SageMaker in AWS, and Cloud AI in Google. The extra layer—the systems of intelligence—can be seen in the following diagram:

Figure 5.3: Simple representation of the systems of intelligence layer

To be clear: systems of record or engagement don't say anything about the type of underlying resource. It can be anything from a physical server to a **virtual machine (VM)**, a container, or even a function. Systems of record or engagement only say something about the functionality of a specific resource.

Application principles

Data doesn't stand on its own. If we look at TOGAF once more, we'll see that data and applications are grouped into one architectural phase, known as phase C, which is the information systems architecture. In modern applications, one of the main principles of applications is that they have a data-driven approach, following the recommendation of Steven Spewak's enterprise architecture planning. Spewak published his book *Enterprise Architecture Planning* in 1992, but his approach is still very relevant, even—and perhaps even more—in multi-cloud environments.

Also mentioned in Spewak's work: the business mission is the most important driver in any architecture. That mission is data-driven; enterprises make decisions based on data, and for that reason, data needs to be relevant, but also accessed and usable. These latter principles are related to the applications disclosing the data to the business. In other words, applications need to safeguard the quality of the data, make data accessible, and ensure that data can be used. Of course, there can be—and there is—a lot of debate regarding, for instance, the accessibility of data. The sole relevant principle for an application architecture is that it makes data accessible. To whom and on what conditions are both security principles.

In multi-cloud, the storage data changes, but also the format of applications. Modern applications are usually not monolithic or client-server-based these days, although enterprises can still have a large base of applications with legacy architectures. Cloud-native apps are defined with roles and functions and build on the principles of code-based modularity and the use of microservices. These apps communicate with other apps using APIs or even triggers that call specific functions in other apps. These apps don't even have to run on the same platform; they can be hosted anywhere. Some architects tend to think that monolithic applications on mainframes are complex, so use that as a guideline to figure out how complex apps in multi-cloud can get.

However, a lot of architectural principles for applications are as valid as ever. The technology might change, but the functionality of an application is still to support businesses when it comes to rendering data, making it accessible, and ensuring that the data is usable.

Today, popular principles for applications are taking the specific characteristics of cloud-native technology into consideration. Modern apps should be enabled for mobility, be platform-independent using open standards, support interoperability, and be scalable. Apps should enable users to work with them at any time, anywhere.

One crucial topic is the fact that the requirements for applications change at almost the speed of light: users demand more and more from apps, so they have to be designed in an extremely agile way so that they can adopt changes incredibly fast in development pipelines. Cloud technology does support this: code can easily be adapted. But this does require that the applications are well-designed and documented, including in runbooks.

Infrastructure and technology principles

Finally, we get to the real technology: machines, wires, nuts, and bolts. Here, we're talking about virtual nuts and bolts. Since data is stored in many places in our multi-cloud environment and applications are built to be cloud-native, the underlying infrastructure needs to support this. This is phase D in TOGAF, the phase in architecture development where we create the target technology architecture, which comprises the platform's location, the network topology, the infrastructure components that we will be using for specific applications and data stores, and the system interdependencies. In multi-cloud, this starts with drafting the landing zone: the platform where our applications and data will land.

One of the pitfalls of this is that architects create long, extensive lists with principles that infrastructure and technology should adhere to, all the way up to defining the products that will be used as a technology standard. However, a catalog with products is part of a portfolio. Principles should be generic and guiding, not constraining. In other words, a list of technology standards and products is not a principle. A general principle could be about bleeding edge technology: a new, non-proven, experimental technology that imposes a risk when deployed in an environment because it's still unstable and unreliable.

Other important principles for infrastructure can be that it should be scalable (scale out, up, and down) and that it must allow micro-segmentation. We've already talked about the Twelve-Factor App, which sets out specific requirements for the infrastructure. These can be used as principles. The principles for the Twelve-Factor App were set out in 2005, but as we already concluded in *Chapter 2, Collecting Business Requirements*, they are still very accurate and relevant.

The Twelve-Factor App sets the following three major requirements for infrastructure:

- The app is portable between different platforms, meaning that the app is platform-agnostic and does not rely on a specific server or system's settings.
- There's little to no difference between the development stage and the production stage of the app so continuous development and deployment are enabled. The platform that the app is deployed on should support this (meaning that everything is basically code-based).

- The app supports scaling up without significant changes needing to be made to the architecture of the app.

In the next section, we will discuss the principles for usability and processes. We will also touch upon the transition and transformation to cloud environments.

Principles for processes

The last group of principles is concerned with processes. This is not about the **IT System Management (ITSM)** processes, but about the processes of deployment and automation in multi-cloud. One of the principles in multi-cloud is that we will automate as much as we can. This means that we will have to define all the tasks that we would typically do manually in an automated workflow. If we have a code-only principle defined, then we can subsequently set a principle that states that we must work from the code base or master branch. If we fork the code and we do have to make changes to it, then a principle is that altered code can only be committed back to the master code if it's tested in an environment that is completely separated from acceptance and production. This is related to the life cycle process of our environment.

So, processes here focus more on our way of working. Today, a lot of companies are devoted to agile and DevOps. If that's the defined way of working, then it should be listed as a principle; for example, an organization embraces agility in its core processes. This can be done through the **Scaled Agile Framework (SAFe)** or the Spotify model. Following that principle, a company should also define the teams, their work packages, and how epics, features, product backlogs, and so on are planned. However, that's not part of the principle anymore. That's a consequence of the principle.

As with all principles, the biggest pitfall is making principles too complex. Especially with processes, it's important to really stick to describing the principle and not the actual process.

We have discussed the architecture principles and why we are doing architecture. The next step is to define the components of the architecture. This is the topic of the next section.

Creating the architecture artifacts

The hierarchy in documents that cover the architecture starts with the enterprise architecture. It's the first so-called architecture artifact. The enterprise architecture is followed by the high-level design and the low-level design, which covers the various components in the IT landscape. We will explore this in more detail in the following sections. Keep in mind that these sections are merely an introduction to the creation of these artifacts. You will find samples of these artifacts at `https://publications.opengroup.org/i093`, where you can download a ZIP file containing relevant templates.

Creating a business vision

Creating a business vision can take years, but it's still a crucial artifact in architecture. It sets out what the business wants to achieve. This should be a long-term outlook since it will drive architectural decisions. Though cloud environments enable the agile deployment of services, deployment should never become ad hoc.

A business vision focuses on the long-term goals in terms of finance, quality of services/products, sustainability of the business, and, above all, the potential growth of the business and market domains that it's targeting. The business vision is the input for the enterprise architecture. It's the only document that will not be produced by architects, although the enterprise architect might be an important stakeholder that gives their view on the vision. After all, the vision must be realistic and obtainable. In other words, the architecture must be able to support the vision and help achieve its goals.

Enterprise architecture

The enterprise architecture is the first document that will be written by architects. Typically, this is the deliverable that is created by a team of architects, led by the enterprise or business architect. They will work together with domain architects. The latter can also be a cloud architect or an architect specialized in cloud-native development. The enterprise architecture describes the business structures and processes and connects these to the use of data in the enterprise. This data drives the business in the enterprise. In essence, enterprise architecture bridges business and IT.

Principles catalog

This document lists all the architecture principles that have to be applied to any architecture that will be developed. We discussed this in detail in the first section of this chapter, *Defining architecture principles for multi-cloud*. Principles are assembled per architecture domain.

Requirements catalog

This document lists all the requirements that a business has issued in order to achieve its goals since these are set out in the business vision. Going from a business vision to a requirements catalog is a long haul, so there are intermediate steps in creating the enterprise architecture and the principles catalog. From there, business functionality must be translated into requirements regarding the use of data and application functionality. Since not everything is known in detail at this stage, the catalog also contains assumptions and constraints. At the end, the catalog holds a list of building blocks that represent solutions to the business requirements.

High-level design

This is not an official TOGAF document. TOGAF talks about a solution concepts diagram. In practice, a lot of people find it hard to read just a diagram and grasp the meaning of it. A high-level design provides the solution concepts and includes the rationales of why specific concepts have been chosen to fulfill the requirements. Typically, a high-level design is created per architecture domain: data, application, and technology. Cloud concepts are part of each of these domains. Networking, computing, and storage are concepts that fit into the technology design. Data logics and data streams are part of the data design. Application functions must be described in the design for applications.

Low-level design

This document contains the nitty-gritty details per building block. Low-level designs for data comprise data security and data transfer. Application designs contain the necessary software engineering diagrams and distribution patterns. Technology designs hold details on the core and boundaries (networks and security), including lists of used ports; IP plan and communication protocols; platform patterns and segmentation processing units (VMs, containers, and so on); storage division; interfaces; and so on.

One note that has to be made here is that in *Chapter 4*, *Service Design for Multi-Cloud*, we agreed that we would work with everything as code. So, does it make sense to have everything written out in documents that are stored in some cabinet or drawer, never to be looked at again? Still, documenting your architecture is extremely important, but we can also have our documentation in wikis that can easily be searched through and directly linked to the related code that is ready to be worked with or even deployed. In today's world of multi-cloud, DevOps, and CI/CD pipelines, this will be the preferred way of working.

Working in DevOps pipelines and having documentation in wikis enforces the fact that the cycle of creating and maintaining these artifacts never stops. Code and wikis can easily be maintained and are more agile than chunky documents. Keep in mind that artifacts will constantly be updated. This is the ground principle of continuous architecture (reference: *Continuous Architecture*, by Murat Erder and Pierre Pureur, 2016). Continuous architecture doesn't focus on solutions, for a good reason.

In multi-cloud, there are a zillion solutions and solution components (think of PaaS) already available and a zillion more on their way. Continuous architecture focuses on the quality of the architecture itself and describes how to design, build, test, and deploy solutions, as in DevOps pipelines. Other than this, it has a strong focus on continuous validation of the architecture, which is something we will explore in the last section of this chapter—that is, *Validating the architecture*.

Planning transition and transformation

We have done a lot of work already. Eventually, this should all add up to an architecture vision: a high-level view of the end state of our architecture and the objective of that architecture. However, an architecture is more than just a description or a blueprint of that end state. An architecture should also provide a roadmap—that is, a guide on how we will reach that end state. To be honest, there's nothing new under the sun here. On the contrary, this is how IT programs are typically run: it's all about transition and transformation.

Let's assume that our end state is full cloud adoption. This means that the business has all of its IT systems, data, and applications in the cloud. Everything is code-based and automated, deployed, and managed from CI/CD pipelines. We've adopted native cloud technology such as containers and serverless functions. In *Chapter 2, Business Acceleration Using a Multi-Cloud Strategy*, we defined this as the dynamic phase, but that's a technical phase. The dynamic phase can be part of the end state of our architecture. However, we need to be absolutely sure that this dynamic technology does fit the business needs and that we are ready to operate this environment in the end state. We will refer to this end state as the **Future Mode of Operation** (**FMO**).

How do we get to this FMO? By starting at the beginning: the current situation, the **Current Mode of Operation** (**CMO**), or the **Present Mode of Operation** (**PMO**). A proper assessment of the existing landscape is crucial to get a clear, indisputable insight into the infrastructure, connections, data, and applications that a business has in its IT environment. From there, we can start designing the transition and transformation to the FMO.

If we combine the methodology of Spewak—the model that we discussed under the section about *Application principles*—with CMO-FMO planning, the model will look as follows:

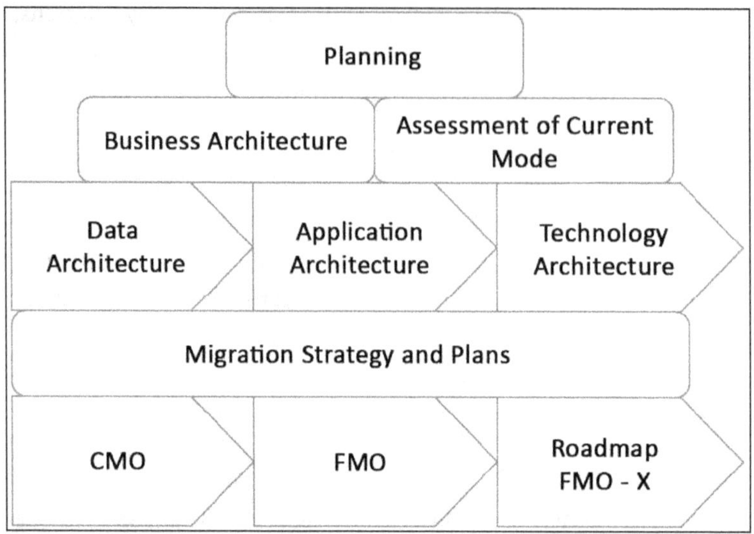

Figure 5.4: Spewak's Enterprise Architecture model plotted with transition planning

If we don't change anything in our application and we simply move it to a public cloud using IaaS or bare-metal propositions, then we can talk about transition. The technology phase would be the standard phase. Transition just means that we move the workloads, but we don't change anything at all in terms of the technology or services. However, we are also not using cloud technologies to make our environment more agile, flexible, or cost-efficient. If we want to achieve that, we will need to make a transformation: we need to change the technology under the data and applications. This is a job that needs to be taken care of through architecture. Why are we changing? What are we changing? How are we changing applications? And also, how do we revert changes if things don't work as planned; that is, what is our fallback solution?

There's one debate that needs discussing in terms of transition and transformation. As explained already, transition means that we are not changing the underlying technology and services. We just move an environment from A to B, as it is. But is that true when we are shifting environments to a public cloud? Moving an application to Azure, AWS, or GCP always implies that we are changing something: either the underlying platform or the services.

By moving an application to a major public cloud, the services will almost definitely change. We are introducing a third party to our landscape: a public cloud provider. Hence, we are introducing an agreement to our landscape. That agreement comprises terms and conditions on how our applications will be hosted in that public cloud. This is something the architecture should deal with in a clear change management process. We will cover change management in the next section.

Change management and validation as the cornerstone

We are working under architecture from this point onward. This implies that the changes that are made to the systems in our environment are controlled from the architecture. Sometimes, these changes have an impact on the architecture itself, where we will need to change the architecture. In multi-cloud environments, that will actually happen a lot.

Cloud platforms are flexible in terms of use and thus our architecture can't be set in stone: it needs to allow improvements to be made to the environments that we have designed, thereby enabling these improvements to be documented and embedded in the architecture. Improvements can be a result of fixing a problem or mitigating an issue with enhancements. Either way, we have to make sure that changes that are the result of these improvements can be validated, tracked, and traced. Change management is therefore crucial in maintaining the architecture.

Since we have already learned quite a bit from TOGAF, we will also explore change management from this angle: phase H. Phase H is all about change management: keeping track of changes and controlling the impact of changes on the architecture. But before we dive into the stages of proper change management, we have to identify what type of changes we have in IT. Luckily, that's relatively easy to explain since IT organizations typically recognize two types: standard and non-standard changes. Again, catalogs are of great importance here.

Standard changes can be derived from a catalog. This catalog should list changes that have been foreseen from the architecture as part of standard operations, release, or life cycle management. A standard change can be to add a VM. Typically, these are quite simple tasks that have either been fully automated from a repository and the code pipeline or have been scripted. Non-standard changes are often much more complex. They have not been defined in a catalog or repository, or they consist of multiple subsequent actions that require these actions to be planned.

In all cases, both with standard and non-standard changes, a request for change is the trigger for executing change management. Such a request has a trigger: a drive for change. In change management for architecture, the driver for change always has a business context: what problem do we have to solve in the business? The time to market for releasing new business services is too slow, for instance. This business problem can relate to not being able to deploy systems fast enough, so we would need to increase deployment speed. The solution could lie in automation—or designing systems that are less complex.

That is the next step: defining our architecture objectives. This starts with the business objective (getting services to market faster) and defining the business requirements (we need faster deployment of systems), which leads to a solution concept (automatic deployment). Before we go to the drawing board, there are two more things that we must explore.

Here, we need to determine what the exact impact of the change will be and who will be impacted: we need to assess who the stakeholders are, everyone who needs to be involved in the change, and the interests of these people. Each stakeholder can raise concerns about the envisioned change. These concerns have to be added to the constraints of the change. Constraints can be budgetary limits but also timing limits: think of certain periods where a business can't absorb changes.

In summary, change management to architecture comprises the following:

1. Request for change
2. The request is analyzed through change drivers within the business context
3. Definition of business objectives to be achieved by the change
4. Definition of architecture objectives
5. Identifying stakeholders and gathering their concerns
6. Assessment of concerns and constraints of the change

These steps have to be documented well so that every change to the architecture can be validated and audited. Changes should be retrievable at all times. Individual changes in the environment are usually tracked via a service, if configured. However, a change can comprise multiple changes within the cloud platform. We will need more sophisticated monitoring to do a full audit trail on these changes, to determine who did what. But having said that, it remains of great importance to document changes with as much detail as possible.

Validating the architecture

You might recognize this from the process where we validate the architecture of software. It is very common to have an architecture validation in software development, but any architecture should be validated. But what do we mean by that and what would be the objective? The first and most important objective is quality control. The second objective is that improvements that need to be made to the architecture need to be considered. This is done to guarantee that we have an architecture that meets our business goals, addresses all the principles and requirements, and can be received for continuous improvement.

Validating the architecture is not an audit. Therefore, it is perfectly fine to have the first validation procedure be done through a peer review: architects and engineers that haven't been involved in creating the architecture. It is also recommended to have an external review of your cloud architecture. This can be done by cloud solutions architects from different providers, such as Microsoft, AWS, and Google. They will validate your architecture against the reference architectures and best practices of their platforms, such as the Well-Architected Framework. These companies have professional and consultancy services that can help you assess whether best practices have been applied or help you find smarter solutions for your architecture. Of course, an enterprise would need a support agreement with the respective cloud provider, but this isn't a bad idea.

The following is what should be validated at a minimum:

- **Security**: Involve security experts and the security officer to perform the validation process.
- **Discoverability:** Architects and developers must be sure that services can find each other, especially when the architecture is defined by microservices, for instance, using container and serverless technology.
- **Scalability**: At the end of the day, this is what multi-cloud is all about. Cloud environments provide organizations with great possibilities in terms of scaling. But as we have seen in this chapter, we have to define scale sets, determine whether applications are allowing for scaling, and define triggers and thresholds, preferably all automated through auto-scaling.
- **Availability and robustness**: Finally, we have to validate whether the availability of systems is guaranteed, whether the backup processes and schemes are meeting the requirements, and whether systems can be restored within the given parameters of RTO and RPO.

In summary, validating our architecture is an important step to make sure that we have completed the right steps and that we have followed the best practices.

Summary

In the cloud, it's very easy to get started straight away, but that's not a sustainable way of working for enterprises. In this chapter, we've learned that, in multi-cloud, we have to work according to a well-thought-out and designed architecture. This starts with creating an architecture vision and setting principles for the different domains such as data, applications, and the underlying infrastructure. Quality attributes are a great help in setting up the architecture.

With these quality attributes, we have explored topics that make architecture for cloud environments very specific in terms of availability, scalability, discoverability, configurability, and operability. If we have designed the architecture, we have to manage it. If we work under the architecture, we need to be strict in terms of change management. Finally, it's good practice to have our architectural work validated by peers and experts from different providers.

With this, we have learned how to define enterprise architecture in different cloud platforms by looking at the different stages of creating the architecture. We have also learned that we should define principles in various domains that determine what our architecture should look like. Now, we should have a good understanding that everything in our architecture is driven by the business and that it's wise to have our architecture validated.

Now, we are ready for the next phase. In the next chapter, we will design the landing zones using the Well-Architected Framework and introduce BaseOps: the basic operations of multi-cloud.

Questions

1. Name at least four quality attributes that are discussed in this chapter.
2. What would be the first artifact in creating the architecture?
3. What are the two types of changes?

Further reading

- The official page of The Open Group Architecture Framework: `https://www.opengroup.org/togaf`.
- *Enterprise Architecture Planning*, by Steven Spewak, John Wiley & Sons Inc.

6

Controlling the Foundation Using Well-Architected Frameworks

This chapter of the book is all about the basic operations, or BaseOps for short, in multi-cloud. We're going to explore the foundational concepts of the major cloud providers and learn about the basics, starting with managing the landing zone—the foundation of any cloud environment. Before a business can start migrating workloads or developing applications in cloud environments, they will need to define that foundation. Best practices for landing zones include using enterprise-scale landing zones in Azure, AWS Landing Zone, and landing zones and projects in Google Cloud and Alibaba Cloud.

In this chapter, we will cover the following topics:

- Understanding BaseOps and the foundational concepts
- Building the landing zone with Well-Architected and cloud adoption principles
- BaseOps architecture patterns
- Managing the landing zone using policies
- Understanding the need for demarcation

Understanding BaseOps and the foundational concepts

BaseOps might not be a familiar term to all, although it is simple to interpret: **basic operations**. In cloud environments, these are more often referred to as **cloud operations**. BaseOps is mainly about operating the cloud environment in the most efficient way possible by making optimal use of the cloud services that major providers offer on the different layers: network, compute, storage, but also PaaS and SaaS.

The main objective of BaseOps is to ensure that cloud systems are available to the organization and that these can safely be used to do the following:

- Monitor network capacity and appropriately route traffic.
- Monitor the capacity of compute resources and adjust this to the business requirements.
- Monitor the capacity of storage resources and adjust this to the business requirements.
- Monitor the availability of resources, including health checks for backups and ensuring that systems can be recovered when required.
- Monitor the perimeter and internal security of systems, ensuring data integrity.
- Overall, manage systems at the agreed-upon service levels and use **Key Performance Indicators** (**KPIs**), as agreed upon by the business.
- Assuming that the systems are automated as much as possible, part of BaseOps is also being able to monitor and manage the pipeline.

At the end of the day, this is all about the quality of service. That quality is defined by service levels and KPIs that have been derived from the business goals. BaseOps must be enabled to deliver that quality via clear procedures, skilled people, and the proper tools.

We have already explored the business reasons regarding why we should deploy systems in cloud environments: the goal is to have flexibility, agility, but also cost efficiency. This can only be achieved if we standardize and automate. All repetitive tasks should be automated. Identifying these tasks and monitoring whether these automated tasks are executed in the right way is part of BaseOps. The automation process itself is development, but a key reason we should have DevOps in the first place is so that we can execute whatever the developer invents. Both teams have the same goal, for that matter: protect and manage the cloud systems according to best practices.

We can achieve these goals by executing the activities mentioned in the following sections. The Well-Architected Frameworks are a good methodology to capture the requirements of a cloud implementation.

Defining and implementing the base infrastructure—the landing zone

This is by far the most important activity in the BaseOps domain. It's really the foundation of everything else. The landing zone is the environment on the designated cloud platform where we will host the workloads, the applications, and the data resources. The starting principle of creating a landing zone is that it's fully provisioned through code. In other words, the landing zone contains the building blocks that form a consistent environment where we can start deploying application and data functionality, as we discussed in *Chapter 4, Service Designs for Multi-Cloud*, where we talked about scaffolding.

First, let's define what a landing zone is. Landing zones are the place to start building environments in public clouds. Let's use this analogy: if we want to land an aircraft, we need an airfield. The airfield is much more than just a long strip of tarmac where the plane actually lands. We also need some form of governance such as air traffic controllers who guide the plane toward the landing strip and decide in what order planes can land to keep it all safe.

That's exactly what landing zones in the cloud are: a place to land workloads. As we discussed in the previous chapter, the public cloud providers have **Cloud Adoption Frameworks** (**CAFs**) and **Well-Architected Frameworks** (**WAFs**) to help build the right landing zone, based on the requirements of the workloads that are supposed to land in the public cloud.

We could also compare landing zones to the foundations of a house. An architect must know how the house will look before an engineer can start calculating how strong and what shape the foundations must be on which the house stands. This implies that we must know what sort of workloads we will host on the landing zone in the public cloud before we start designing and building the landing zone. The architect must at a minimum have an idea of what the cloud platform will be used for in terms of applications and type of data. There have to be business requirements to create the right landing zone. These requirements are translated into standards and guardrails.

Defining standards and guardrails for the base infrastructure

The base infrastructure typically consists of networking and environments that can host, compute, and store resources. You could compare this with the **Hyperconverged Infrastructure** (**HCI**), which refers to a physical box that holds compute nodes, a storage device, and a switch to make sure that compute nodes and storage can communicate. The only addition that we would need is a router that allows the box to communicate with the outside world. The cloud is no different: the base infrastructure consists of compute, storage nodes, and switches to enable traffic. The major difference with the physical box is that, in the cloud, all these components are code.

But as we have already learned, this wouldn't be enough to get started. We also need an area that allows us to communicate from our cloud to the outside and to access our cloud. Next, we will need to control who accesses our cloud environment. So, a base infrastructure will need accounts and a way to provision these accounts in a secure manner. You've guessed it: even when it comes to defining the standard and policies for setting up a base infrastructure, there are a million choices to make. Landing zone concepts make it a lot easier to get started fast.

As a rule of thumb, the base infrastructure consists of five elements:

- Network
- Compute nodes
- Storage nodes
- Accounts
- Defense (security)

The good news is that all cloud providers agree that these are the base elements of an infrastructure. Even better, they all provide code-based components to create the base infrastructure. From this point onward, we will call these components building blocks. The issue is that they offer lots of choices in terms of the different types of building blocks and how to deploy them, such as through blueprints, templates, code editors, command-line programming, or their respective portals.

The Well-Architected Frameworks provide good guidance on making the right choices.

Building the landing zone with Well-Architected and Cloud Adoption principles

AWS explains it well on their website: the landing zone is a Well-Architected environment that is scalable and secure. The Well-Architected Framework provides prescriptive guidance on how to design and build the landing zone. In *Chapter 4, Service Designs for Multi-Cloud*, we discussed the various pillars of the framework. In this section, we will see how the guidelines are used in designing and building the landing zones.

All cloud providers offer services that enable the design and implementation of landing zones, using the principles of the Cloud Adoption and Well-Architected Frameworks. The terms might change per cloud provider, but the overall principles are all the same in Azure, AWS, GCP, and Alibaba. Having said that, the commercial offerings are different per provider. We will discuss the propositions per provider.

Enterprise-scale in Azure

Microsoft calls the landing zone the "plumbing" that must be in place before customers can start hosting workloads in Azure. From the CAF, the landing zone contains best practices for scalability, governance, security, networking, and identity management. In 2020, Microsoft introduced enterprise-scale landing zones. This is a service that offers prescriptive architectures and design principles. Enterprise-scale delivers a ready-to-go landing zone for eight architecture domains: Microsoft themselves call it an "accelerator." Landing zones provide many implementation options built around a set of common design areas. So, it's an actionable resource. You can deploy the templates provided by Microsoft:

- **Enterprise Agreement (EA)** enrollment and Azure Active Directory tenants
- Identity and access management
- Management group and subscription organization
- Network topology and connectivity
- Management and monitoring
- Business continuity and disaster recovery
- Security, governance, and compliance
- Platform automation and DevOps

Standardized Microsoft Azure offerings and services are included in these architectures, including **Privileged Identity Management (PIM)**, web application firewalls, and Azure Virtual WAN for connectivity. Enterprise-scale sets up a number of subscriptions in Azure that provide all the foundational services to start hosting workloads on the platform:

- Identity subscription with Azure Key Vault
- Management subscription for monitoring and logging
- Connectivity subscription with Virtual WAN, and hub-and-spoke models
- Sandbox subscription for testing applications and services
- Landing zone subscription for production workloads with applications

These are all part of the subscription with the root management group and subscription organization. The following diagram shows a high-level representation of an enterprise-scale landing zone:

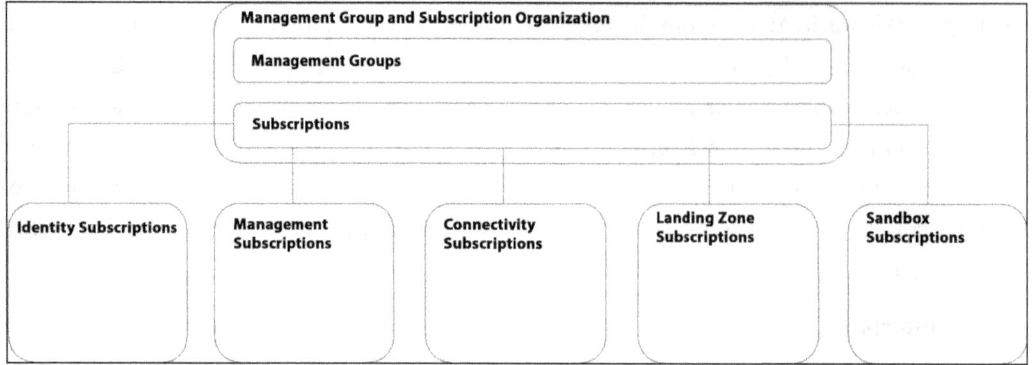

Figure 6.1: High-level representation of an enterprise-scale landing zone in Azure

In the enterprise-scale landing zone, there is an area for "landing zone subscriptions." Those are the ones for applications or workloads.

More information about enterprise-scale landing zones in Azure can be found at `https://learn.microsoft.com/en-us/azure/cloud-adoption-framework/ready/landing-zone/`.

AWS Landing Zone

AWS offers AWS Landing Zone as a complete solution, based on the `Node.js` runtime. Like Azure, AWS offers numerous solutions so that you can set up an environment. All these solutions require design decisions. To save time in getting started, AWS Landing Zone sets up a basic configuration that's ready to go. To enable this, AWS Landing Zone deploys the so-called AWS **Account Vending Machine** (**AVM**), which provisions and configures new accounts with the use of single sign-on.

AWS uses accounts. AWS Landing Zone comprises four accounts that follow the CAF of AWS:

- **Organization account**: This is the account that's used to control the member accounts and configurations of the landing zone. It also includes the so-called manifest file in the S3 storage bucket. The manifest file sets parameters for region and organizational policies. The file refers to AWS CloudFormation, a service that we could compare to ARM in Azure. CloudFormation helps with creating, deploying, and managing resources in AWS, such as EC2 compute instances and Amazon databases. It supports Infrastructure as Code.

- **Shared services account**: By default, Landing Zone manages the associated accounts through **SSO**, short for **single sign-on**. The SSO integration and the AWS-managed AD are hosted in the shared services account. It automatically peers new accounts in the VPC where the landing zone is created. AVM plays a big role in this.
- **Log archive account**: AWS Landing Zone uses CloudTrail and Config logs. CloudTrail monitors and logs account activity in the AWS environment that we create. It essentially keeps a history of all actions that take place in the infrastructure that is deployed in a VPC. It differs from CloudWatch in that it's complementary to CloudTrail. CloudWatch monitors all resources and applications in AWS environments, whereas CloudTrail tracks activity in accounts and logs these activities in an S3 storage bucket.
- **Security account**: This account holds the key vault—the directory where we store our accounts—for cross-account roles in Landing Zone and the two security services that AWS provides: GuardDuty and Amazon SNS. GuardDuty is the AWS service for threat detection, the **Simple Notification Service (SNS)** that enables the sending of security notifications. Landing Zone implements an initial security baseline that comprises (among other things) central storage of config files, the configuration of IAM password policies, threat detection, and Landing Zone notifications. For the latter, CloudWatch is used to send out alerts in case of, for example, root account login and failed console sign-in.

The following diagram shows the setup of the landing zone in AWS:

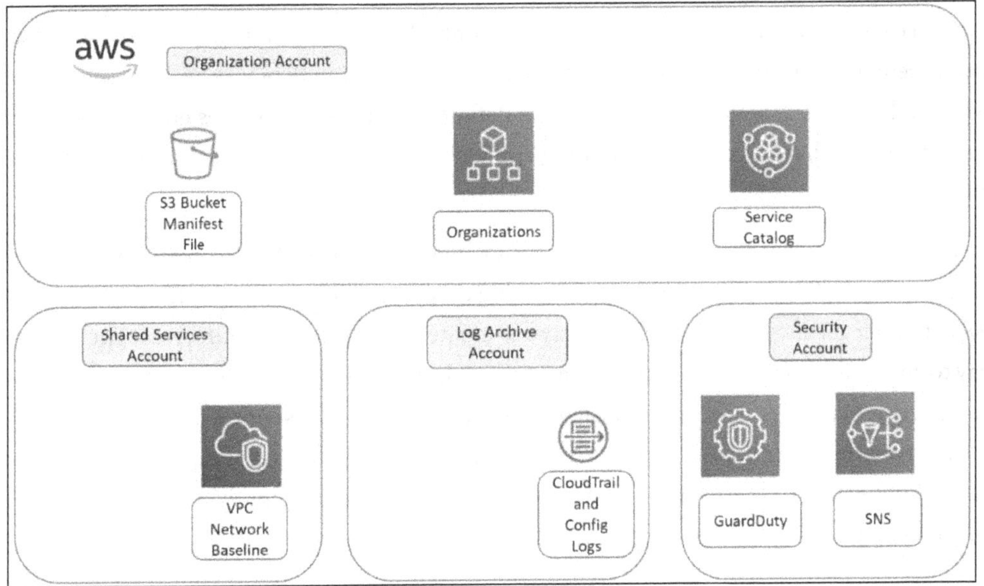

Figure 6.2: The AWS Landing Zone solution

The one thing that we haven't covered in detail yet is AVM, which plays a crucial role in setting up Landing Zone. AVM launches basic accounts in Landing Zone with a predefined network and security baseline. Under the hood, AVM uses `Node.js` templates that set up organization units wherein the previously described accounts are deployed with default, preconfigured settings. One of the components that is launched is the AWS SSO directory, which allows federated access to AWS accounts.

More information about AWS Landing Zone can be found at `https://aws.amazon.com/solutions/aws-landing-zone/`.

Landing zones in Google Cloud

GCP differs a lot from Azure and AWS, although the hub-and-spoke model can also be applied in GCP. Still, you can tell that this platform has a different vision of the cloud. GCP focuses more on containers than on IaaS by using more traditional resources. Google talks about a landing zone as somewhere you are planning to deploy a Kubernetes cluster in a GCP project using GKE, although deploying VMs is, of course, possible on the platform.

In the landing zone, you create a **Virtual Private Cloud (VPC)** network and set Kubernetes network policies. These policies define how we will be using isolated and non-isolated Pods in our Kubernetes environment. Basically, by adding network policies, we create isolated Pods, meaning that these Pods—which hold several containers—only allow defined traffic, whereas non-isolated Pods accept all traffic from any source. The policy lets you assign IP blocks and deny/allow rules to the Pods. The next step is to define service definitions to the Kubernetes environment in the landing zone so that Pods can start running applications or databases. The last step to create the landing zone is to configure DNS for GKE.

As we mentioned previously, Google very much advocates the use of Kubernetes and containers, which is why GCP is really optimized for running this kind of infrastructure. If we don't want to use container technology, then we will have to create a project in GCP ourselves. The preferred way to do this is through Deployment Manager and the `gcloud` command line. You could compare Deployment Manager to ARM in Azure: it uses the APIs of other GCP services to create and manage resources on the platform. One way to access this is through Cloud Shell within the Google Cloud portal, but GCP also offers some nice tools to get the work done. People who are still familiar with Unix command-line programming will find this recognizable and easy to work with.

The first step is enabling these APIs; that is, the Compute Engine API and the Deployment Manager API. By installing the Cloud SDK, we get a command-line tool called `gcloud` that interfaces with Deployment Manager. Now that we have `gcloud` running, we can simply start a project with the `gcloud config set project` command, followed by the name or ID of the project itself; for example, `gcloud config set project [Project ID]`. Next, we must set the region where we will be deploying our resources. It uses the very same command; that is, `gcloud config set compute/region`, followed by the region ID; that is, `gcloud config set compute/region [region]`.

With that, we're done! Well, almost. You can also clone samples from the Deployment Manager GitHub repository. This repository also contains good documentation on how to use these samples.

The following diagram shows a basic setup for a GCP project:

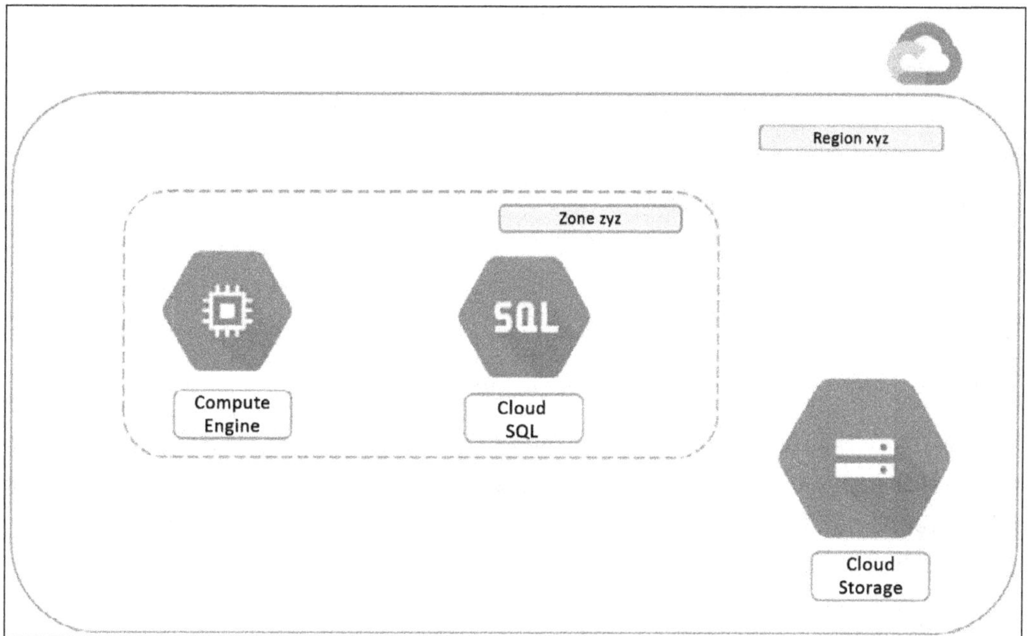

Figure 6.3: Basic setup of a project in GCP, using Compute Engine and Cloud SQL

To clone the GitHub repository for Deployment Manager in your own project, use the `git clone` command `https://github.com/GoogleCloudPlatform/deploymentmanager-samples` or go to `https://github.com/terraform-google-modules/terraform-google-migrate`. There are more options, but these are the commonly used ways to do this.

Landing Zone in Alibaba Cloud

Like AWS and Azure, Alibaba Cloud provides a full service to design and implement a landing zone. It does work a bit differently though: with Alibaba, a customer applies for the service fifteen days before the order is placed. These fifteen days are used to agree on objectives and service requirements. Alibaba collects information that is required to design the solution, including an assessment of the current IT landscape of the customer and the objectives that the customer has to migrate workloads to the cloud.

After the agreement is settled, the customer pays for the solution design and implementation. Alibaba Cloud offers three flavors: basic, standard, and advanced. The pricing is fixed. Basic is charged at 75,000 USD, standard at 150.000 USD, and advanced is based on the business scenario. Prices are for one-off payments.

The Landing Zone service includes account management, financial management, network planning, security management, resource management, and compliance auditing. The latter, compliance auditing, is only provided in standard and advanced.

The landing zone is only implemented after the acceptance of the solution by the customer.

More information about applying for the Landing Zone service in Alibaba Cloud can be found at `https://www.alibabacloud.com/services/alibaba-cloud-landing-zone`.

With that, we have created landing zones in three major cloud platforms and by doing so, we have discovered that, in some ways, the cloud concepts are similar, but that there are also some major differences in the underlying technology. Now, let's explore how we can manage these landing zones using policies, as well as how to orchestrate these policies over the different platforms.

Landing zones in Oracle Cloud Infrastructure

A landing zone in OCI is a pre-defined, automated infrastructure deployment that helps organizations quickly set up a secure, scalable, and efficient cloud environment. Like the other concepts that we have discussed, it's essentially a best-practice blueprint for cloud infrastructure that includes account structure, networking, security, and governance policies. Using a landing zone can help organizations avoid common pitfalls and accelerate their cloud adoption journey.

Once we have defined our requirements, we can create a landing zone environment using OCI's Landing Zone service. This service provides a wizard-driven interface that will guide us through the process of creating a landing zone. The service offers choices in various landing zone patterns, such as a standard landing zone, a multi-account landing zone, or a regulated landing zone.

After creating the landing zone environment, we can customize it to meet specific needs. This includes configuring **Virtual Cloud Networks (VCNs)**, subnets, security controls, and IAM policies. We can also define governance policies and automation using tools such as Terraform.

OCI provides various monitoring and management tools, such as the OCI console, APIs, and a CLI, as well as third-party tools such as Terraform, Ansible, and Chef.

BaseOps architecture patterns

A way to define a reference architecture for your business is to think outside-in. Think of an architecture in terms of the circles that can be seen in *Figure 6.4*. The circle just within the intrinsic security zone is the business zone, where all the business requirements and principles are gathered. These drive the next inner circle: the solutions zone. This is the zone where we define our solutions portfolio. For example, if the business has a demand for analyzing large sets of data (a business requirement), then a data lake could be a good solution.

The solution zone is embedded between the business zone on the outer side and the platform zone on the inner side. If we have, for instance, Azure as our defined platform, then we could have Azure Data Factory as part of a data lake solution. The principle is that from these platforms, which can also be third-party PaaS and SaaS platforms, the solutions are mapped to the business requirements. By doing so, we create the solutions portfolio, which contains specific building blocks that make up each solution.

The heart of this model—the most inner circle—is the integration zone, from where we manage the entire ecosystem in the other, outer circles.

Security should be included in every single layer or circle. Due to this, the boundary of the whole model is set by the intrinsic security zone:

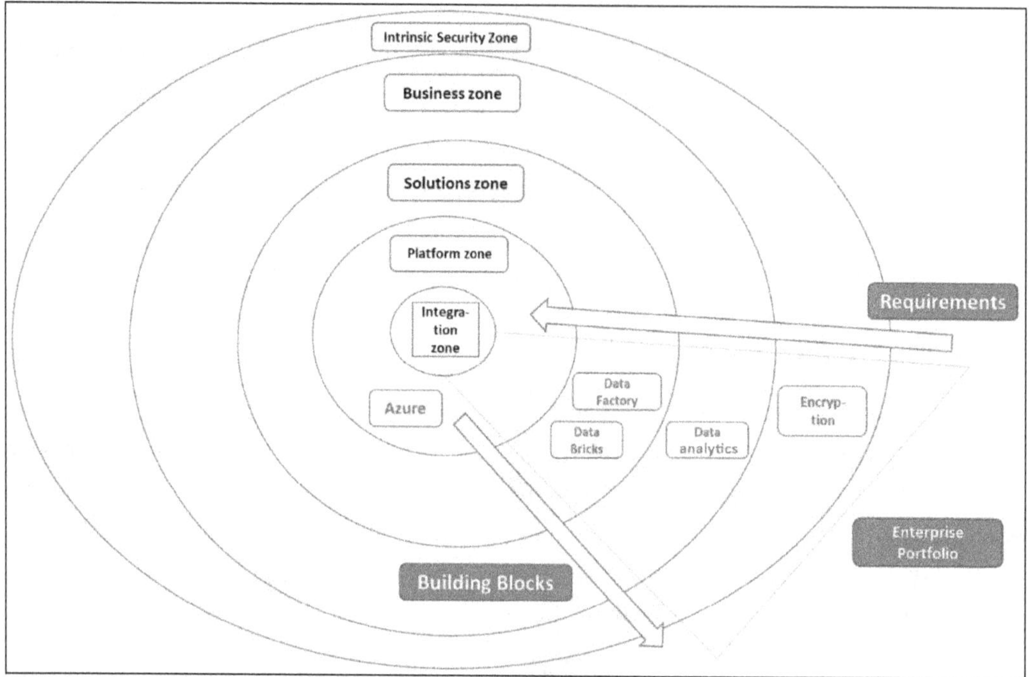

Figure 6.4: Circular model showing the layers of the enterprise portfolio

The preceding diagram shows this model with an example of a business requiring data analytics, with Data Factory and Databricks as solutions coming from Azure as the envisioned platform. The full scope forms the enterprise portfolio.

This will define the pattern for the architecture, working from the outside in. Next, we can start detailing the foundation architecture with patterns for infrastructure in the cloud. The basics are the same for every public cloud.

For a network, we will have to manage, at a minimum, the following:

- Provisioning, configuring, and managing virtual networks (vNets, VPCs, subnets, internet-facing public zones, and private zones)
- Provisioning and managing routing, **Network Address Translation** (**NAT**), **Network Access Control** (**NAC**), **Access Control Lists** (**ACLs**), and traffic management
- Provisioning and managing load balancing, network peering, and network gateways for VPNs or dedicated connections

- Provisioning and managing DNS
- Network monitoring
- Detecting, investigating, and resolving incidents related to network functions

For compute, we will have to manage, at a minimum, the following:

- Provisioning, configuring, and the operations of virtual machines. This often includes managing the operating system (Windows, various Linux distributions, and so on)
- Detect, investigate, and resolve incidents related to the functions of virtual machines
- Patch management
- Operations of backups (full, iterative, and snapshots)
- Monitoring, logging, health checks, and proactive checks/maintenance

Do note that compute in the cloud involves more than virtual machines. It also includes things such as containers, container orchestration, functions, and serverless computing. However, in the landing zone, these native services are often not immediately deployed. You might consider having the container platform deployed as part of the base infrastructure. Remember that, in the cloud, we do see a strong shift from VMs to containers, so we should prepare for that while setting up our landing zone.

In most cases, this will include setting up a Kubernetes cluster. In Azure, this is done through **Azure Kubernetes Service** (**AKS**), where we create a resource group that will host the AKS cluster. AWS offers its own cluster service through **Elastic Kubernetes Service** (**EKS**). In GCP, this is **Google Kubernetes Engine** (**GKE**). The good news is that a lot of essential building blocks, such as Kubernetes DNS, are already deployed as part of setting up the cluster. Once we have the cluster running, we can start deploying cluster nodes, Pods (a collection of application containers), and containers. In terms of consistently managing Kubernetes platforms across multi-cloud platforms, there are multiple agnostic solutions that you can look at, such as Rancher or VMware's Tanzu Mission Control.

For storage, we will have to manage, at a minimum, the following:

- Provisioning, configuring, and the operations of storage, including disks for managed virtual machines
- Detecting, investigating, and resolving incidents related to the functions of storage resources
- Monitoring, logging, health checks on local and different redundant types of storage solutions (zone, regional, globally redundant), and proactive checks/maintenance, including capacity checks and adjustments (capacity allocation)

Next, we will have to manage the accounts and make sure that our landing zone—the cloud environment and all its building blocks—is secure. Account management involves creating accounts or account groups that need access to the cloud environment. These are typically created in Active Directory.

In the final section of this chapter, *Understanding the need for demarcation*, we will take a deeper look at admin accounts and the use of global admin accounts. Security is tightly connected to account, identity, and access management, but also to things such as hardening (protecting systems from outside threats), endpoint protection, and vulnerability management. From day 1, we must have security in place on all the layers in order to prevent, detect, assess, and mitigate any breaches. This is part of SecOps. *Section 4* of this book is all about securing our cloud environments.

Managing the base infrastructure

After we have defined and deployed a landing zone in a public cloud platform, there are quite a number of building blocks that we will have to manage from that point onward. In this section, we will discuss how we can manage these building blocks. Firstly, we need to be able to connect to the landing zone, and that requires connectivity.

Implementing and managing connectivity

One of the most used technologies in cloud technology is the **VPN**, the **virtual private network**. In essence, a VPN is a tunnel using the internet as a carrier. It connects from a certain IP address or IP range to the IP address of a gateway server in the public cloud.

Before we get into this, you have to be aware of what a public cloud is. If you as a business deploy services in Azure, AWS, GCP, OCI, Alibaba Cloud, or any other public cloud, you are extending your data center to that cloud. It therefore needs a connection between your data center and that extension in the public cloud. The easiest and probably also the most cost-efficient way to get that connection fast is through a VPN. The internet is already there, and all you would have to do in theory is assign IP addresses or the IP range that is allowed to communicate to that extension, creating a tunnel. That tunnel can be between an office location (site) or from just one user connecting to the cloud. The latter is something we refer to as a point-to-site VPN.

In the public cloud itself, that connection needs to terminate somewhere, unless you want all resources to be opened up for connectivity from the outside. That is rarely the case and it's certainly not advised. Typically, a business would want to protect workloads from direct and uncontrolled outside connections. When we're setting up VPNs, we need to configure a zone in the public cloud with a gateway where the VPN terminates.

From the gateway, the traffic can be routed to other resources in the cloud, using routing rules and tables in the cloud. It works the same way as in a traditional data center where we would have a specific connectivity zone or even a **Demilitarized Zone** (**DMZ**) as a zone in the cloud environment that users connect to before they get to the actual systems. The following architecture shows the basic principle of a VPN connection to a public cloud:

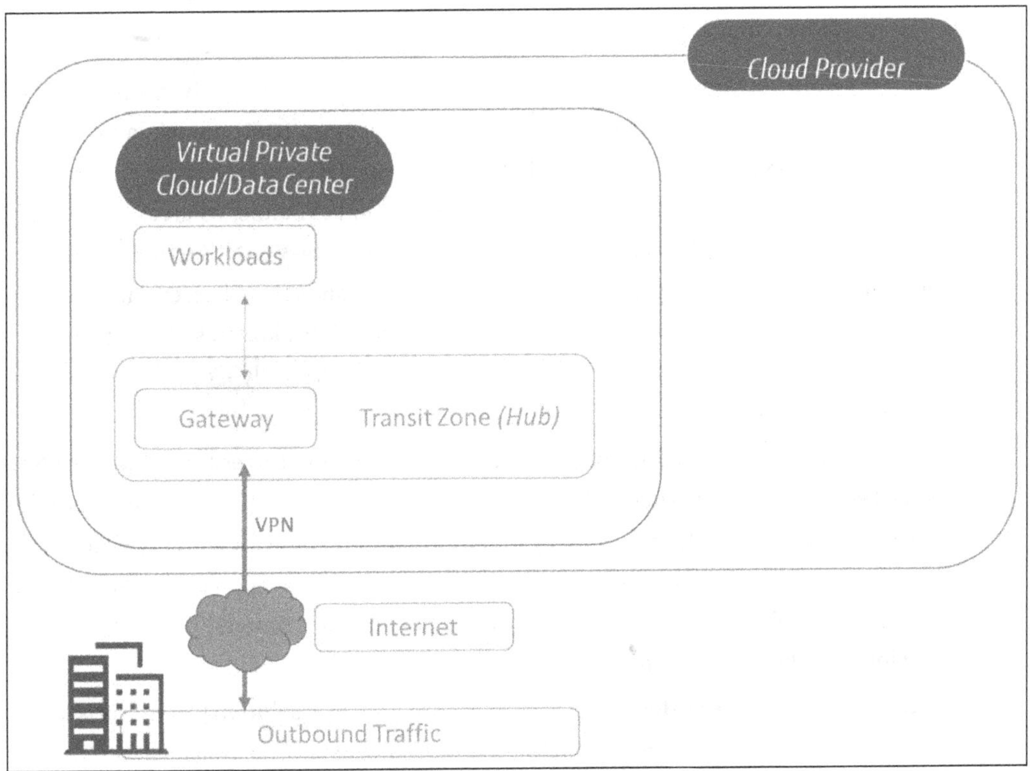

Figure 6.5: The basic architecture of VPN connectivity

There are other ways to connect to a public cloud. VPN tunnels use the internet as a carrier. For several reasons, companies are often not very keen to have their traffic sent over the internet, not even when it's through secured, encrypted tunnels. A more stable, predictable, and even more secure solution is a direct connection between the router or firewall in your on-premises environment and the cloud platform that you as a business use. Down to the core, this solution involves a cable from an on-premises network device straight into a gateway device that routes directly to services that your business purchases from the cloud. The leading cloud providers all offer direct connectivity solutions. Partner interconnects to the different clouds are also available through colocations such as Equinix, Digital Realty, and Interxion.

Implementing Azure ExpressRoute

Azure offers ExpressRoute as a direct connection from your network to your environments in Azure. ExpressRoute provides a private connection to Microsoft cloud services: Azure, Microsoft 365, and Dynamics. The offering comprises four different deployment types: cloud exchange co-location, point-to-point Ethernet, any-to-any IPVPN, and ExpressRoute Direct, which provides dual 100 Gbps or 10 Gbps connections supporting active/active connectivity:

- **Point-to-point Ethernet**: This type of connection provides connectivity on layer 2 or managed on layer 3. In the **Open Systems Interconnection** (**OSI**) model, layer 2 is the data link layer and layer 3 is the network layer. The main difference is that layer 3 provides routing and also takes care of IP routing, both static and dynamic, whereas layer 2 only does switching. Layer 3 is usually implemented when intra-**VLAN** (**Virtual Local Area Network**) is involved. Simply explained, layer 3 understands IP addresses and can route traffic on an IP base, whereas layer 2 does not understand IP addresses. Having said that, a point-to-point Ethernet ExpressRoute connection will typically—or is at least recommended to—be on layer 3.
- **Any-to-any IPVPN**: With this deployment, ExpressRoute integrates the **Wide-Area Network** (**WAN**) of the company with Azure, extending the on-premises environment with a virtual data center in Azure—making it literally one environment with the virtual data center as a *branch office*. Most companies will have network connectivity over **MPLS** (**Multiprotocol Label Switching**), provided by a telecom provider. ExpressRoute connects this network over layer 3 to Azure.
- **Azure cloud exchange co-location**: This will be the preferred solution for enterprises that host their systems in a co-location. If that co-location has a cloud exchange, you can use the connections from this exchange. Typically, these will be managed layer 3 connections. The hosted environment is connected to the exchange—often in so-called "meet-me rooms" or racks—and from the exchange, the connection is set up to Azure.

ExpressRoute lets customers connect directly to Microsoft's network through a pair of 10 or 100 Gbps ports. Bandwidths are offered from 50 Mbps up to 10 Gbps.

Implementing AWS Direct Connect

The direct connection from your on-premises network to AWS is very appropriately called AWS Direct Connect. It connects your router or firewall to the Direct Connect service in your AWS Region. The connection goes from your own router to a router of your connectivity partner and from there, it gets directly linked to the Direct Connect endpoint.

That endpoint connects to a **Virtual Private Gateway** (**VPG**) in your AWS **Virtual Private Cloud** (**VPC**) or to the AWS services.

Implementing Google Dedicated Interconnect

Google offers Dedicated Interconnect. The principle is the same as with Direct Connect from AWS: from the router in the on-premises network, a direct link is established to the Google peering edge; this is done in a co-location facility where Google offers these peering zones. From there, the connection is forwarded to a cloud router in the GCP environment.

Dedicated Interconnect is offered as a single 10 Gbps or 100 Gbps link, or a link bundle that connects to the cloud router. Multiple connections from different locations or different devices to Google require separate interconnects.

Implementing Alibaba Express Connect

Alibaba Cloud offers Express Connect. Like the solutions that we discussed already, Express Connect also provides a dedicated, physical connection from a customer's data center to Alibaba Cloud. The service is available worldwide.

Implementing direct connectivity in OCI

In OCI, there are several options available for direct connectivity to OCI resources. These options are designed to provide high-bandwidth, low-latency connectivity between on-premises data centers, colocation facilities, or other cloud providers.

FastConnect is a dedicated, private connectivity option that provides high-bandwidth and low-latency connections between on-premises data centers or colocation facilities and OCI. FastConnect provides dedicated, private connections up to 100 Gbps and allows you to bypass the public internet.

An alternative is VPN Connect: This option provides an encrypted and secure connection between an on-premises network and a VCN in OCI. VPN Connect uses the internet to establish a secure connection, making it a more cost-effective option compared to FastConnect. VPN Connect is available in different configurations, such as site-to-site VPN and remote access VPN, to suit different connectivity needs.

Lastly, we mention Oracle Interconnect as an option to set up private connectivity between OCI and other cloud providers, such as Azure and AWS. Oracle Interconnect provides dedicated, private connections up to 10 Gbps.

Accessing environments in public clouds

Once we have connectivity, we need to design how we access the resources in the cloud. We don't want admins to access every single resource separately but through a single point of access so that we can control access on that single instance. Steppingstone or bastion servers are a solution for this.

Steppingstone servers are sometimes referred to as jump servers. Azure, AWS, and GCP call these servers bastion servers. The idea is to allow access to one or more network segments through a select number of machines, in this case, the jump server(s). This not only limits the number of machines that can access those network segments, but also the type of traffic that can be sent to those network segments. It's highly recommended and best practice to deploy these jump or steppingstone servers in a virtual network in the cloud.

Defining and managing infrastructure automation tools and processes

In the cloud, we work with code. There's no need to buy physical hardware anymore; we simply define our hardware in code. This doesn't mean we don't have to manage it. To do this in the most efficient way, we need a master code repository. This repository will hold the code that defines the infrastructure components, as well as how these components have to be configured to meet our principles in terms of security and compliance. This is what we typically refer to as the desired state.

Azure, AWS, and Google offer native tools to facilitate infrastructure and configuration as code, as well as tools to automate the deployment of the desired state. In Azure, we can work with Azure DevOps, GitHub Actions, and Azure Automation, all of which work with **Azure Resource Manager (ARM)**. AWS offers CloudFormation, while Google has Cloud Resource Manager and Cloud Deployment Manager. These are all tied into the respective platforms, but the market also offers third-party tooling that tends to be agnostic to these platforms. We will explore some of the leading tools later in this chapter.

For source code management, we can use tools such as GitHub, Azure DevOps, AWS CodeCommit, and GCP Cloud Source Repositories.

Defining and implementing monitoring and management tools

We've already discussed the need for monitoring. The next step is to define what tooling we can use to perform these tasks. Again, the cloud platforms offer native tooling: Azure Monitor, Application Insights, and Log Analytics; AWS CloudTrail and CloudWatch; and Google Cloud's operations suite (formerly Stackdriver). And, of course, there's a massive set of third-party tools available, such as Splunk and Nagios. These latter tools have a great advantage since they can operate independently of the underlying platform. This book won't try to convince you that tool A is preferred over tool B; as an architect, you will have to decide what tool fits the requirements—and the budget, for that matter.

Security is a special topic. The cloud platforms have spent quite some effort in creating extensive security monitoring for their platforms. Monitoring is not only about detecting; it's also about triggering mitigating actions. This is especially true when it comes to security where detecting a breach is certainly not enough. Actually, the time between detecting a vulnerability or a breach and the exploit can be a matter of seconds, which makes it necessary to enable fast action. This is where **SIEM** comes into play: **security incident and event management**. SIEM systems evolve rapidly and, at the time of writing, intelligent solutions are often part of the system.

An example of this is Azure Sentinel, an Azure-native SIEM and **SOAR (Security Orchestration, Automation and Response)** solution: it works together with Azure Security Center, where policies are stored and managed, but it also performs an analysis of the behavior it sees within the environments that an enterprise hosts on the Azure platform. Lastly, it can automatically trigger actions to mitigate security incidents. For instance, it can block an account that logs in from the UK one minute and from Singapore the next—something that wouldn't be possible without warp-driven time traveling. As soon as it monitors this type of behavior, an action might be triggered to block the user from the system.

In other words, monitoring systems become more sophisticated and developments become as fast as lightning.

Supporting operations

Finally, once we have thought about all of this, we need to figure out who will be executing all these tasks. We will need people with the right skills to manage our multi-cloud environments. As we have said already, a truly T-shaped engineer or admin doesn't exist. Most enterprises end up with a group of developers and operators that all have generic and more specific skills.

Some providers refer to this as the **Cloud Center of Excellence** (**CCoE**), and they mark it as an important step in the cloud journey or cloud adoption process of that enterprise. Part of this stage would be to identify the roles this CCoE should have and get the members of the CCoE on board with this. The team needs to be able to build and manage the environments, but they will also have a strong role to fulfill to evangelize new cloud-native solutions.

Just as a reading tip, please have a look at an excellent blog post on forming a CCoE by Amazon's Enterprise Strategist Mark Schwartz: `https://aws.amazon.com/blogs/enterprise-strategy/using-a-cloud-center-of-excellence-ccoe-to-transform-the-entire-enterprise/`.

In this section, we have learned what we need to cover to set up our operations in multi-cloud. The next step is managing the landing zones.

Managing the landing zones using policies

When we work in cloud platforms, we work with code. Everything we do in the cloud is software-and code-defined. This makes cloud infrastructure very agile, but it also means that we need some strict guidelines for how we manage the code, starting with the code that defines our landing zone or foundation environment. As with everything in IT, it needs maintenance. In traditional data centers and systems, we have maintenance windows where we can update and upgrade systems. In the cloud, things work a little differently.

First, the cloud providers apply maintenance whenever it's needed. There's no way that they can agree upon maintenance windows with thousands of customers spread across the globe. They simply do whatever needs to be done to keep the platform healthy and ready for improvements and the release of new features. Enterprises don't want to be impacted by these maintenance activities, so they will have to make sure that their code is always safe.

The next thing we need to take into account is the systems that the enterprise has deployed on the platform, within their own virtual cloud or project. These resources also need maintenance. If we're running VMs, we will need to patch them every now and then. In this case, we are patching code. We want to make sure that, with these activities, administrators do not accidentally override certain security settings or, worse, delete disks or any critical code that is required for a specific function that a resource fulfills. This is something that we must think about from the very start when setting up the landing zones. From that point onward, we must start managing them. For that, we can use policies and management tooling.

In this section, we have set up the landing zones. In the next section, we'll learn how to manage them.

Managing basic operations in Azure

When we look at Azure, we must look at a service called **Test-Driven Development** (**TDD**) for landing zones in Azure. Be aware that TDD is not an Azure service in itself: it's an agile practice used in software development that can be applied to the development of landing zones in the cloud. TDD is particularly known in software development as it aims to improve the quality of software code. As we have already discussed, the landing zone in Azure is expanded through the process of refactoring, an iterative way to build out the landing zone. Azure provides a number of tools that support TDD and help in the process of refactoring the landing zone:

- **Azure Policy**: This validates the resources that will be deployed in Azure against the business rules. Business rules can be defined as cost parameters or thresholds, as well as security parameters such as checking for the hardening of resources or consistency with other resources. For instance, they can check if a certain ARM template has been used for deployment. Policies can also be grouped together to form an initiative that can be assigned to a specific scope, such as the landing zone. A policy can contain actions, such as denying changes to resources or deploying after validation. Azure Policy offers built-in initiatives that can be specifically used to execute TDD: it will validate the planned resources in the landing zone against business rules. A successful validation will result in a so-called definition of done and, with that, acceptance that resources may be deployed.
- **Azure Blueprints**: With Blueprints, you can assemble role assignments, policy assignments, ARM templates, and resource groups in one package so that they can be reused over and over again in case an enterprise wants to deploy multiple landing zones in different subscriptions. Microsoft Azure offers various blueprint samples, including policies for testing and deployment templates. The good thing is that these can easily be imported through Azure DevOps so that you have a CI/CD pipeline with a consistent code repository right from the start.
- **Azure Graph**: Azure Landing Zone is deployed based on the principle of refactoring. So, in various iterations, we will be expanding our landing zone. Since we are working according to the principles of TDD, this means that we must test whether the iterations are successfully implemented, that resources have been deployed in the right manner, and that the environments have interoperability. For these tasks, Azure offers Graph. It creates test sets to validate the configuration of the landing zone. Azure Graph comes with query samples, since it might become cumbersome to get started with the settings and coding that Graph uses.

- **Azure Quickstart Templates**: If we really want to get going fast, we can use Quickstart Templates, which provide default settings for the deployment of the landing zone itself and its associated resources.
- **Azure Bicep**: This is a **Domain-Specific Language** (**DSL**) that uses declarative syntax to deploy Azure resources. It was introduced in 2020. In a Bicep file, you define the infrastructure you want to deploy to Azure and then use that file throughout the development lifecycle. The file can be used to deploy infrastructure repeatedly, but always in a consistent manner. The big advantage is that developers and administrators only have to manage the Bicep file that is used to deploy resources. Bicep files have an easier syntax than **JSON** (**JavaScript Object Notation**) code. The Bicep syntax directly communicates with **Azure Resource Manager** (**ARM**) so that resources are always implemented in the most optimized way. In a way, Bicep is comparable to Terraform, a tool that provides an automated way to deploy **Infrastructure as Code** (**IaC**) too.

More information on test-driven development in Azure Landing Zone can be found at `https://docs.microsoft.com/en-us/azure/cloud-adoption-framework/ready/considerations/azure-test-driven-development`.

Managing basic operations in AWS

AWS offers CloudFormation guardrails. This is a very appropriate name since they really keep your environment on the rails. Guardrails come with four principal features for which policies are set in JSON format. To create policies, AWS offers Policy Generator. In Policy Generator, you define the type of policy first and then define the conditions, meaning when the policy should be applied. The following policies are in scope:

- **Termination protection**: Here, AWS talks about stacks and even nested stacks. Don't get confused—a stack is simply a collection of AWS resources that can be managed as one unit from the AWS Management Console. An example of a stack can be an application that comprises a frontend server, a database instance using an S3 bucket, and network rules. Enabling termination protection prevents that stack from being deleted unintentionally. Termination protection is disabled by default, so you need to enable it, either from the management console or by using command-line programming.
- **Deletion policies:** Where termination protection has entire stacks as the scope, deletion policies target specific resources. To enable this, you must set `DeletionPolicy` attributes within the CloudFormation templates.

Now, this policy comes with a lot of features. For instance, the policy has a retain option so that whenever a resource is deleted, it's still kept as an attribute in your AWS account. You can also have CloudFormation take a snapshot of the resource before it gets deleted. Keep in mind that deletion policies are set per resource.

- **Stack policies**: These policies are set to define actions for a whole stack or group of resources. Stack policies are a type of access control policy that aims to protect critical infrastructure resources by preventing updates to them. Stack policies are applied to AWS CloudFormation stacks, which are templates that define the infrastructure resources that make up an application. Stack policies work by specifying a set of rules that restrict the types of updates that can be made to specific resources within a stack. For example, we might use a stack policy to prevent updates to a production database or to restrict changes to your load balancer configuration.
- **IAM policies**: These policies define access controls; that is, who is allowed to do what and when? Access controls can be set with fine granularity for whole stacks, specific resource groups, or even single resources and only allow specific tasks to define the roles that users can have. In other words, this is the place where we manage **Role-Based Access Control** (**RBAC**). The last section of this chapter, *Understanding the need for demarcation*, is all about IAM and the separation of duties.

One important service that must be discussed here is AWS Control Tower. In most cases, enterprises will have multiple accounts in AWS, corresponding with divisions or teams in the enterprise. On an enterprise level, we want to keep the AWS setup consistent, meaning that guardrails are applied in the same way in every account. Control Tower takes care of that.

Control Tower is a service that creates a landing zone with AWS Organizations and helps in setting up multi-account governance, based on the AWS best practices. These best practices obviously comply with the CAF and the WAF. So, with Control Tower you set up a landing zone based on best-practice blueprints from AWS Organizations and apply the guardrails to this landing zone. Blueprints contain policies for identity management, federated access to accounts, centralized logging, automated workflows for the provisioning of accounts, and network configurations.

Next, the administrator gets a dashboard in Control Tower that shows how accounts and guardrails are applied.

More information on guardrail policies in AWS can be found at `https://aws.amazon.com/blogs/mt/aws-cloudformation-guardrails-protecting-your-stacks-and-ensuring-safer-updates/`. Information about Control Tower is provided at `https://aws.amazon.com/controltower/?control-blogs.sort-by=item.additionalFields.createdDate&control-blogs.sort-order=desc`.

Managing basic operations in GCP

As we have seen, GCP can be a bit different in terms of the public cloud and landing zones. This originates from the conceptual view that Google has, which is more focused on container technology using Kubernetes. Still, GCP offers extensive possibilities in terms of setting policies for environments that are deployed on GCP. In most cases, these policies are comprised of organizations and resources that use IAM policies:

- **Organizations**: In GCP, we set policies using constraints. A constraint is an attribute that is added to the service definition. Just as an example, we'll take the Compute Engine service that deploys VMs to our GCP project. In Compute Engine projects, logging in for operating systems is disabled by default. We can enable this and set a so-called Boolean constraint, named after George Boole, who invented this type of logic as an algebraic system in the nineteenth century: a statement or logical expression is either true or false. In this case, we set Compute Engine to `true`. Next, we must set a policy that prevents that login from being disabled. The command in gcloud + code style is: `constraints/compute.requireOsLogin`. A lot of policies and constraints in GCP work according to this principle.

More on organization policy constraints in GCP can be found at `https://cloud.google.com/resource-manager/docs/organization-policy/org-policy-constraints`.

- **Resource policies**: Cloud IAM policies set access controls for all GCP resources in JSON or YAML format. Every policy is defined by bindings, an audit configuration, and metadata. This may sound complex, but once you understand this concept, it does make sense. First, let's look at bindings. Each binding consists of a member, a role, and a condition. The member can be any identity. Remember what we said previously: in the cloud, basically, everything is an identity. This can be users, but also resources in our cloud environment that have specific tasks so that they can access other resources and have permission to execute these tasks.

> Thus, a member is an identity: a user, a service account, a resource, or a group of resources. The member is bound to a role that defines the permission that a member has. Finally, we must determine under what condition a member may execute its role and what constraints are valid. Together, this makes a binding.

However, the binding is only one part of the policy. We also have an AuditConfig to log the policy and the metadata. The most important field in the metadata is `etag`. The `etag` field is used to guarantee that policies are used in a consistent way across the various resources in the project. If a policy is altered on one system, the `etag` field makes sure that the policies stay consistent. Inconsistent policies will lead resource deployments to fail.

Policies can have multiple bindings and can be set on different levels within GCP. However, be aware that there are limitations. As an example, GCP allows a maximum of 1,500 members per policy. So, do check the documentation thoroughly, including the best practices for using policies.

Extensive documentation on Cloud IAM policies in GCP can be found at `https://cloud.google.com/iam/docs/policies`.

Managing basic operations in Alibaba Cloud

Managing resources in Alibaba Cloud is done best through the Cloud Governance Center service. Alibaba recommends setting this up with a corporate account or the RAM user: the Resource Access Management user. Note that is not RAM as in random access memory. RAM in Alibaba Cloud is related to resource access.

The RAM user gets permission to access the various resources in Alibaba Cloud. Next, the resource structure is defined by setting up the resource directory and the management accounts. The resource directory holds the resource folders with, for instance, the **Elastic Compute Service (ECS)** instances and the storage accounts. Also, shared service and logging accounts are defined.

Administrators are advised to use Account Factory, where the baselines—the guardrails—for identities, permissions, networking, and security are set. Cloud Governance Center offers a way to centrally manage these items and make sure that resources are configured in a consistent manner.

This completes this section, in which we have learned how to create policies by enabling the **Basic Operations (BaseOps)** of our landing zones in the different clouds. The next section talks about orchestrating policies in a true multi-cloud setup, using a single repository.

Managing basic operations in OCI

To start managing resources in **Oracle Cloud Infrastructure** (**OCI**), we need to create an OCI account and obtain the necessary credentials to access the OCI console and APIs.

First, we need an OCI account to access the OCI console and APIs. This account provides access to the OCI services and resources that we need to manage our infrastructure in OCI. Next, we create user accounts for each person who will be managing resources in OCI. User accounts are associated with a specific OCI identity domain and can be granted specific permissions to access and manage resources within that domain.

We also need to define groups. We can create groups within the OCI identity domain to organize users based on their roles and responsibilities. Groups can be used to simplify permissions management and make it easier to grant or revoke access to resources.

Resources are organized in compartments: logical containers for resources in OCI. They provide a way to organize and isolate resources based on their business purpose or application. We must create compartments within the OCI tenancy and assign permissions to users and groups to access and manage resources within those compartments.

Finally, we must set up an API signing key. To access the OCI APIs, we need to generate an API signing key for each user account. This key is used to authenticate API requests and authorize access to specific resources.

Understanding the need for demarcation

Typically, when we talk about demarcation in cloud models, we refer to the matrix or delineation of responsibility: who's responsible for what in IaaS, PaaS, and SaaS computing? The following diagram shows the very basics of this matrix—the shared responsibility model:

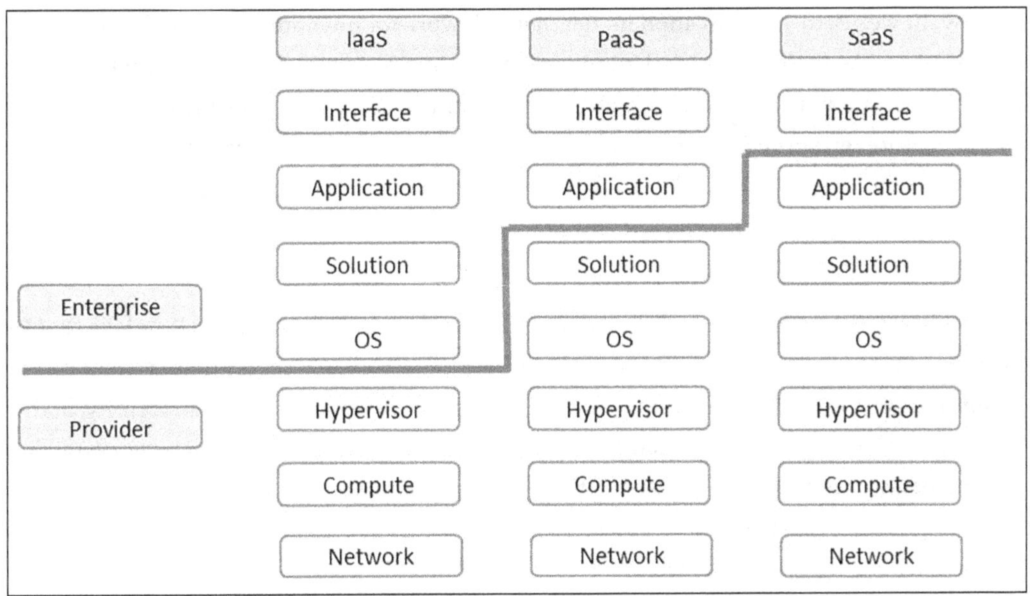

Figure 6.6: Shared responsibility model in cloud deployment

However, we need a much more granular model in multi-cloud. We have been discussing policies throughout this chapter and by now, we should have concluded that it's not very easy to draw some sharp lines when it comes to responsibilities in our multi-cloud environment. Just look at the solution stack: even in SaaS solutions, there might be certain security and/or compliance policies that the solution needs to adhere to. Even something such as an operating system might already be causing issues in terms of hardening: are monitoring agents from a PaaS provider allowed or not? Can we run them alongside our preferred monitoring solution? Or will that cause too much overhead on our systems? In short, the world of multi-cloud is not black and white. On the contrary, multi-cloud has an extensive color scheme to work with.

So, how do we get to a demarcation model that will work for our enterprise? Well, that's architecture. First, we don't need global admins all over our estate. This is a major pitfall in multi-cloud. We all know such cases: the database administrator that needs global admin rights to be able to execute certain actions or, worse, solutions that require service accounts with such roles. It's global admin galore. Do challenge these requests and do challenge software providers—or developers, for that matter—when it comes to why systems would need the highest possible access rights in the environment.

That's where it starts: policies. In this case, a good practice is the **Policy of Least Privilege** (**PoLP**). This states that every identity is granted the minimum amount of access that is necessary to perform the tasks that have been assigned to that identity. Keep in mind that an identity, in this case, doesn't have to be a user: it can be any resource in the environment. When we are talking about users, we refer to this as **Least-Privileged User Account or Access** (**LPUA**). PoLP helps in protecting data as data will only be accessible when a user or identity is explicitly granted access to that data. But there are more reasons to adhere to this policy. It also helps in keeping systems healthy as it minimizes risks or faults in systems. These faults can be unintended or the result of malicious conduct. We should follow the rule of least privilege at all times. We will discuss this in more detail in *Chapter 15, Implementing Identity and Access Management*, which is all about identity and access management.

Regarding this very first principle, there are a few more considerations that need to be made at this stage. These considerations translate into controls and, with that, into deliverables that are part of BaseOps, since they are absolutely part of the foundational principles in multi-cloud. The following table shows these controls and deliverables:

Control	**Deliverable**
A policy document is available and active that describes how user/admin accounts are generated, maintained, and disposed of throughout their lifecycle.	Policy and approvals
An RBAC authorization matrix is available that describes the access delegation from data or system owners.	Authorization matrix
User accounts are created by following established approval procedures that adhere to LPUA.	List of user accounts
Periodic checks are performed to ensure the continuous validity of accounts; for example, the account is needed and in active use, the correct role has been applied, and so on.	Checklist

Unauthorized access (attempts) to system resources is logged in an audit trail and periodically reported to and reviewed by the CISO/Security Manager.	Report of unauthorized access attempts

Table 6.1 - Cloud controls and associated deliverables

Demarcation and separation of duties are very strongly related to identity and access management. We will learn more about that in the forthcoming sections about security.

Summary

In this chapter, we have designed and set up our landing zones on the different major cloud platforms. We have learned that the foundational principles might be comparable, but the actual underlying implementation of the landing zone concepts differs. We studied these various concepts, such as enterprise-scale in Azure and Control Tower in AWS.

We also learned how to work with patterns to define the components of a landing zone, starting with connectivity. VPN connections are still the most commonly used way to connect to public clouds, but enterprises might also choose direct dedicated connections. In this chapter, the various direct connectivity offerings of Azure, AWS, GCP, Alibaba Cloud, and OCI have been discussed. As soon as we have connectivity, we can start managing workloads in the cloud. The main concern for administrators is to keep resources and configurations consistent in different deployments.

Finally, we learned that there's a need for a redundant demarcation model in multi-cloud. This all adds up to the concept of BaseOps: getting the basics right.

Part of getting the basics right is making sure that our environments are resilient and performing well. That's what we will be discussing in the next chapter, where we will concentrate on risks and how to mitigate these.

Questions

1. A basic infrastructure in the cloud consists of five major domains, three of which are network, compute, and storage. What are the other two domains?
2. What solution does Azure offer to provide scaling of landing zones?
3. AWS offers a service that enables the central management of guardrails for resources. What's the name of this service?
4. What does RAM stand for in Alibaba Cloud?

Further reading

- *Azure for Architects*, by Ritesh Modi, Packt Publishing
- *Architecting Cloud Computing Solutions*, by Kevin L. Jackson, Packt Publishing

7

Designing Applications for Multi-Cloud

An important topic in any multi-cloud architecture is the resilience and performance of the applications that we will host in the cloud. We will have to decide on the type of solution that fits the business requirements and mitigates the risks of environments not being available, not usable, or not secured. Solutions and concepts such as 12-factor apps, PaaS, and SaaS must be considered as strategies.

In this chapter, we will gather and validate business requirements for resilience and performance. We will learn how to use tools to build resiliency in cloud platforms and how to optimize the performance of our application environments using advisory tools and support plans.

We will cover the following topics in this chapter:

- Architecting for resilience and performance
- Starting with business requirements
- Using the principles of the 12-factor app
- Accelerating application design with PaaS
- Designing SaaS solutions
- Performance KPIs in a public cloud—what's in it for you?
- Optimizing your multi-cloud environment
- Use case: creating solutions for business continuity and disaster recovery

Architecting for resilience and performance

Resilience and performance are two important factors in application architecture. Cloud platforms offer great tools to enhance both resilience and performance. Typically, these factors are included in the Well-Architected Frameworks that we discussed in the previous chapter. But what do we mean by resilience and performance?

We can use the definition that AWS uses for resilience in its Well-Architected Framework: *the capability to recover when stressed by load, attacks, and failure of any component in the workload's components*. To simplify this a bit: if any component of a system fails, how fast and to what extent can it recover, preferably to its original state, without loss of data. Obviously, an organization will need to test that, and that's where Chaos Monkey comes in as a testing strategy. In *Chapter 18, Developing for Multi-Cloud with DevOps and DevSecOps*, we will discuss testing strategies in more detail since they are part of the DevSecOps Maturity Model.

Defining the resilience of systems starts with understanding the criticality to the business of that system and what service levels need to be applied. That begins with having a clear understanding of business requirements. But it's important to realize one major thing: resilience and performance are not strictly technical issues. It's also about operational responsibility, which is a shared responsibility across all teams that build and manage applications.

Starting with business requirements

In *Chapter 6, Controlling the Foundation Using Well-Architected Frameworks*, we talked a little bit about things such as availability, backup, and disaster recovery. In this chapter, we will take a closer look at the requirements and various solutions that cloud platforms offer to make sure that your applications are available, accessible, and, most of all, safe to use. Before we dive into these solutions and the various technologies, we will have to understand what the potential risks are for our business if we don't have our requirements clearly defined.

In multi-cloud, we recognize risks at various levels, again aligning with the principles of enterprise architecture.

Understanding data risks

The biggest risk concerning data is ambiguity about the ownership of the data. This ownership needs to be regulated and documented in contracts, as well as with the cloud providers. International and national laws and frameworks such as the **General Data Protection Regulation** (**GDPR**) already define regulations in terms of data ownership in Europe, but nonetheless, be sure that it's captured in the service agreements as well. First, involve your legal department or counselor in this process.

We should also make a distinction between types of data. Is it business data, metadata, or data that concerns the operations of our cloud environments? In the latter category, you can think of the monitoring logs of the VMs that we host in our cloud. For all these kinds of data, there might be separate rules that we need to adhere to in order to be compliant with laws and regulations.

We need to know and document where exactly our data is. Azure, AWS, and GCP have global coverage and will optimize their capacity as much as they can by providing resources and storage from the data centers where they have that capacity. This can be a risk. For example, a lot of European countries specify that specific data cannot leave the boundaries of the European Union (EU). In that case, we will need to ensure that we store data in cloud data centers that are in the EU. So, we need to specify the locations that we use in the public cloud: the region and the actual country where the data centers reside.

Recently, the US implemented a new law protecting European citizens from unauthorized usage of data. The possibility that privacy-sensitive data from European citizens could be tapped by American companies—including the major cloud providers Microsoft, AWS, and GCP, which provide a lot of services to European companies and individuals—was an immense legal issue. In September of 2022, US President Joe Biden signed a bill stating that an independent institute—not related to the US government—has to judge whether European data is well-enough protected from unauthorized usage. The debate about this shows how complicated this matter is when data is collected, used, and shared across global cloud platforms.

We also need to ensure that when data needs to be recovered, it's recovered in the desired format and in a readable state. Damaged or incomplete data is the risk here. We should execute recovery tests on a regular basis and have the recovery plans and the actual test results audited. This is to ensure that the data integrity is guarded at all times. This is particularly important with transaction data. If we recover transactions, we need to ensure that all the transactions are recovered, but also that the transactions are not replicated during the recovery procedure. For this, we also need to define who's responsible for the quality of the data, especially in the case of SaaS.

To help with structuring all these requirements, a good starting point would be to create a model for data classification. Classification helps you decide what type of solution needs to be deployed to guarantee the confidentiality, integrity, and availability of specific datasets. Some of the most commonly used data categories are public data, confidential company data, and personal data.

Understanding application risks

The use of SaaS is becoming increasingly popular. Many companies have a strategy that prefers SaaS before PaaS before IaaS. In terms of operability, this might be the preferred route to go, but SaaS does come with a risk. In SaaS, the whole solution stack is managed by the provider, including the application itself. A lot of these solutions work with shared components, and you have to combat the risk of whether your application is accessible to others or whether your data can be accessed through these applications. A solution to mitigate this risk is to have your own application runtime in SaaS.

One more risk that is associated with the fact that the whole stack—up until the application itself—is managed by the provider is that the provider can be forced out of business. At the time of writing, the world is facing the coronavirus pandemic, and a lot of small businesses are really struggling to survive. We are seeing businesses going down, and even in IT, it's not always guaranteed that a company will keep its head above water when a severe crisis hits the world. Be prepared to have your data safeguarded whenever a SaaS provider's propositions might be forced to stop the development of the solution or, worse, to close down the business.

Also, there's the risk that the applications fail and must be restored. We have to make sure that the application code can be retrieved and that applications can be restored to a healthy state.

Understanding technological risks

We are configuring our environments in cloud platforms that share a lot of components, such as data centers, the storage layer, the compute layer, and the network layer. By configuring our environment, we merely **silo**. This means that we are creating a separate environment—a certain area on these shared services. This area will be our virtual data center. However, we are still using the base infrastructure of Azure, AWS, GCP, or any other cloud.

Even the major cloud providers can be hit by outages. It's up to the enterprise to guarantee that their environments will stay available, for instance, by implementing redundancy solutions when using multiple data centers, zones, or even different global regions.

Observability and monitoring are a must, but they don't have any added value if we're looking at only one thing in our stack or at the wrong things. Bad monitoring configuration is a risk. As with security, the platforms provide their customers with tools, but the configuration is up to the company that hosts its environment on that platform.

Speaking of security, one of the biggest risks is probably weak security. Public clouds are well protected as platforms, but the protection of your environment always remains your responsibility. Remember that clouds are a wonderful target for hackers since they're a platform hosting millions of systems. That's exactly the reason why Microsoft, Amazon, and Google spend a fortune securing their platforms. Make sure that your environments on these platforms are also properly secured and implement endpoint protection, hardening of systems, network segmentation, firewalls, and vulnerability scanning, as well as alerting, intrusion detection, and prevention. You also need to ensure you have a view of whatever is happening on the systems.

However, do not overcomplicate things. Protect what needs to be protected but keep it manageable and comprehensible. The big question is, what do we need to protect, to what extent, and against which costs? This is where gathering business requirements begins.

Although we're talking about technological, application, and data risks, at the end of the day, it's about business requirements. These business requirements drive data, applications, and technology, including risks. So far, we haven't answered the question of how to gather these business requirements.

The main goal of this process is to collect all the relevant information that will help us create the architecture, design our environments, implement the right policies, and configure our multi-cloud estate as the final product. Now, this is not a one-time exercise. Requirements will change over time, and especially in the cloud era, the demands and, therefore, the requirements are constantly changing at an ever-increasing speed. So, gathering requirements is an ongoing and repetitive process.

How do we collect the information we need? There are a few key techniques that we can use for this:

- **Assessment**: A good way to do this is to assess whether we're assessing resilience and performance from the application layer. What does an application use as resources, and against which parameters? How are backups scheduled and operated? What are the restore procedures? Is the environment audited regularly, what were the audit findings, and have these been recorded, scored, and solved? We should also include the end-user experience regarding the performance of the application and under what conditions, such as office rush hour, when the business day starts, and normal hours.
- **Stakeholder interviews**: These interviews are a good way to understand what the business need is about. We should be cautious, though. Stakeholders can have different views on things such as what the business-critical systems are.

- **Workshops**: These can be very effective for drilling down a bit deeper into the existing architectures, the design of systems, the rationales behind demands, and the requirements, while also giving us the opportunity to enforce decisions since all stakeholders will ideally be in one room. A risk of this is that discussions in workshops might become too detailed. A facilitator can help steer this process and get the desired result.

Once we have our requirements, then we can map to the functional parameters of our solution. A business-critical environment can have the requirements that it needs to be available 24/7, 365 days a year. The system may hold transactions where every transaction is worth a certain amount of money. Every transaction that's lost means that the company is losing money. The systems handle a lot of transactions every minute, so every minute of data loss equals an amount of real financial damage. This could define the **recovery point objective** (**RPO**)—the maximum amount of data loss the company finds acceptable—which should be close to 0. This means that we have to design a solution that is highly available, redundant, and guarded by a disaster recovery solution with a restore solution that guarantees an RPO of near 0—possibly a solution that covers **data loss prevention** (**DLP**).

Is it always about critical systems, then? Not necessarily. If we have development systems wherein a lot of developers are coding, the failure of these systems could actually trigger a financial catastrophe for a company. The project gets delayed, endangering the time to market of new services or products, and the developers will sit idle, but the company will still have to pay them. It's always about the business case, the risks a company is willing to accept, and the cost that the company is willing to pay to mitigate these risks.

Using the principles of the 12-factor app

The 12-factor app is a methodology for building modern apps, ready for deployment on cloud platforms, abstracted from server infrastructure. The major benefit of applying the 12-factors is that developers can create apps that are not only scalable, but also reliable since the app is self-contained. Dependencies are clearly defined and isolated from the app. This includes the dependencies on the underlying infrastructure.

Next, it allows for a consistent way of managing code through a single code base and a defined, declarative way of deployment.

The format of the 12-factor app is based on the book *Patterns of Enterprise Application Architecture and Refactoring* by Martin Fowler, published by Addison-Wesley Professional, 2005. Although it was written in 2005—a long time ago in terms of cloud technology—it is still very relevant in terms of its architecture. In cloud-native development, the 12-factor principles are widely used.

The twelve factors are as follows:

1. **Code base**: One code base is tracked in revision control, with many deployments (including Infrastructure as Code).
2. **Dependencies**: Explicitly declare and isolate dependencies.
3. **Config**: Store config in the environment (Configuration as Code).
4. **Backing services**: Treat backing services as attached resources. Think of databases and queues.
5. **Build, release, run**: Strictly separate build and run stages (pipeline management, release trains).
6. **Processes**: Execute the app as one or more stateless processes.
7. **Port binding**: Export services via port binding.
8. **Concurrency**: Scale out via the process model.
9. **Disposability**: Maximize robustness with fast startup and graceful shutdown.
10. **Dev/prod parity**: Keep development, staging, and production as similar as possible.
11. **Logs**: Treat logs as event streams.
12. **Admin processes**: Run admin/management tasks as one-off processes.

More on these principles can be found at `https://12factor.net`. We will find out that a number of these twelve principles are also captured in different frameworks and embedded in the governance of multi-cloud environments, such as DevOps and **Site Reliability Engineering** (**SRE**).

Accelerating application design with PaaS

PaaS is the abbreviation for **Platform as a Service**. The first question that we must answer then is: what is a platform? Opinions and definitions differ in this case. The best way to define a platform would be a service or collection of services that enable teams and organizations to speed up the development and delivery of software. To put it very simply: PaaS makes life easy for software teams.

With the use of PaaS, the teams do not need to bother about a wide variety of services that they need to build before they can start with the software product that really adds business value. In essence, PaaS is an abstraction layer on top of the infrastructure. Let's have another look at the diagram that we discussed in the previous chapter, where we talked about demarcation:

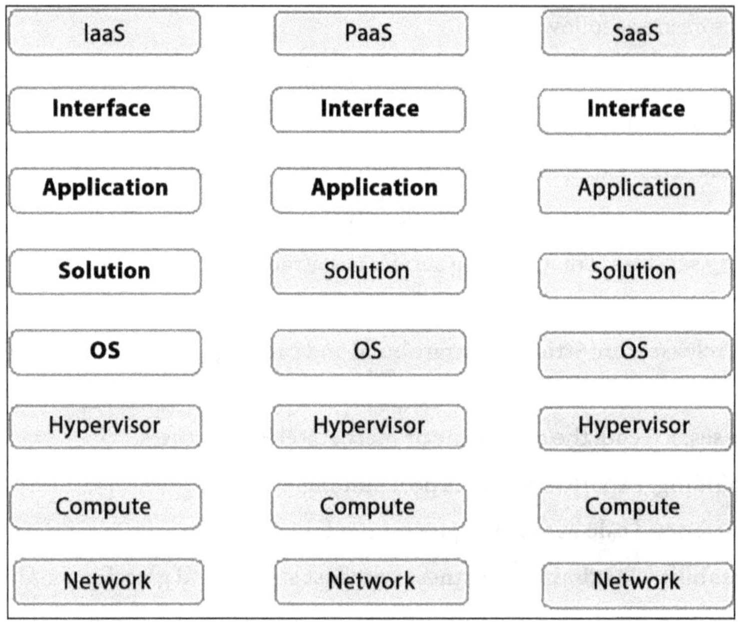

Figure 7.1: The differences in responsibilities between IaaS, PaaS, and SaaS

In PaaS everything up until the operating system is taken care of by the platform. Developers only have to write code for the applications.

The container hosting platform Kubernetes fits the definition. Kubernetes manages the infrastructure for containers that are used to host applications. Developers only need to worry about the application code. The code is packaged into containers that land on the Kubernetes clusters. Kubernetes, as a platform, makes sure that the underlying infrastructure is available. The major advantages of using PaaS are:

- **Limit complexity**: PaaS is on a higher abstraction layer than IaaS.
- **Efficient onboarding for applications**: A few bloggers will call this the "paved way for shipping stuff." There's no need to configure infrastructure: we only need to build the application; the platform does the rest.
- **Consistency**: Since everything is taken care of by the platform, the underlying infrastructure remains consistent across the various development teams.

- **Compliancy**: Compliancy is built into the platform services. Guardrails are implemented and managed at the platform level.

The use of PaaS can accelerate the development of applications, but there's a downside. Platforms require heavy standardization, and this may lead to restrictions for the developers of applications.

Designing SaaS solutions

The market for **Software as a Service (SaaS)** has been growing with massive numbers over the past five years. It's expected that the market for **SaaS** will grow to over 600 billion US dollars during the year 2023.

SaaS as a strategy makes sense if a company wants to expand its business fast. SaaS products allow customers access to applications over the internet without needing to buy a physical product that they need to install themselves on a device. SaaS products are centrally hosted on cloud platforms, where the product is fully managed by the developers. The customer is assured of the newest releases because of the central hosting and management model.

SaaS is extremely customer-centric: the customer simply subscribes to the service, gets access, and, next, the service is ready to use on the device of the customer's choice. Hence, developing SaaS products has become sort of the holy grail for many enterprises.

The major cloud providers offer toolkits to develop SaaS functionality on their platforms. AWS calls this the SaaS Factory, and Azure has the SaaS Development Kit. All these services aim to help companies transform monolithic applications into modern web-based applications that can be delivered as SaaS. At re:Invent 2020, the main convention of AWS, SaaS Boost was released. In 2022 version 2 was released. SaaS Boost provides out-of-the-box and ready-to-use core components to build software. It uses CloudFormation to dynamically create and deploy resources to run application code, using container infrastructure with **Elastic Container Services (ECS)** and Lambda serverless functions to orchestrate workflows between the various components.

The magic word in SaaS is *easy*. Subscribing and onboarding to SaaS should be easy and extremely user-friendly. The user interface, typically using a web interface, must be accessible and easy to use. Most important: SaaS must be available. Think of Netflix and Spotify: these services are always "on." They can't allow for major disruptions in the services. These environments are built as micro-services, allowing parts of the services to be continuously updated and upgraded in CI/CD pipelines, without causing downtime for the customers. Keep in mind that customers pay for these services in the form of subscription fees. Major outages will seriously cause trust issues with customers.

SaaS applications must be designed with the customer and the customer experience permanently in mind. The formula for SaaS is *Great User Experience + Great User Interface = Great Product.*

Performance KPIs in a public cloud—what's in it for you?

As we mentioned in the previous section, performance is a tricky subject and, to put it a bit more strongly, if there's one item that will cause debates in service-level agreements, it's going to be performance. In terms of KPIs, we need to be absolutely clear about what performance is, in terms of measurable objectives.

What defines performance? It's the user experience. What about how fast an application responds and processes a request? Note that *fast* is not a measurable unit. A lot of us can probably relate to this: a middle-aged person may think that an app on their phone responding within 10 seconds is fast, while someone younger may be impatiently tapping their phone a second after they've clicked on something. They have a relative perception of *fast*. Thus, we need to define and agree on what is measurably fast. One thing to keep in mind is that without availability, there's nothing to measure, so resilience is still a priority.

What should we measure? Let's take VMs in the cloud as an example since VMs are still widely used in cloud. Some key metrics that we must take into consideration are:

- **CPU and memory**: All cloud providers offer a wide variety of VM instance sizes. We should look carefully at what instance is advised for specific workloads. For instance, applications that run massive workflow processes in memory require a lot of memory in the first place. For example, SAP S4/HANA instances can require up to 32 GB of RAM or more. For these workloads, Azure, AWS, and GCP offer large, memory-optimized instances that are next to complete SAP solutions. If we have applications that run heavy imaging or rendering processes, we might want to look at specific instances for graphical workloads that use GPUs. So, it comes down to the right type of instance and infrastructure, as well as the sizing. You can't blame the provider for slow performance if we choose a low-CPU, low-memory machine underneath a heavy-duty application. Use the advisor tools to fulfill the best practices.

- **Responsiveness:** How much time does it take for a server or a service to get a response to a request? There are a lot of factors that determine this. To start with, it's about the right network configuration, routing, and dependencies in the entire application stack. It does matter whether we connect through a low bandwidth VPN or a high-speed dedicated connection. And it's also about load balancing. If, during peak times, the load increases, we should be able to deal with that. In the cloud, we can scale out and up, even in a fully automated way. In that case, we need proper configuration for the load balancing solution.
- **IO throughput**: This is about throughput rates on a server or in an environment. Throughput is a measure of **Requests Per Second** (**RPS**), the number of concurrent users, and the utilization of the resources, such as servers, connections, firewalls, and load balancers. One of the key elements in architecture is sizing. From a technological perspective, the solution can be well-designed, but if the sizing isn't done correctly, then the applications may not perform well or be available at all. The advisor tools that we will discuss in this chapter provide good guidance in terms of setting up an environment, preparing the sizes of the resources, and optimizing the application (code) as such.

The most important thing about defining KPIs for performance is that all stakeholders—business, developers, and administrators—have a mutual understanding of what performance should look like and how it should be measured.

Optimizing your multi-cloud environment

Cloud providers offer advisor tools we can use to optimize environments that are hosted on their platforms. In this section, we will briefly look at these different tools and how we can use them.

Optimizing environments using Azure Advisor

Like AWS, Azure offers a tool to help optimize environments, called Azure Advisor. Azure Advisor is a service that helps in analyzing our Azure environments and making recommendations around the pillars of the Well-Architected Framework. Next to this, we should evaluate the support plans.

Let's start with the support plans. Azure offers four types of plans: basic, developer, standard, and professional direct. The basic plan is free, while the most expensive one is professional direct. However, you can't compare this to the enterprise plan of AWS. Every provider offers free and paid services—the difference per provider is which services are free or must be paid for.

Azure Advisor comes at no extra cost. It provides recommendations on the five pillars of WAF: reliability, security, performance, costs, and operational excellence. The dashboard can be launched from the Azure portal, and it will immediately generate an overview of the status of the resources that we have deployed in Azure, as well as recommendations for improvements. For resiliency, Azure Advisor, for example, checks whether VMs are deployed in an availability set, thereby remediating fault tolerance for VMs. Be aware of the fact that Azure Advisor only advises that the actual remediation needs to be done by the administrator. It offers a recommended action for a recommendation to implement it. A simple interface will open that enables you to implement the recommendation or refer you to documentation that assists you with implementation. We could also automate this with Azure Policy and Azure Automation, but there's a good reason why Azure Advisor doesn't already do this. Remediation actions might incur extra costs, and we want to stay in control of our costs. If we automate through Policy and Automation, that's a business decision and architectural solution that will be included in cost estimations and budgets.

On the other hand, Azure Advisor does provide us with some best practices. In terms of performance, we might be advised to start using managed disks for our app, do a storage redesign, or increase the sizes of our VNets. It's always up to the administrator to follow up, either through manual tasks or automation.

Next to Advisor, we will use Azure Monitor to guard our resources and Azure Service Health to monitor the status of Azure services. Specific to security monitoring, Azure offers Microsoft Defender for Cloud, previously known as Azure Security Center, and the Azure-native **Security Information and Event Manager** (**SIEM**) tool known as Sentinel. Most of these services are offered on a pay-as-you-go basis: you pay for the ingestion and retention of data that they collect.

More information on Azure Advisor can be found at `https://learn.microsoft.com/en-us/azure/advisor/advisor-overview`.

Using Trusted Advisor for optimization in AWS

In all honesty, getting the best out of AWS—or any other cloud platform—is really not that easy. There's a good reason why these providers have an extensive training and certification program. The possibilities are almost endless, and the portfolios for these cloud providers grow bigger every day. We could use some guidance while configuring our environments. AWS provides that guidance with **Trusted Advisor**. This tool scans your deployments, references them against best practices within AWS, and returns recommendations. It does this for cost optimization, security, performance, fault tolerance, and service limits.

Before we go into a bit more detail, there's one requirement we must fulfill in order to start using Trusted Advisor: we have to choose a support plan, although a couple of checks are for free, such as a check on **Multi-Factor Authentication** (**MFA**) for root accounts and IAM use. Also, checks for permissions on S3 (storage) buckets are free. Note that basic support is included at all times, including AWS Trusted Advisor on seven core checks, mainly focusing on security. Also, the use of the Personal Health Dashboard is included in basic support.

Support plans come in three levels: developer, business, and enterprise. The latter is the most extensive one and offers full 24/7 support on all checks, reviews, and advice on the so-called Well-Architected Framework, as well as access to AWS support teams. The full service does come at a cost, however. An enterprise that spends one million dollars on AWS every month would be charged around 70 USD per month on this full support plan. This is because AWS typically charges the service against the volumes that a customer has deployed on the platform. The developer and business plans are way lower than that. The developer plan can be a matter of as little as 30 USD per month, just to give you an idea. Note that prices change all the time.

However, this full support does include advice on almost anything that we can deploy in AWS. The most interesting parts, though, are the service limits and performance. The service limits perform checks on volumes and the capacity of a lot of different services. It raises alerts when 80 percent of a limit for that service has been reached, and it then gives advice on ways to remediate this, such as providing larger instances of VMs, increasing bandwidth, or deploying new database clusters. It strongly relates to the performance of the environment: Trusted Advisor checks the high utilization of resources and the impact of this utilization on the performance of those resources.

The full checklist for Trusted Advisor can be found at `https://aws.amazon.com/premiumsupport/technology/trusted-advisor/best-practice-checklist/`.

One more service that we should mention in this section is the free Personal Health Dashboard, which provides us with very valuable information on the status of our resources in AWS. The good news is that not only does it provide alerts when issues occur and impact your resources, but it also guides us through remediation. What's even better is that the dashboard can give you proactive notifications when planned changes might affect the availability of resources. The dashboard integrates with AWS CloudWatch, but also with third-party tooling such as Splunk, ServiceNow, and Datadog.

Optimizing GCP with Cloud Trace and Cloud Debugger

GCP offers two interesting features in terms of optimizing environments: Cloud Trace and Cloud Debugger. Both can be accessed from the portal. From this, you can tell that GCP comes from the cloud-native and native apps world.

Cloud Trace is really an optimization tool: it collects data on latency from the applications that you host on instances in GCP, whether these are VMs, containers, or deployments in App Engine or the native app environment in GCP. Cloud Trace measures the amount of time that elapses between incoming requests from users or other services and the time the request is processed. It also keeps logs and provides analytics so that you can see how performance evolves over a longer period of time. Cloud Trace uses a transaction client that collects data from App Engine, load balancers, and APIs. It gives us good insight into the performance of apps, dependencies between apps, and ways to improve performance.

Cloud Trace doesn't only work with GCP assets but with non-Google assets too. In other words, we can use Cloud Trace in AWS and Azure as a REST API using JSON.

Cloud Debugger is another tool, and it's used to debug code in apps that you run in GCP. Debugger will analyze the code while the application is running. It does this by taking a snapshot of the code, although you can use it on the source code as well. It integrates with versioning tools such as GitHub. Debugger supports the most commonly used programming languages in code apps, at least when it runs in containers on GKE. In this case, Java, Python, Go, Node.js, Ruby, PHP, and .NET Core are supported. In Compute Engine, .NET Core is not supported at the time of writing.

Cloud Trace and Cloud Debugger are part of the operations suite of GCP and are a charged service.

More information on Cloud Trace can be found at `https://cloud.google.com/trace/docs/overview/`. Documentation on Cloud Debugger can be found at `https://cloud.google.com/debugger/docs`.

Optimizing in OCI

In OCI, we will find a number of tools that will help us in optimizing workloads.

Oracle Cloud Infrastructure Monitoring provides real-time performance metrics, logs, and alerts to help optimize the performance of OCI resources. It monitors the health and performance of resources such as compute instances, load balancers, and databases.

Another tool is Oracle Cloud Infrastructure Advisor, which analyzes OCI resources and provides recommendations to help optimize your infrastructure. It offers suggestions for improving performance, reducing costs, and enhancing security.

Next to this, OCI also offers Oracle Cloud Infrastructure Resource Manager, which helps to automate the creation, deployment, and management of resources. It offers a simple way to manage infrastructure as code and ensure consistency across your environment.

Use case: creating solutions for business continuity and disaster recovery

Now that we have gathered the business requirements, identified the risks, and considered our application strategy, including the usage of PaaS and SaaS, we can start thinking about solutions and align these with the requirements. The best way to do this is to create a matrix with the systems, the requirements for resilience, and the chosen technology to get the required resilience. The following table shows a very simple example of this, with purely fictional numbers:

System or system group/category	Business level	RTO	RPO	Solution
Application X	Critical	<2 hours	<2 hours	Standby (failover) systems, disaster recovery
Application Y	Important	>2 hours <8hours	>2 hours <8 hours	Full daily backup, increments with snapshots
Application Z	Non-critical	<48 hours	<48 hours	Weekly backup, daily incrementals

Table 7.1: Example of a table for required resilience KPIs

Resilient systems are designed in such a way that they can withstand disruptions. Regardless of how well the systems might be designed and configured, sooner or later, they will be confronted with failures and, possibly, disruptions. Resilience is, therefore, often associated with quality attributes such as redundancy and availability.

Creating backups in the Azure cloud with Azure Backup and Site Recovery

Azure Backup works with the principle of snapshots. First, we must define the schedule for running backups. Based on that schedule, Azure will start the backup job. During the initial execution of the job, the backup VM snapshot extension is provisioned on the systems in our environment.

Azure has extensions for both Windows and Linux VMs. These extensions work differently from each other: the Windows snapshot extension works with Windows **Volume Shadow Copy Services (VSS)**. The extension takes a full copy of the VSS volume. On Linux machines, the backup takes a snapshot of the underlying system files.

Next, we can take backups of the disks attached to the VM. The snapshots are transferred to the backup vault. By default, backups of operating systems and disks are encrypted with Azure Disk Encryption. The following diagram shows the basic setup for the Azure Backup service:

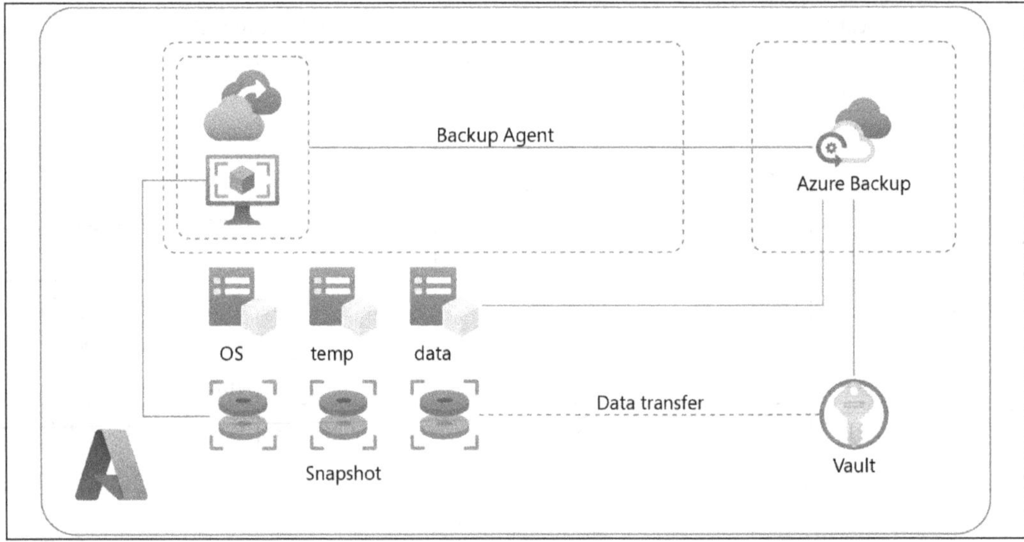

Figure 7.2: High-level overview of the standard backup components in Azure

We can create backups of systems that are in Azure, but we can also use Azure Backup for systems that are outside the Azure cloud.

Backing up non-Azure systems

Azure Backup can be used to create backups of systems that are not hosted in Azure itself. For that, it uses different solutions. **Microsoft Azure Recovery Services (MARS)** is a simple solution to do this. In the Azure portal, we have to create a Recovery Services vault and define the backup goals.

Next, we need to download the vault credentials and the agent installer that must be installed on the on-premises machine or machines that are outside Azure. With the vault credentials, we register the machine and start the backup schemes. A more extensive solution is **Microsoft Azure Backup Server (MABS)**. MABS is a real VM—running Windows Server 2016 or 2019—that controls the backups within an environment, in and outside Azure. It can execute backups on a lot of different systems, including SQL Server, SharePoint, and Exchange, but also VMware VMs—all from a single console.

MABS, like MARS, uses the recovery vault, but in this case, backups are stored in a geo-redundant setup by default. The storage replication option allows you to choose between geo-redundant storage and locally redundant storage. The following diagram shows the setup of MABS:

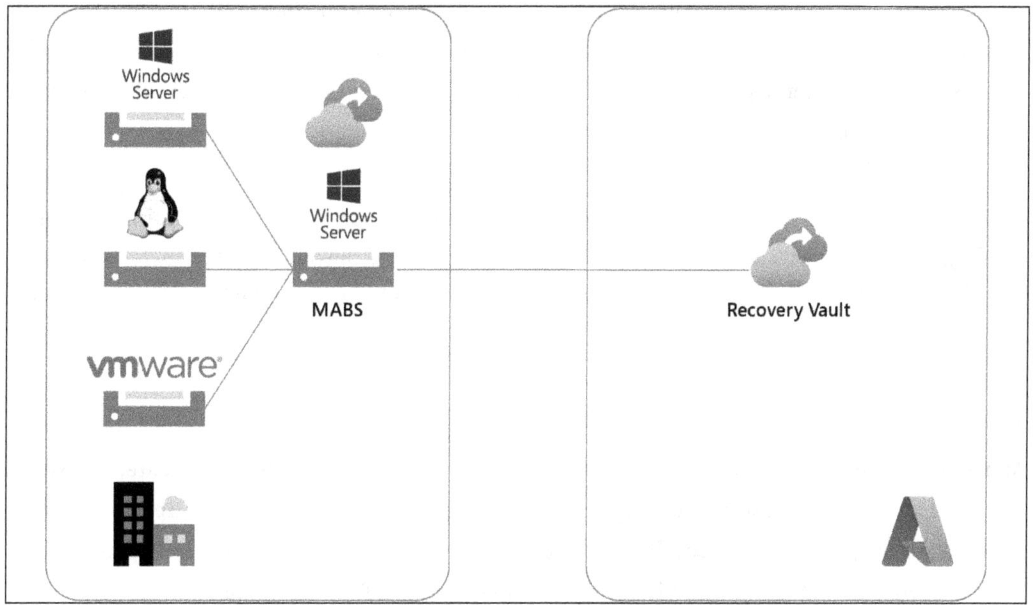

Figure 7.3: High-level overview of the setup for Microsoft Azure Backup Server

Documentation on the different backup services that Azure provides can be found at `https://docs.microsoft.com/en-us/azure/backup/`.

Before we dive into the recovery solutions for Azure and the other cloud providers, we will discuss the generic process of disaster recovery. Disaster recovery has three stages: detect, response, and restore. We must have monitoring in place that is able to detect whether critical systems are failing and that they are not available anymore.

It then needs to trigger actions to execute mitigating actions, such as failover to standby systems that can take over the desired functionality and ensure that business continuity is safeguarded. The last step is to restore the systems back to the state that they were in before the failure occurred. In this last step, we also need to make sure that the systems are not damaged in such a way that they can't be restored.

Recovery is a crucial element in this process. However, recovery can mean that systems are completely restored back to their original state where they still were fully operational, but we can also have a partial recovery where only the critical services are restored, and, for example, the redundancy of these systems must be fixed at a later stage. Two more options are **cold standby** and **warm standby**. With cold standby, we will have systems that are reserved that we can spin up when we need them. Until that moment, these systems are in shut down modus. In warm standby, the systems are running but not yet operational in production modus. Warm standby servers are much faster to get operational than cold standby servers, which merely have reserved capacity available.

Donald Firesmith wrote an excellent blog post about resilience for the Software Engineering Institute of Carnegie Mellon University. You can find it at `https://insights.sei.cmu.edu/sei_blog/2019/11/system-resilience-what-exactly-is-it.html`.

Understanding Azure Site Recovery

Azure Site Recovery (**ASR**) offers a solution that helps set up disaster recovery in Azure. In essence, ASR replicates and orchestrates the replication of VMs (between regions in Azure, on-premises to Azure, or on-premises to secondary on-premises). If the primary location where you host the environments becomes unavailable because of an outage, ASR will execute a failover to the secondary location, where the copies of your systems are. As soon as the primary location is back online again, ASR will execute the failback to that location again.

A bit of bad news is that this is not as simple as it sounds. You will need to design a recovery plan and assess whether workloads can actually failover from the application and data layer. Then, probably the trickiest part is getting the network and boundary security parameters right: think of switching routes, reserved IP addressing, DNS, and replicating firewall rules.

Azure has solutions for this as well, such as DNS routing with traffic manager, which helps with DNS switching in case of a failover, but still, it takes some engineering and testing to get this in place.

The last thing that really needs serious consideration is what region you will have the secondary location in. A lot of Azure regions do have dual zones (data centers), but there are some regions that only have one zone, and you will need to choose another region for failover. Be sure that you are still compliant in that case. To ensure resiliency, a minimum of three separate availability zones are present in all availability zone-enabled regions.

The following diagram shows the basic concept of ASR. It's important to remember that we need to set up a cache storage account in the source environment. During replication, changes that are made to the VM are stored in the cache before being sent to storage in the replication environment:

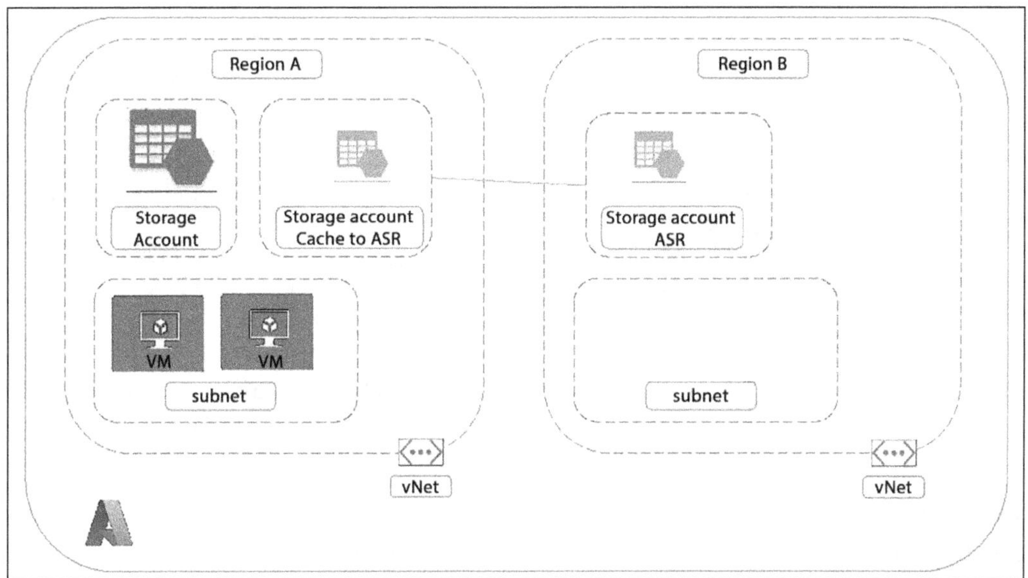

Figure 7.4: High-level overview of ASR

More information on ASR can be found at `https://docs.microsoft.com/en-us/azure/site-recovery/`.

With that, we have covered Azure. In the next section, we will look at backup and disaster recovery in AWS and GCP.

Working with AWS backup and disaster recovery

In this section, we will explore the backup and disaster recovery solutions in AWS. We will learn how to create backups based on policies and on tags. We will also look at the hybrid solution for AWS.

Creating policy-based backup plans

As in Azure, this starts with creating a backup plan that comprises backup rules, frequency, windows, and defining and creating the backup vault and the destination where the backups should be sent. The backup vault is crucial in the whole setup: it's the place where the backups are organized and where the backup rules are stored. You can also define the encryption key in the vault that will be used to encrypt the backups. The keys themselves are created with AWS **Key Management Service (KMS)**. AWS provides a vault by default, but enterprises can set up their own vaults.

With this, we have defined a backup plan, known as backup policies in AWS. These policies can now be applied to resources in AWS. For each group of resources, you can define a backup policy in order to meet the specific business requirements for those resources. Once we have defined a backup plan or policy and we have created a vault, we can start assigning resources to the corresponding plan. Resources can be from any AWS service, such as EC2 compute, DynamoDB tables, **Elastic Block Store (EBS)** storage volumes, **Elastic File System (EFS)** folders, **Relational Database Services (RDS)** instances, and storage gateway volumes.

Creating tag-based backup plans

Applying backup plans or policies to resources in AWS can be done by simply tagging the plans and the resources. This integration with tags makes it possible to organize the resources and have the appropriate backup plan applied to these resources. Any resource with a specific tag will then be assigned to the corresponding backup plan. An example, if we have set out policies for business-critical resources, we can define a tag that says `BusinessCritical` as a parameter for classifying these resources. If we have defined a backup plan for `BusinessCritical`, every resource with that tag will be assigned to that backup plan.

The following diagram shows the basic concept of AWS Backup:

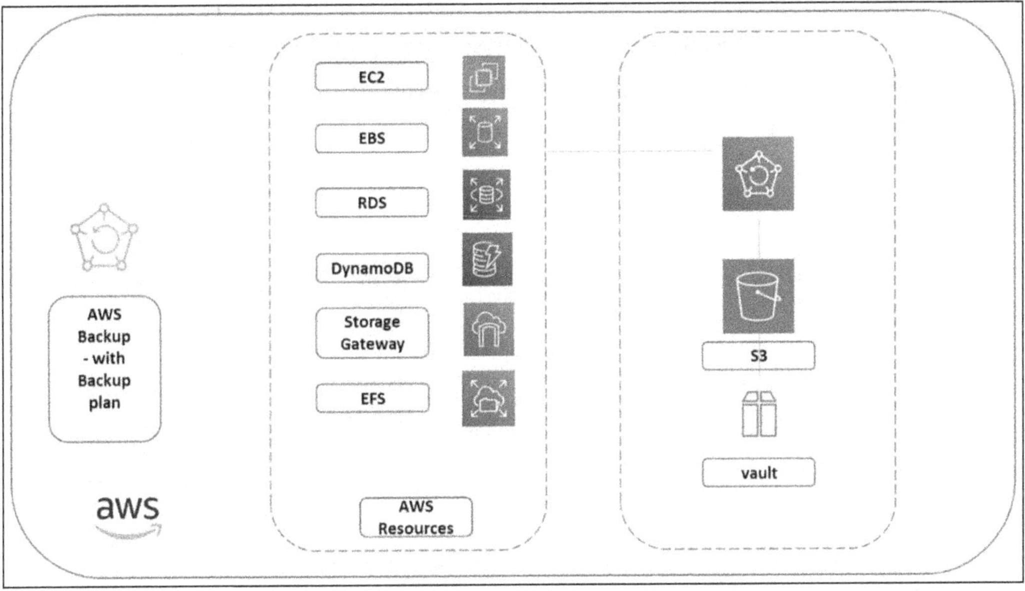

Figure 7.5: High-level overview of AWS Backup

Similar to Azure, we can also create backups of systems that are outside of AWS using the hybrid backup solution of AWS. We'll describe this in the next section.

Hybrid backup in AWS

AWS calls backing up resources in AWS a native backup, but the solution can be used for on-premises workloads too. This is what AWS calls hybrid backup. For this, we'll have to work with the AWS Storage Gateway. We can compare this to MABS, which Microsoft offers.

In essence, the on-premises systems are connected to a physical or virtual appliance over industry-standard storage protocols such as **Network File System (NFS)**, **Server Message Block (SMB)**, and **internet Small Computer System Interface (iSCSI)**. The appliance—the storage gateway—connects to the AWS S3 cloud storage, where backups can be stored. You can use the same backup plans for hybrid backup that you do for the native backup. The following diagram shows the principle of hybrid backup:

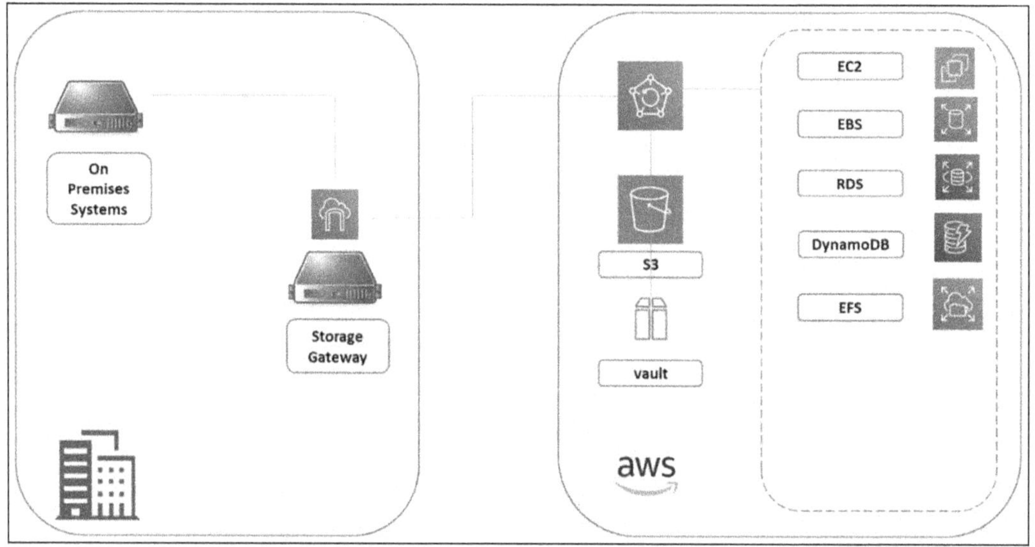

Figure 7.6: High-level overview of hybrid backup in AWS

Now, let's learn about the disaster recovery options available in AWS.

AWS disaster recovery and cross-region backup

AWS allows us to perform cross-region backups, meaning that we can make backups according to our backup plans and replicate these to multiple AWS regions. However, this occurs in the data layer. We can do this for any data service in AWS like RDS, EFS, EBS, and Storage Gateway volumes. So, with Storage Gateway included, it also works for data that is backed up on-premises.

Next to this, AWS also has another proposition that's a **Business Continuity and Disaster Recovery (BCDR)** solution: CloudEndure **disaster recovery (DR)**. This solution doesn't work with snapshots but keeps target systems for DR continuously in sync with the source systems with continuous data protection. By doing this, they can even achieve sub-second recovery points and barely lose any data. CloudEndure supports a lot of different systems, including physical and virtualized machines, regardless of the hypervisor. It also supports enterprise applications such as Oracle and SAP.

This principle is shown in the following diagram:

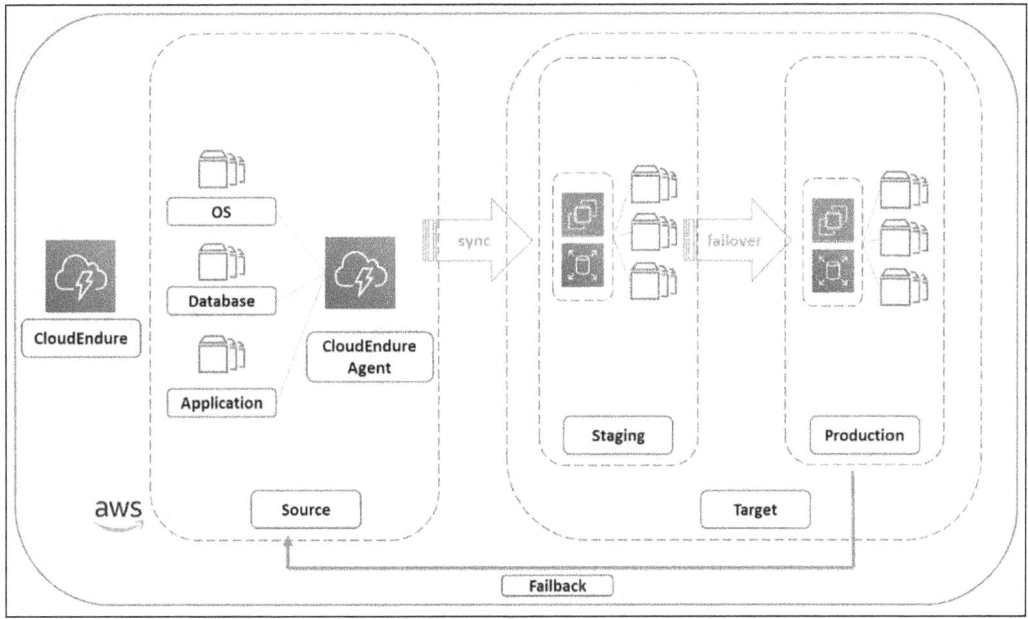

Figure 7.7: High-level overview of the CloudEndure concept in AWS

CloudEndure uses agents on the source systems and a staging area in AWS where the system duplicates are stored on low-cost instances. In the case of a failover, the target DR systems are booted and synced from this staging area. The failback is done from the DR systems in AWS.

More information on AWS Backup can be found at `https://docs.aws.amazon.com/aws-backup/latest/devguide/whatisbackup.html`. Documentation on CloudEndure can be found at `https://aws.amazon.com/cloudendure-disaster-recovery/`.

Creating backup plans in GCP

GCP also uses snapshot technology to execute backups. The first snapshot is a full backup, while the ones that follow are iterative and only back up the changes that have been made since the last snapshot. If we want to make a backup of our data in GCP Compute Engine, we have to create a persistent disk snapshot. It's possible to replicate data to a persistent disk in another zone or region, thus creating geo-redundancy and a more robust solution.

As with AWS and Azure, you will first have to design a backup plan, or in GCP, a snapshot schedule so that backups are taken at a base frequency. Next, we have to set the storage location. By default, GCP chooses the region that is closest to the source data. If we want to define a solution with higher availability, we will need to choose another region ourselves where we wish to store the persistent disks.

Be aware that, in GCP, we work with constraints. If we have defined a policy with a constraint that data can't be stored outside a certain region, and we do pick a region outside of that boundary, the policy will prevent the backup from running.

GCP proposes to flush the disk buffers prior to the backup as a best practice. You don't need to stop the applications before snapshots are taken, but GCP does recommend this so that the application stops writing data to disk. If we stop the application, we can flush the disk buffer and sync the files before the snapshot is created. For Unix programmers, this will be very familiar since GCP lets you connect with SSH to the disk and `sudo sync` to execute the synchronization process. All of this is done through the command-line interface.

But what about Windows? We can run Windows-based systems in GCP, and we can take backups of these systems. GCP uses VSS for this, which is the Volume Shadow Copy Services of Windows. Before we do that, GCP recommends unmounting the filesystem and then taking the snapshots. We can use PowerShell for this.

More information on backups in Compute Engine from GCP can be found at `https://cloud.google.com/compute/docs/disks/create-snapshots`. Specific documentation and how-to information about creating snapshots of Windows systems can be found at `https://cloud.google.com/compute/docs/instances/windows/creating-windows-persistent-disk-snapshot`.

Disaster recovery planning

GCP lets us define the RTO and RPO first when planning a DR strategy. Based on the requirements, we define the building blocks within GCP to fulfill the strategy. These building blocks comprise Compute Engine and Cloud Storage. In Compute Engine, we can define how we want our VMs to be deployed and protected from failures. The key components here are persistent disks, live migration, and virtual disk import. We discussed the creation of persistent disks as part of backing up in GCP. Live migration keeps VMs up by migrating them into a running state to another host. Virtual Disk Import lets you import disks from any type of VM to create new VMs in Compute Engine.

These newly created machines will then have the same configuration as the original machines. The supported formats are **Virtual Machine Disk** (**VMDK**), which is issued by VMware, and **Virtual Hard Drive** (**VHD**), which is used by Microsoft and RAW.

As you can tell, GCP does not offer a predefined solution for DR. And again, there's much more focus on containers with GKE. GKE has some built-in features that you can use to create highly available clusters for containers. Node auto-repair checks the health status of cluster nodes: if nodes don't respond within 10 minutes, they get automatically remediated. If we're running nodes in a multi-region setup, GKE offers multi-cluster ingress for Anthos as a load-balancing solution between the clusters. This solution was previously offered as **Kubernetes Multicuster Ingress** (**Kubemi**). For all of this, we do need a solution to route the traffic across the GCP regions and to make sure that DNS is pointing to the right environments. This is done through Cloud DNS using Anycast, the Google global network, and Traffic Director.

More information on disaster recovery planning in GCP can be found at `https://cloud.google.com/solutions/dr-scenarios-planning-guide`.

Google suggests looking at third-party tools when we have to set up a more complex DR infrastructure solution. Some of the tools that are mentioned in the documentation include Ansible with Spinnaker and Zero-to-Deploy with Chef on Google Cloud.

Creating backups in OCI

OCI offers several options for creating backups, including the Backup service, Database backup, Object Storage, and Compute Instance backup.

The first solution to look at is the OCI Backup service, as a fully managed, centralized backup solution for all resources that we host in OCI. It offers automatic backup policies and retention policies and allows you to create on-demand backups of resources. The service supports backups for a variety of OCI resources, including compute instances, block volumes, file storage, and database services.

We can use OCI Database backup specific for the databases that we run in OCI. This is a built-in backup and recovery feature. The service offers several backup types, including full, incremental, and automated backups, and allows us to schedule backups and retention policies. The backup files are stored in OCI Object Storage. Restoring a database can be done to any point in time within the backup retention period.

If we have deployed OCI Compute instances, we can create manual backups using the OCI Console or the OCI CLI. The backup creates a boot volume backup, which includes the instance's boot volume and any attached Block Volumes. Now, we can restore the backup to a new instance or to the same instance and restore the entire backup or individual files.

More information on backups and disaster recovery in OCI can be found at `https://www.oracle.com/database/technologies/high-availability/secure-backup.html`. Specific details on implementing a DR strategy in OCI can be found at `https://www.oracle.com/cloud/backup-and-disaster-recovery/`.

We have discussed backup solutions from the cloud provider themselves. There's a risk in doing this: we are making our businesses completely dependent on the tools of these providers. From an integration perspective, this may be fine, but a lot of companies prefer to have their backups and DR technology delivered through third-party tooling. Reasons to do this can sprout from compliance obligations but also from a technological perspective. Some of these third-party tools are really specialized in these types of enterprise cloud backup solutions, can handle many more different types of systems and data, and can be truly cloud agnostic. Examples of such third-party tools include Cohesity, Rubrik, Commvault, and Veeam.

The most important thing is to have a good strategy. This starts with defining the RPO and RTO: RPO and RTO are critical metrics that define how much data you can afford to lose and how quickly you need to recover your services. These metrics are specific to your business needs and should be defined before designing your DR strategy. Possible strategies are:

- Active-Active: In an active-active configuration, you have multiple active environments in different regions. This configuration provides high availability and can be used to distribute traffic across regions. However, it can be more complex to manage than other configurations.
- Active-Passive: In an active-passive configuration, you have a primary environment in one region and a secondary environment in another region. The secondary environment is only activated in the event of a disaster. This configuration is simpler to manage but may have longer RTOs.

Lastly, we need to replicate the data and design the failover tactic, including the DNS failover, in case we need to execute our DR strategy.

This concludes the chapter where we have learned to architect and deploy applications to the cloud, making use of the various native services that cloud providers offer to host these applications and making sure that applications perform well and are resilient. In the next chapter, we will discuss the data in cloud.

Summary

In this chapter, we discussed the definitions of resilience and performance. Companies want their applications to be available and to perform well: it defines the customer experience. We discussed various concepts that developers can use to build performant and resilient applications, using cloud-native tools as much as possible in 12-factor apps, PaaS, and SaaS.

We also learned how to optimize our environments using different advisory tools that cloud providers offer. We then learned how to identify risks in the various layers: business, data, applications, and technology. We studied the various methods we can use to mitigate these risks.

One of the biggest risks is that we "lose" systems without the ability to retrieve data from backups or without the possibility of failover to other systems. For real business-critical systems, we might want to have disaster recovery, but at a minimum, we need to have proper backup solutions in place. Various backup and disaster recovery solutions that are available in the major cloud platforms have been explored.

We talked about infrastructure and applications a lot in the past chapters. But applications only make sense if there's data that these applications can use. In a connected world, the amount of data that is available to us is massive. How do we design platforms that can handle these amounts of data and make it usable for specific business goals? The next chapter is all about data platforms in the cloud.

Questions

1. What do the terms RPO and RTO stand for?
2. What tool would you use to capture failures in application code that's running in Google Cloud?
3. Cloud providers offer toolkits to develop SaaS applications. Name the service that AWS offers to build and host SaaS applications.
4. True or false: We can use the backup solutions in Azure and AWS for systems that are hosted on-premises too.

Further reading

- *Reliability and Resilience on AWS*, by Alan Rodrigues, Packt Publishing
- *Architecting for High Availability on Azure*, by Rajkumar Balakrishan, Packt Publishing

Join us on Discord!

Read this book alongside other users, cloud experts, authors, and like-minded professionals. Ask questions, provide solutions to other readers, chat with the authors via. Ask Me Anything sessions and much more.

Scan the QR code or visit the link to join the community now.

`https://packt.link/cloudanddevops`

8

Creating a Foundation for Data Platforms

If there's one big advantage of the public cloud, it's that platforms that utilize it can hold massive amounts of data. That is one of the main reasons that companies use cloud technology. In the cloud, they can collect data from a variety of sources and start analyzing this data to improve their business strategy, products, and services. The first question that companies will ask themselves is: where do we start with building data platforms?

In this chapter, we will discuss the basic architecture of data lakes and consider the various solutions that the major cloud providers offer. We will also look at the challenges that come with collecting and analyzing vast amounts of data, including the phenomenon that is called data gravity, since it's hard to transport these large amounts of data across different platforms. A solution to overcome this challenge is data mesh.

We will cover the following topics in this chapter:

- Choosing the right platform for data
- Building and sizing a data platform
- Designing for portability and interoperability
- Overcoming the challenges of data gravity
- Managing the foundation for data lakes

Choosing the right platform for data

It is a cliché, but nonetheless, it's very true: data is the new gold. It is with good reason that in enterprise architecture frameworks, data is named as the first thing that a business must consider, analyzing what data it should use and how to gain optimal benefits from that data. No business can operate without data; it needs data to gain insights into markets and the demands of its customers. It needs data to drive the business.

You will find the term **data-driven** in almost every cloud assessment study. What does data-driven mean? A company makes decisions based on the analysis of data. Intuition or decision based on previous experience is ruled out. Every action is supported by the analysis of data.

To enable a data-driven business, we need one thing: the data itself, typically in vast amounts and preferably in (near) real time. The collection of data is prerequisite number one. Prerequisite number two is that this data must be accessible. So, we need accurate, relevant data that is available for data analytics. These are the key requirements for data and to become a data-driven organization.

To enable capturing data, storing it, cleansing the data, and preparing it for analysis through data analytics, companies need data platforms. Unfortunately, there's no magic formula or golden nugget that will get an instant data platform. Let's first define what a platform is.

A data platform is first of all a central repository where a company captures data. Data that is scattered among a variety of databases and other sources is very hard to analyze. Hence, we try to collect all the relevant data in one repository. From that single repository, data analysts can start processing the data. In this process, logical collections are transformed into datasets, including the cleansing of data; next, the algorithms are defined to mine the data and produce valuable outcomes for the business.

Defining and designing a data platform that can do this must take architectural layers into account, regardless of the specific cloud platform that is used. These layers are:

- **Discovery**: Where are the data sources and is the data accessible?
- **Observability**: This is mainly about the quality of the data. Is the data recent and accurate? Has all the relevant data been collected?
- **Ingestion**: Moving data from one place to the other, for instance, raw data from the data lake to the data warehouse for further processing.
- **Storage**: The "physical" place where data is stored. Think of AWS S3 or Azure blob storage.
- **Modeling**: Building the data models.

- **Analytics**: The usage of cleaned data to run metrics against, aiming to get meaningful results out of data.

A data platform is often referred to as a data lake, which is nothing other than a massive storage location that holds raw data from multiple sources. This can be all sorts of data, from files to streaming data. Typically, these data lakes use object storage such as blob storage in Azure or S3 in AWS. Data lakes can be built on-premises, but more common is the use of public cloud providers where data lakes are configured in storage clusters. To distribute large datasets across these clusters, technology such as Apache Hadoop is used. A basic architecture of Hadoop, using HDFS, or the Hadoop Distributed File System, is shown in *Figure 8.1*. MapReduce is a technology that is used for applications to process vast amounts of data in parallel across nodes in the cluster.

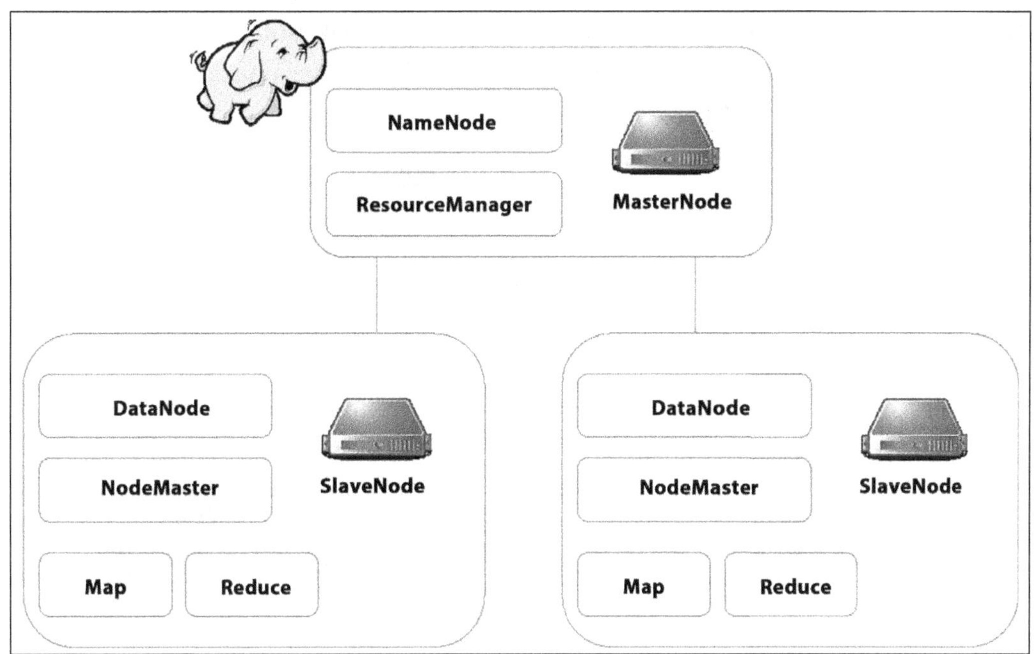

Figure 8.1: High-level architecture of Hadoop cluster

What's the difference between data lakes and data warehouses? A data lake holds raw, unstructured data, whereas warehouses offer structured data. Typically, datasets that are extracted from data lakes are inserted into warehouses for querying and analytics.

A combination of a data lake and a data warehouse is a **lakehouse**. With a lakehouse, additional, formatted structures are implemented on top of the data lake. This was initially pushed by Databricks, but since then, more technologies that provide lakehouse solutions have entered the market. Examples are Delta Lake and Apache Iceberg.

Now, let's explore the various solutions that cloud providers offer for creating data platforms.

Azure Data Lake and Data Factory

The full name for the service is Azure Data Lake Storage Gen2. The reason to have this very specific name is that the service is nothing less than a solution built on top of Azure storage. So, you can use the blob API or **Azure Data Lake Storage (ADLS)**. Keep in mind that not all features of blob storage are available in ADLS at this time.

Data Lake Gen2 adds file system semantics to blob storage so that petabytes of data can be organized in objects and files with a hierarchical structure of directories. This enables easier access to various types of data. But: it's still raw data.

Azure also provides solutions to start assembling and analyzing datasets. This solution is **Data Factory**. With Data Factory, analysts can process data using **extract-transform-load** (ETL) or **extract-load-transform** (ELT) as a code-free service running in Azure. It allows for building data pipelines that cover the different stages of data processing:

- **Ingest**: Collecting datasets
- **Control and data flow**: Designing data workflows in pipelines
- **Schedule**: To run data processes at specified times or as a trigger when new data is ingested
- **Monitor**: Tracking the activities in the pipelines

A high-level overview of the Data Factory architecture is presented in *Figure 8.2*.

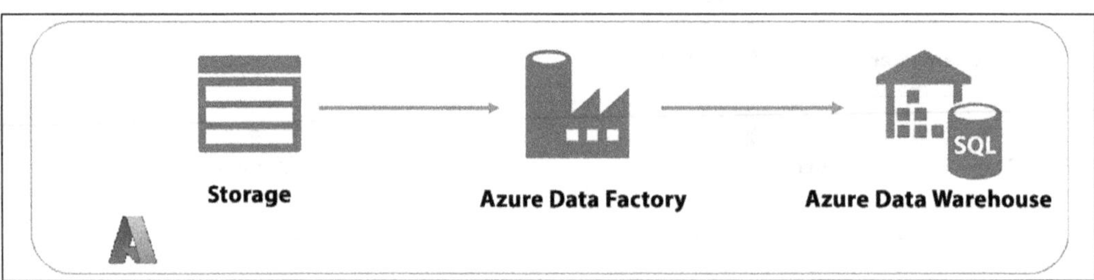

Figure 8.2: High-level architecture of Azure Data Factory

To explain it in very simple terms: Data Factory allows for connecting to data sources, collecting data, processing that data, and preparing it for analytics. The outcomes of analysis might be presented to tools such as Microsoft Power BI. **Azure Data Factory (ADF)** starts with ingesting, then preparing/transforming/analyzing, and next publishing the results to data stores, ready for other tools (such as Power BI) to consume them. ADF supports dozens of data stores (see `https://learn.microsoft.com/en-us/azure/data-factory/connector-overview`).

At Ignite 2022, Microsoft introduced Azure Data Explorer: a fully managed big data analytics platform, supporting the analysis of massive amounts of data in near real time. It also offers ingestion support for various data sources, including AWS S3 and OpenTelemetry Metrics, Logs, and Traces. It's one development that shows how important the market for big data is to cloud providers.

One other service that must be mentioned here is Azure Synapse, which connects enterprise data warehousing with big data analytics. Azure Synapse brings together SQL technologies used in enterprise data warehousing, Spark technologies used for big data, Data Explorer for log and time series analytics, pipelines for data integration and ETL/ELT, and deep integration with other Azure services such as Power BI, Cosmos DB, and Azure **ML (Machine Learning)**. For data governance, Microsoft offers Microsoft Purview, which provides a unified data governance service that helps you manage your on-premises, multi-cloud, and **software-as-a-service (SaaS)** data.

AWS Data Lake and Redshift

The AWS Data Lake solution consists of various components. The raw data is stored on top of S3 storage. Part of the solution is Amazon DynamoDB, a NoSQL database service that is fully managed by AWS and offers continuous backups and automated replication across regions to make the service resilient. Import and export tools are provided too with DynamoDB.

One final element in the AWS Data Lake proposition is AWS Glue. The name of the service has been chosen accurately since it really glues together the various components. AWS Glue performs the ETL process: discovering and preparing the datasets—or data catalog—into DynamoDB where the data can be analyzed. Both DynamoDB and Glue are serverless services. The analytics service is provided through OpenSearch, previously Elasticsearch. The Lambda serverless function is used as a message trigger to start the process.

Figure 8.3 shows a high-level architecture.

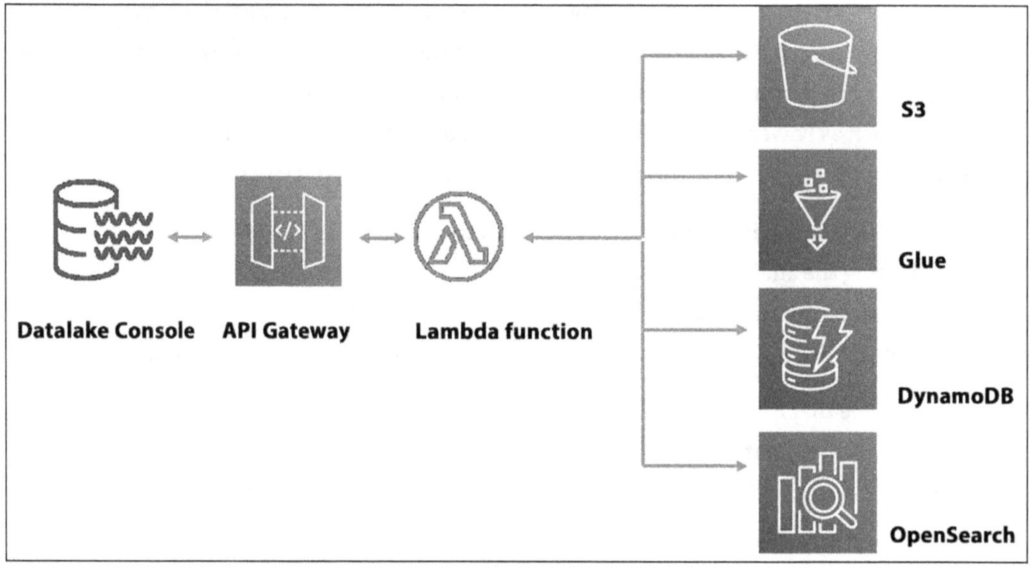

Figure 8.3: High-level architecture of AWS Data Lake

In the context of big data, a different AWS service is often mentioned: Amazon Redshift. This is a data warehouse based on PostgreSQL. Redshift allows for 16 petabytes of—structured—data on one cluster.

Google's data lake and BigLake

The foundation of Google's data lake is Google Cloud's Cloud Storage, comparable to Azure blob and AWS S3. Cloud Storage holds objects and files up to 5 TB per item. Also comparable to other cloud providers is the fact that Cloud Storage comes in different classes with various service levels in terms of availability and performance, which is reflected in the pricing. Exploring raw data in Cloud Storage can be done with Cloud Datalab and Cloud Dataprep. These tools will allow you to view data and determine if data is relevant, for instance.

The offloading of data into a data warehouse is done through BigQuery, a fully managed serverless service that even includes built-in machine learning capabilities to perform real-time and predictive analysis. *Figure 8.4* shows a simple architecture for GCP's data lake solution.

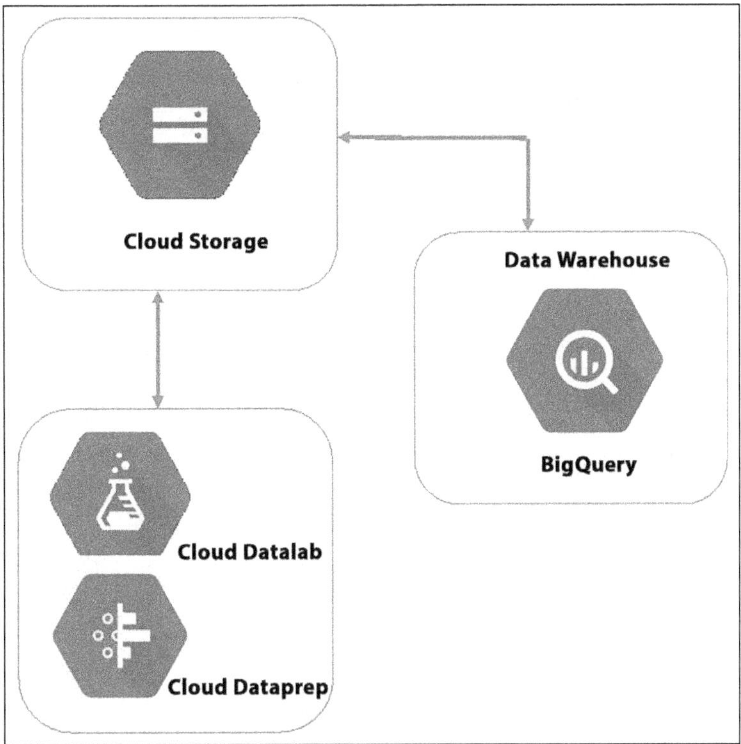

Figure 8.4: High-level architecture of Google's data lake solution

BigQuery Omni is a multi-cloud variant of this solution that also allows for data analytics using data that is stored in AWS and Azure.

BigLake, which was introduced in 2022, takes the idea of multi-cloud a bit further. BigLake provides a single-pane-of-glass view and a storage engine across various platforms, meaning that it "combines" data sitting in Cloud Storage, AWS S3, and Azure Data Lake Storage Gen2 as if it were one big lake. Data engineers would only have to work in one console. Next, BigLake offers fine-grained security controls.

Alibaba Cloud Lakehouse

A common way to start data engineering and data analytics is to offload data from a data lake to a data warehouse. The warehouse is separated from the lake, and tools such as Apache Hive are used to request and transport data between sources and processing platforms.

Alibaba Cloud Lakehouse takes a different approach, by integrating the warehouse with the lake, allowing data to flow between the two platforms. The data in the lake and the warehouse are seamlessly integrated, including the metadata. The data lake uses Alibaba's cloud storage **Object Storage Service (OSS)**, comparable to S3 or blob, but the warehouse is built on a solution called MaxCompute. This solution implements a unified storage access layer that supports HDFS, the file system of Hadoop, and OSS, all in read/write modus. *Figure 8.5* is a high-level presentation of the architecture.

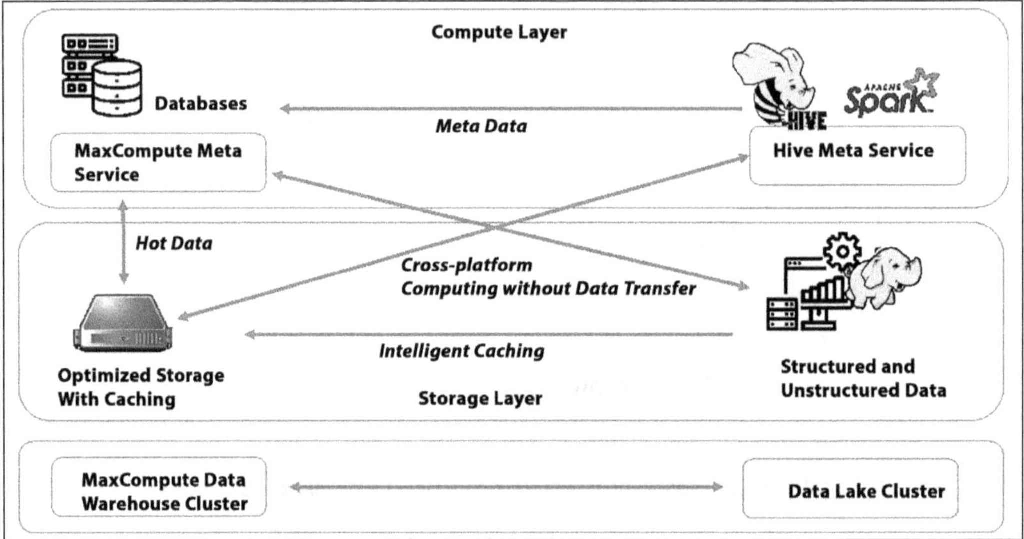

Figure 8.5: High-level architecture of Alibaba Cloud Lakehouse

A good blog post on Alibaba's solution can be found at `https://alibabatech.medium.com/data-lakes-or-data-warehouses-cd5122ba7634`.

MaxCompute still uses Apache Hive and Spark to map databases in the data lake to projects in MaxCompute, the data warehouse. So, although it's presented as one system, the architecture still resembles the common architecture to process flows between lakes and warehouses.

Oracle Big Data Service

There's one more solution that deserves mentioning. As a company that finds its origin in processing data with database solutions, it makes sense that Oracle has developed a solution for big data, accurately called **Oracle Big Data Service**.

The solution is built on top of **Oracle Cloud Infrastructure** (**OCI**) and contains the deployment of fully managed Hadoop clusters to form a data lake. This is Oracle's own version of Hadoop, called **Oracle Distribution of Apache Hadoop** (**ODH**).

It's a complete service to get customers up and running fast, including a toolbox with various ETL and analytics tools such as the analytics engine Spark, the non-relational database HBase, the data warehouse software Hive, and the configuration tool ZooKeeper.

Building and sizing a data platform

As with every service that we deploy in the cloud, we need a foundation to build the platform on. Hence, building a landing zone that can hold raw data is the first step. This landing zone should be an environment that serves only one purpose: to capture raw data. It's recommended to build this landing zone separate from core IT systems. It should be scalable but low cost, since it will hold a lot of data. The issue with keeping data is that it might increase the cloud bill exponentially. Data storage comes at a very low price per unit of data, but the catch is that we need a lot of these small units.

It is important to implement governance from the start. This includes defining and implementing guardrails for the classification of data and tagging.

Once the landing zone has been established, data analysts can start using the data lake as a sandbox environment. This is the second stage. Analysts can start building prototypes of data models and work with the raw data that is collected in the data lake. They can also test various tools to find what will work best and give the most benefit to the business.

There will be a moment when datasets have been defined and tools selected. This is the time to start integrating the datasets with other business data. This is the process of **ETL** or **ELT**: **extract-transform-load** or **extract-load-transform**. From the raw data in the data lake, the required datasets are collected, extracted, and loaded in the enterprise data warehouses. Since only relevant data is extracted, the sizing of the data warehouse doesn't need to be increased. A lot of data will remain in the data lake, built on cheaper storage. Tools such as the Data Factory and OpenSearch will help in optimizing queries.

Now, the data lake is a core component of the IT infrastructure and the business of an enterprise. Strong governance is an absolute requirement in operating the data lake and the data flows that connect the data lake with data warehouses and various data models. These models will generate output to applications and enable detailed insights into the business itself, the efficiency and performance of business-supporting operations, and the markets the business operates in.

We have now defined the four stages of building and sizing the data lake. It's summarized in *Figure 8.6*.

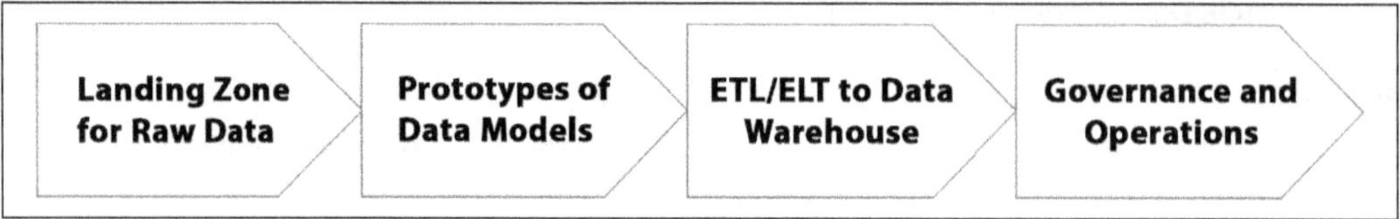

Figure 8.6: Four stages of building and managing data platforms

There are some best practices to keep in mind in building and sizing a data platform. Number one to keep in mind is that you have to understand the business in order to collect the right, relevant data. There must be a clear objective for implementing a data lake: what business problem are we solving by querying vast amounts of data? And what sort of data would we really need as a business?

Second: know what the data is about. Data must be recognized and that can be achieved by tagging the data. Hence, metadata is crucial. There's always the risk of a data swamp, that is, companies and their employees drowning in data. One way to solve this is data cleansing, which is as important a process as ingesting, scheduling, and monitoring. Depending on the data classification, data must be cleaned. This is also important in terms of controlling costs.

Lastly: security must be priority number one. Keep data safe by implementing authentication, authorization, and encryption, both for data in transit and for data at rest. Remember what we said earlier in this chapter: data is the new gold.

Designing for interoperability and portability

Portability and interoperability should be driven by use and business cases—not purely for the sake of portability or interoperability. In IT systems, there are four levels that define the portability of systems: data, applications, platforms, and infrastructure, following the **Architecture Development Method** (**ADM**) of TOGAF.

- Data represents information in such a form that it can be processed by computers. Data is stored in storage that is accessible to computers.
- An application is software that performs actions that are triggered by business requests.
- Platforms support applications.
- Infrastructure is a collection of computation, storage, and network resources. Computation can also refer to cloud computing including VMs, containers, and serverless functions.

One important note that we have to make at this point is that cloud computing causes a "blurring" effect in the demarcation of infrastructure, platforms, and applications. Think of PaaS and SaaS, where PaaS includes the infrastructure layer and SaaS is a fully integrated stack of application software, the platform, and the underlying infrastructure. In PaaS and SaaS, the demarcation of the various architecture layers is not as defined anymore as in traditional IT architecture.

However, the aim of TOGAF in achieving portability is to create abstract architecture layers in IT systems. Thus: data is separated from the applications and applications are separated from the technology. Portability is a requirement in itself but subsequently brings requirements along which systems should adhere to. To summarize:

- The architecture vision includes the portability of applications and interoperability.
- Requirements are collected to fulfill the vision.
- Business architecture defines the business processes where portability and interoperability are a requirement.
- Information systems architecture defines how portability and interoperability are achieved on application and data levels.
- Technology architecture defines how technology is supporting portability and interoperability.

In theory, this would allow for data and application architecture that is not dependent on specific technology layers. This is also the starting point for microservices; microservices-based applications are built as a collection of highly decoupled services that handle a single action. Each service is independently built, deployed, and monitored. The use of microservices architecture is important to create portability.

Portability is enabled through decoupling: applications are decoupled from data, data is decoupled from storage, and application code is decoupled from the underlying infrastructure including networks. Interoperability is enabled through standardization, preferably with open standards.

Portability is defined by abstraction between the basic infrastructure (often referred to as the landing zone; please refer to *Chapter 6, Controlling the Foundation Using Well-Architected Frameworks*), the configurations of the infrastructure components, the data, and the application. Again, we're using the definitions by The Open Group as the industry standard. The definitions for portability are provided below:

- **Data portability** is essentially about reusing data across various applications. Data portability is likely the most difficult component in achieving portability. The structure of the data is often designed to fit a particular form of application processing, and a significant transformation is needed to produce data that can be handled by a different product. This is separate from the data carrier, the database technology, or the storage layer hosting the database. For example, PostgreSQL as a database will run on different platforms. That doesn't mean that the data in the database can be used by different applications without transforming this data.
- Application portability is about reusing application components across various computing platforms. These can be traditional on-premises platforms, but also cloud platforms including PaaS. Examples of the latter are managed database services such as **Relational Database Service (RDS)** in AWS and Azure SQL Database. Portability requires a standard interface exposed by the supporting platform. This must enable the application to use the service discovery and communication between the platforms, as well as providing access to the platform capabilities that support the application directly. This same principle applies to interoperability.
- For platform portability, we can think of reusing platform components across clouds and on-premises infrastructure. Kubernetes as an underlying platform for containers hosting applications is an example. Kubernetes can be deployed on various platforms, including public cloud infrastructure and on-premises machines that sit in privately owned datacenters. But virtualization is also an example: virtual machines can be transferred between different platforms if the machine images are portable. Note: this is different from managing a variety of machines from one console. For example, Azure Arc allows administrators to bring non-Azure machines and Kubernetes from other clouds and on-premises under the control of Azure and manage these machines from the Azure console. The non-Azure machines, however, stay where they are; the images of these machines are not transferred to other instances. This technology leverages the possibilities for interoperability between platforms, but it doesn't make environments portable.

It's important to realize that cloud (native) services do not necessarily contribute to portability and interoperability.

Portability is one of an application's non-functional requirements. An application should be as portable as possible, not tied to a specific infrastructure or platform. We can achieve this through the abstraction of layers and decoupling services. To summarize, if we approach portability from the perspective of defining, designing, and applying microservices, we can obtain the highest level of portability.

Cloud portability is the ability to move applications and data from one cloud to another with minimal disruption. Cloud portability enables the migration of cloud services from one cloud provider to another or between a public cloud and a private cloud. If we're designing and implementing data and applications to be compliant with cloud portability, then we are really following a multi-cloud strategy. A multi-cloud strategy could be having the data lake in one cloud and having apps (with their data) in other clouds. But architects must implement the principle of decoupling into their designs from the start.

Decoupling services come with challenges, especially in refactoring existing landscapes and applications. The biggest challenge: interoperability. But when do we speak about interoperability? Products, systems, or organizations are interoperable if they can work together without restrictions. Services can discover each other, connect, and communicate in a coordinated way, without interference from the end user. If we decouple systems, applications, and data in a microservices architecture, this becomes a challenge. We will have to make sure that all services can find each other (discoverability) and know how they can communicate with each other.

Best practices to achieve interoperability in multi-cloud architectures include:

- Homogenous virtualization technology, including container orchestration platforms
- Standardized protocols for authorization and authentication
- Use of standard **Application Programming Interfaces (APIs)**

Data synchronization is a particular issue when components in different clouds or internal resources work together, whether or not they share the same protocols or even when they are absolutely identical. Copies of the same data are kept in different systems and the challenge is to keep this data synchronized, and with that, consistent. If data is stored in different clouds, this can become a challenge. Data must first of all be discoverable and accessible, but to keep it consistently in sync, it also requires high-speed connectivity to overcome issues in latency. Items that we must consider in synchronization of data are:

- Management of the master data sources: where is the single source of truth?
- Management of data at rest and data in transit across platforms
- Data visibility
- Access management, authentication, and authorization

Full interoperability includes the continuous, dynamic discovery of infrastructure, data sources, and application components, communicating with other components, at run time.

Overcoming the challenges of data gravity

Applications don't just hold data; they also produce a lot of data that they share with other applications. Data will attract new data and services in other applications. As data accumulates, more and more applications and services will use it. Data and applications are attracted to each other, as in the law of gravity. To keep it short and simple: the amounts of data will grow, either autonomously or, more likely, because data sources will be connected to other data sources.

In addition to the strategic advantage of having access to this data, this also presents a major challenge. Databases are becoming so large that it becomes almost impossible to move the data. This can lead to a situation where companies are tied to a certain location or provider to hold that data. In addition, companies that use each other's data and services must stay close to each other in order to provide good service. By keeping data physically close together, it can be exchanged quickly without end users experiencing slow processes—the effect that we call **latency**.

Data gravity forces companies and their architects to consider a variety of topics, data privacy being one of them. This will increasingly become an issue since data will need to be accessible across different platforms and applications. At the time of writing, there's a debate going on in several Western-European countries where universities, healthcare institutions, and governmental bodies are storing data in public clouds that are US-owned and managed. One of the main arguments: once you have the data in one of these clouds, you will never get it out of it. And: the US government might force American companies to grant insights into the data that they store, despite agreements between the US and the European Union.

The challenge of data gravity forces companies to think about where they keep their data, how they connect to other data sources, and what data they can use and/or share. Successful business processes must be able to bring data together and bring the user and the applications to the data. Since enormous amounts of data are hard to move, we will have to architect solutions where we use the data where it is, acknowledging that data is decentralized by default. If we take that as a principle, then there's no need to collect every single piece of data in a central data store.

Again, this will inevitably lead to debates about data privacy. But if we want to create global solutions for the major challenges of our time—think of climate change and the accessibility of healthcare around the world—then we need data. We can try to transfer all this data to one repository, or design applications in such a way that we can use this data in a safe way, wherever the data is.

As a result of technology such as the cloud, IoT, and analytics, data is created everywhere. It is produced in smartphones, buildings, homes, and even across entire cities. Companies and other institutions must be prepared for this. A decentralized infrastructure can address the challenges of data gravity by providing the right coverage, capacity, and connectivity, quickly bringing users, networks, and clouds to that data. Cloud providers are already prepared for this by implementing global coverage of their services. With edge computing and 5G, we will also have the technology to access this data swiftly.

Introducing the principles of data mesh

One solution that we must mention here is the principle of data mesh, an architecture principle that was introduced by Zamak Deghani.

The paper about the data mesh architecture can be found at `https://martinfowler.com/articles/data-monolith-to-mesh.html`.

Data mesh is an architecture that allows applications, and with that the user, to access data without transporting this data first to a data lake. Data mesh utilizes the principles of decentralized data sources. The challenge in data mesh is still how to get the right data to the application where it can be processed, without moving that data. This can be done through data streaming.

Oracle provides a managed data mesh service called **Oracle Cloud Infrastructure** (**OCI**) GoldenGate. GoldenGate connects to data sources, replicates the data, cleans the data, and delivers this curated data through streaming data pipelines to data consumers. GoldenGate responds to triggers using Apache Kafka: as soon as events occur that call for action in the data chain, GoldenGate will start replicating and streaming relevant raw data in real time, different from traditional ETL tools that transport data in batches.

In short, GoldenGate is the bridge between data producers and consumers. A bridge enables a continuous stream of traffic, in this case, data traffic.

The major cloud providers all provide their own solutions for data mesh. Azure Synapse Analytics also utilizes Kafka or Event Hub to capture data events and next collects and publishes data to data products. A similar solution can be built with AWS Lake Formation and AWS Glue, providing solutions for data discovery, analytics, and publishing data. The data mesh solution that Google offers is Dataplex, which became generally available in early 2022.

Managing the foundation for data lakes

Data engineers design, build, and manage the data pipelines, but the foundation of the data lake and data warehouse is the specific landing zone for the data platform. Typically, landing zones in the cloud are operated by cloud engineers who take care of the compute, storage, and network resources.

Looking at the management of data platforms, we can distinguish various roles:

- Data architect or engineer: The architect and data engineer are often combined in one role. This role is responsible for the design, development, and deployment of the data pipelines. The engineer must have extensive knowledge of ETL or ELT principles and technologies, making sure that data from sources gets collected and transformed into usable datasets in data warehouses or other data products where the data can be further analyzed. Data also needs to be validated, which is a required skill of the engineer too. In essence, the engineer makes sure that the data that is ingested into warehouses is of good quality and usable for analytics.
- Business and data analyst: The main question that a business or data analyst must answer is, what data can be used for the business? What data provides the required insights for the business? The business ambitions, goals, and targets are translated into metrics through which data is analyzed.
- Data scientist: Where the business or data analyst defines the metrics for the data, the data scientist makes sure that the right data sources are discovered, and the appropriate data is made available to the analysts. The data scientist is crucial in finding the right data that the analysts can use to proof the metrics and provide insights that enable companies to fulfill their business strategy.
- Cloud operator: Data must be hosted somewhere, on big machines running in the cloud, for instance. These machines holding all the data must be accessible, performant, and secured from attacks. The design of the landing zone, the operations of the hosting instances, the network configuration, and security guardrails are the responsibility of cloud operators, who keep the data platform running in an optimal and highly secure way.

There's one aspect that we have not mentioned here so far, but that does play a significant role in the management of data platforms: cost management. Big data can cost a company big money, hence implementing methods and tools to control the costs of data platforms has become an important task of the architect and the operators.

FinOps helps organizations to first understand what these costs are, how they can control these costs, and how they forecast budgets in the usage and operations of data lakes and warehouses. *Part 3, Controlling Costs in Multi-Cloud Using FinOps*, of this book is all about the FinOps principles.

Summary

In this chapter, we discussed the basic architecture principles of building and managing a data platform. We looked at data lakes that can hold vast amounts of raw data and how we can build these lakes on top of cloud storage. The next step is to fetch the right data that is usable in data models. We must extract, transfer, and load—ETL for short—the datasets into environments where data analysts can work with this data. Typically, data warehouses are used for this.

We studied the various propositions for data operations of the major cloud providers, AWS, Azure, Google Cloud, Alibaba, and Oracle. Next, we discussed the challenges that come with building and operating data platforms. There will be challenges with respect to access to data, accuracy, as well as privacy and compliance. Data gravity is another problem that we must solve. It's not easy to move huge amounts of data across platforms, hence we must find other solutions to work with data on different platforms. Designing for interoperability and portability is therefore a key capability of architects.

There's one question that we didn't answer in this chapter: where does all this data come from? It's largely coming from a growing number of devices that are connected to the internet: the **Internet of Things (IoT)**. That's the topic of the next chapter.

Questions

1. What does the term ETL mean?
2. What would be the first step in building a data platform?
3. True or false: Data lakes are typically built on the common storage layers of major cloud providers such as Azure blob storage and Amazon S3.
4. What does Oracle's GoldenGate do?

Further reading

- *Data Lake for Enterprises*, by Tomcy John and Pankaj Misra, Packt Publishing

9

Creating a Foundation for IoT

Market researchers such as Gartner expect that in 2030, over 30 billion devices will be connected through the **Internet of Things (IoT)**. This would mean a skyrocketing growth from the current number of connected devices: at the time of writing, the number of IoT-connected devices is around 14 billion, which is already an incredible figure. We have connected devices in our homes, in factories, and even in cars. All these devices produce and receive data.

The challenge that companies face is how they can monitor and manage these devices, preferably from one platform. Cloud providers offer centralized IoT platforms as a solution to this challenge. In this chapter, we will study the architectural principles of an IoT ecosystem and discuss how the cloud can help in managing IoT devices. We will explore some of these cloud solutions and also look at crucial elements of IoT, such as connectivity and security.

We will cover the following topics in this chapter:

- Choosing the right platform for IoT
- Monitoring IoT ecosystems
- Designing for connectivity to the cloud
- Connecting IoT with IPv6, LoRa, and 5G

Choosing the right platform for IoT

There's a lot of talk about the IoT, the internet that connects everything: every device, perhaps even a large amount of humans. Let's first define a "thing." Basically, a thing can be anything, but typically, we talk about devices that have a connection to a network through, for example, Bluetooth or Wi-Fi and, through this, a connection to the internet.

A device can be a machine or someone carrying a device with a **Unique Identifier** (**UID**). The device is capable of autonomously transferring data over a network, so without human interference. In literature, this is referred to as **Machine-To-Machine** (**M2M**).

A thing can thus also be a device that is "attached" to a person. You can think of implants that monitor the health status of patients. But you may also think of cars with all sorts of sensors alerting the driver when something is wrong with the car and sending a message to the dealer to book an appointment to fix the problem. All these devices and the applications in the devices use the **Internet Protocol** (**IP**) for communication and transferring data. The IoT has great benefits for both users and companies.

How does it all work? First, we are talking about a massive amount of devices. All of these devices are web-enabled and use the internet to transfer data. They collect the data through sensors and connect to the internet through an IoT gateway or edge computer. The data is collected in a centralized data platform where it is analyzed. The following diagram shows a high-level architecture of IoT environments.

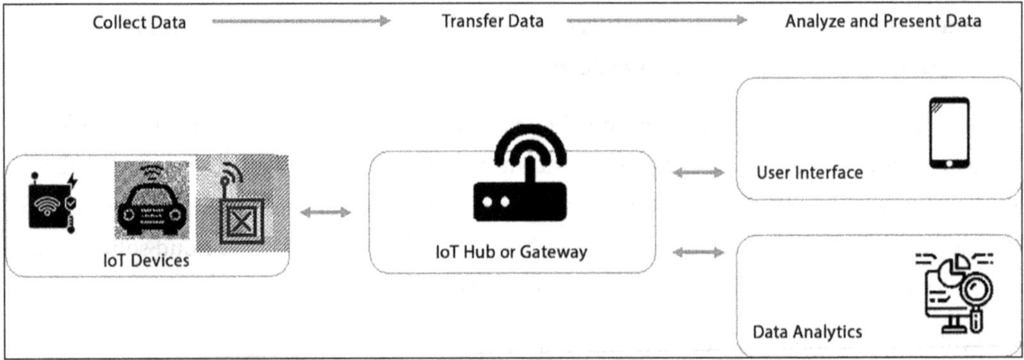

Figure 9.1: High-level architecture of IoT environments

The concept of IoT is used in almost every industry, and in our homes. Think of sensors that automatically turn the lights on when it gets dark.

It's essentially the same concept that is used in smart cities where streetlights automatically turn on when light levels get low enough, just as a very simple example. Sensors in smart cities can also analyze traffic and start rerouting traffic at rush hour, preventing heavy congestion. The data that these sensors collect can be used to analyze patterns and help design new street plans.

There are risks, however. Poorly protected devices are a gold mine for hackers. Once they infiltrate a device, they can virtually hop on to the next device and eventually end up in the main systems holding valuable data.

The number of devices is growing rapidly, leaving companies that have to manage these devices with some real challenges. How do you manage millions of devices? To start with: how do you keep track of these devices?

The public cloud might be of good help in observing, monitoring, and managing IoT ecosystems. In the next sections, we will briefly explore some of these platforms.

Azure IoT Hub

Azure offers **IoT Hub** as a platform to connect devices and receive and process data from these devices. IoT Hub is basically a suite of services, such as over-the-air device updates and integration with Event Grid, IoT Edge, and other Azure services such as Azure Logic Apps, Azure Machine Learning, and Azure Stream Analytics. Device Update allows for group-wise updates of IoT devices, either package-based or image-based. Packages target only specific components of the device, whereas image-based copies the entire updated image to the devices. Device Update was introduced in September 2022 and offers an extensive toolkit to manage devices through IoT Hub, including management, reporting, detailed control of update processes, diagnosis and troubleshooting, and automatic grouping of devices.

It is also worth mentioning IoT Central at this point. This service allows us to develop and manage IoT solutions using an IoT Application **Platform as a Service (aPaaS)**, heavily reducing effort and costs. The simple **User Interface (UI)** lets you connect devices, monitor device conditions, create rules, and manage devices and their data throughout their life cycle in a very comprehensive way.

A good starting point for Azure IoT is `https://azure.microsoft.com/en-us/products/iot-hub/`. More information about IoT Central can be found at `https://learn.microsoft.com/en-us/azure/iot-central/core/overview-iot-central`.

How does IoT Hub communicate system changes to the pool of devices and the applications running on top of these devices? As a message hub, IoT Hub allows for bi-directional communication with the devices. In addition, Azure IoT Hub integrates with Event Grid, allowing more than 500 service endpoints to route events. Event Grid is a serverless broker that is able to communicate events to destinations such as devices. An example would be changing the configuration of an application.

Lastly, IoT Edge must be mentioned. This is a solution where the analysis of IoT data must be closer to the actual devices and data sources. Azure services are wrapped in containers and transferred to the IoT Edge appliances. Workloads such as telemetry but also runtime for artificial intelligence applications are now executed locally on the edge appliance. IoT Edge is also used to send configurations to a specific group of devices. These devices are then connected to a particular edge appliance. The following diagram shows a high-level architecture.

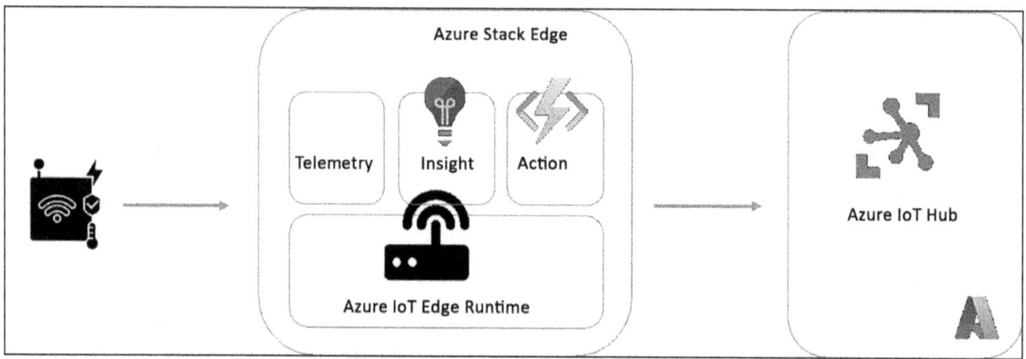

Figure 9.2: Architecture for Azure IoT Edge

Azure IoT Hub includes device management as a standard feature and on top offers a suite of services, including integration with security suites such as Defender for IoT for Sentinel, announced in late 2022. Sentinel is Microsoft's cloud-based **Security Information and Event Management (SIEM)** solution. Defender for IoT, the security endpoint solution for IoT, now integrates into Sentinel, providing visibility of all **Operational Technology (OT)** and IoT devices and network connections, allowing for the fast detection of and response to potential vulnerabilities in the entire ecosystem.

AWS IoT Core, Edge Manager, and IoT Greengrass

AWS IoT Core is comparable with Azure IoT Hub. The service provides secure communication for IoT devices, allowing for the processing of data coming from these devices. However, users will need to add AWS IoT Device Management to IoT Core to enable the remote monitoring and management of devices themselves. In other words, IoT Core lets you connect devices to a central environment where IoT Core serves as a message broker between devices using the lightweight **MQTT (MQ Telemetry Transport)** protocol. *Figure 9.3* shows the principle of IoT Core.

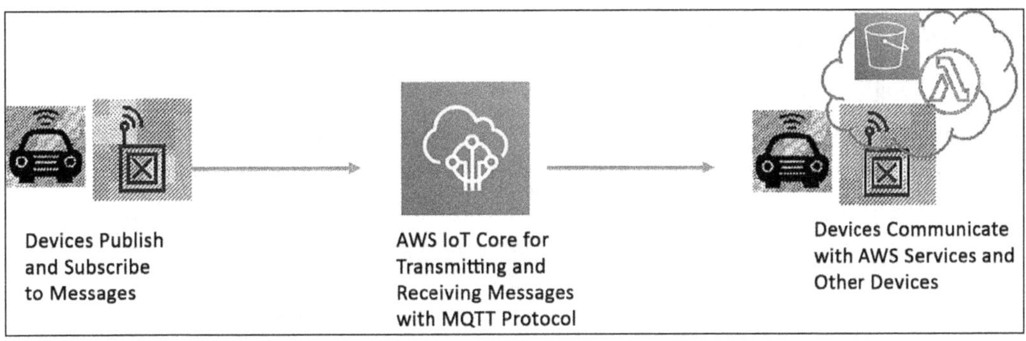

Figure 9.3: Architecture for AWS IoT Core

Like Azure, AWS offers various services that enable the full-stack management of IoT ecosystems. Important services to mention are IoT Greengrass and Edge Manager. The IoT Greengrass cloud service in the AWS cloud helps build, configure, and deploy software to IoT devices. Greengrass communicates with Greengrass client software on edge appliances or core devices. The IoT devices themselves connect to these edge appliances.

The Greengrass architecture makes it easy to manage apps on the devices, operate the devices including anomaly detection, and locally process data. IoT Greengrass is also able to act as a gateway, allowing devices to connect and communicate even when they are not connected to the internet.

The next step would be to predict the behavior of devices and services running on these devices in case of specific events, allowing for the continuous optimization of the services and performance of the devices. This can be done through **Machine Learning (ML)**. AWS offers SageMaker as an ML engine and with SageMaker Edge Manager, the models used for ML can be deployed to IoT devices.

The best way to get started with IoT in AWS is by reading the documentation at `https://aws.amazon.com/iot/`.

In most cases, IoT will be used in industrialized environments with industrial equipment. You can think of sensors in production lines and manufacturing robots. Collecting, modeling, and analyzing industrial data from industrial IoT devices is done through AWS SiteWise.

Be aware that we only listed a few basic services of AWS IoT. Obviously, AWS also offers solutions that can be added to enhance IoT management, including, for instance, IoT Device Defender as endpoint protection for devices and specific industry solutions such as IoT FleetWise for IoT in vehicles.

Google Cloud IoT Core

In 2015, Google introduced two important services for implementing applications across IoT devices. The first one was Brillo, an operating system based on Google's Android, but scrubbed down to make it suitable for low-power devices. Weave was a communication protocol integrated into Brillo that allowed for communication between devices and the cloud. Both Brillo and Weave were rebranded to Android Things, which was depreciated at the beginning of 2022.

What does Google have to offer in terms of IoT nowadays? The main service is Google Cloud IoT Core, a suite of partner-led solutions hosted on the GCP platform. Some examples of these are Aeris and ThingsBoard. Aeris is a software platform for IoT and ThingsBoard is a solution for device management.

Google can integrate these partner solutions with data analytics and machine learning systems such as Cloud Dataflow and BigQuery for data insights.

Alibaba IoT Platform

Alibaba Cloud offers IoT Platform for connecting and managing devices and data. The platform has connection, communication, security, and device management capabilities, including over-the-air updates. IoT Platform is comparable to Azure IoT Hub and AWS IoT, with a full-stack service for entire IoT ecosystems.

Extensive documentation on Alibaba's IoT solutions can be found at `https://www.alibabacloud.com/help/en/iot-platform`.

Monitoring IoT ecosystems

The most important challenge that architects should address in IoT is, not surprisingly, security. The security estate of IoT devices must be monitored continuously. To understand the risks better, we can have a look at the top risks that are listed by the **Open Web Application Security Project (OWASP)**. We've only listed the top five:

- Weak passwords
- Poorly protected network services
- Poorly secured interfaces
- Lack of update mechanisms for security rules and patches
- Use of outdated components

The top risk, however, is the lack of device management and leaving data transfer unmonitored. However, this starts with knowing where the devices are. But knowing where devices sit is not sufficient; we must also know what these devices are doing, what sort of data the devices collect, and how they collect it. Observability is the key principle that any IoT architecture must comply with.

This is already the biggest challenge in IoT. Think of the fact that an IoT ecosystem might consist of hundreds or thousands of devices. A good example might be a water company that has sensors in every pipe that transports water. There are sensors in the production facilities as well, collecting data about the water quality in the various stages. All these sensors collect data about the quality and distribution of water through a complex system of pipes. Important parameters such as water pressure and the functioning of pumps are crucial.

The question that must be raised is: does the company need to monitor every single sensor? Or do we need to determine what is really important to monitor, instead of monitoring and responding to every single error or failure of a sensor? The data that all sensors collect must be accurate and relevant to decide whether the entire system works as designed and predicted. In other words, data analysis is the key to monitoring IoT.

IoT devices collect data, but they also produce data. That data is important to monitor the status of the device itself, including its security posture.

Let's recap the requirements for businesses to monitor IoT ecosystems:

- **IoT hardware**: Typically, IoT devices have a **CPU** (**central processing unit**), memory, and perhaps disks to store data. To ensure that the device is running and performing in an optimal way, we must monitor these components in the device, just as we would do with any other piece of hardware. Monitoring will be about checking that thresholds on usage on the CPU, memory, or disks are not exceeded, which could lead to slow performance or even failures. IoT monitoring tools that have been integrated into IoT suites will be able to send out real-time alerts in case of performance degradation or malfunction of devices.
- **IoT software and applications**: On top of the hardware, there will be software and applications running. This software will vary from operating systems up to small apps or functions that are hosted on lightweight containers with K3s, although K3s itself is developed to run workloads in unattended remote locations. Keep in mind that unattended doesn't mean "not monitored": these are different things. Whatever is running on IoT devices, this software must be monitored too. Monitoring tools must be able to detect issues in the malfunctioning of software since it will prevent devices from working correctly altogether.

- **Data:** As we have discussed, IoT devices collect and transfer data. The devices must be able to send that data to a central environment, such as an IoT hub or gateway. We will be discussing the gateway in a later section of this chapter. Just as important is that devices must be able to receive data, for instance, updates and patches. A non-interrupted flow of data is crucial for the optimal functioning of IoT devices. A glitch in that data flow might already cause severe problems. Monitoring the connectivity by monitoring the data flow between devices and between devices and the central managing environment is an important aspect of the monitoring tool. Interruptions must be detected in real time.
- **Security posture**: Last but not least, IoT devices must be secured and hardened. Hardening is the process of disabling redundant features and/or security risks and encrypting connections. This is to make it as difficult as possible for attackers to gain access to a system. In the occasion that the attacker does get access, it should be as difficult as possible to use obtained data. This also applies to IoT devices. Unnecessary accounts must be deleted, access restricted with least privilege, and connections encrypted. Also, an active policy must be applied for software and firmware updates, with the highest priority for security patches.

Monitoring tools must send alerts whenever security postures are breached. An IoT device is often an open door to a network leading to many other systems with valuable data. This is also true for home IoT: a non-protected door camera can be hacked, providing entry to the home network and computers that are connected to this network—a PC that holds the financial data of the owner, for instance. In companies, the consequences can be even more severe and lead to major data leaks.

The architecture for the IoT ecosystem must cover these topics and address the challenges. Next, we can define how the public cloud can help in providing solutions to overcome these challenges.

Implementing monitoring systems for IoT devices will present several challenges, such as:

- **Observability**: An IoT ecosystem will contain a huge number of distributed devices. This is the first problem that organizations will face: having a complete view of the entire landscape.
- **Detection and response**: It's crucial to detect performance or security issues as quickly as possible to prevent the distribution of the issue through the ecosystem. The speed of detection and response is key. Issues must be detected before critical processes and, with that, services to customers are impacted.

- **Integration**: From time to time, new devices will be added to the landscape. These devices must be integrated into the ecosystem and promptly connected to monitoring systems. The latter is important for security reasons: the new device must be compliant with the security policies. A new device might introduce the risk of vulnerabilities when security policies are not applied instantly.

How can public cloud environments like the ones that we discussed previously help in setting up IoT ecosystems and monitoring these ecosystems? The answer: cloud providers are agnostic and provide a central solution that is not restricted to specific IoT technology or manufacturers of devices. Cloud platforms allow companies to have a single-pane-of-glass view of the entire landscape with central management capabilities.

Solutions such as IoT Hub of Azure, AWS IoT, and Google's IoT Core all provide tools and solutions for:

- The auto-discovery of IoT devices in a network
- Centralized remote device monitoring including alerting
- The centralized monitoring and management of IoT security
- The centralized monitoring and management of connections

The biggest advantage, however, of implementing a cloud solution to manage IoT ecosystems is that the public cloud can offer all of this at speed and scale. The question is: what are the architectural requirements for an IoT system? First, we need to make sure that devices can connect with each other and to our cloud of choice. We will discuss this in the next sections, starting with connecting IoT ecosystems to the cloud.

Designing for connectivity to the cloud

Before we get to monitoring and managing an IoT ecosystem, there's one crucial step that we must take first: connecting the IoT devices to our cloud. Gateways and edge computing are solutions to this.

We can't simply connect thousands of IoT devices one by one to systems in the cloud. First, every connection in itself poses a risk of intrusion and with that the risk of a security breach. The solution to this is to have all connections targeting one system, machine, or instance that sits in front of the cloud environment, before the data of IoT devices is entered into systems that are hosted inside our cloud.

The IoT gateway is such as system. It can be a virtual machine or a service, but in all cases, it serves as the connection point between the cloud and the IoT devices. All data that flows between devices and the cloud environments will have to pass the gateway. The IoT gateway is the central connectivity and data controller. The diagram in *Figure 9.1* shows the principle of the IoT gateway.

Sometimes IoT gateway and edge computing get mixed up, but these are different things. Edge computing is about having compute power at the periphery—the edge—of a network. You could say that it's a small piece of cloud that sits closer to the data source. A typical use case for edge computing is to prevent latency when processing power needs to be extremely close to the data source. Other use cases include privacy and compliance, when data is not supposed to be processed in a central cloud environment.

Edge computers can serve as gateways. Data from, for instance, IoT sensors is collected in edge computers where the data is processed. This is particularly useful when IoT devices are not capable of sending data over large distances, which is true for almost every IoT device. Keep in mind that sending data requires a network and power. IoT devices often only have a small battery. When the device constantly has to send data over long distances, the battery will be worn out fast. Shortening the data distance can save a lot of power. Edge computers used as gateways are a solution. Edge computers will process the data of the devices and also send relevant data to the central cloud environment for further analysis.

Cloud providers offer various solutions for edge computing. The two leaders in this space are Azure Stack Edge and AWS Outposts:

- **Azure Stack Edge**: This is a device that is completely managed by Azure. The appliance can be ordered through the Azure portal. There are four versions of Stack Edge, one with 32 vCPUs (Pro) and one with 40 vCPUs (Pro 2). The 1U rack version can hold up to 4.2 Tb of storage and is also equipped with a **Graphics Processing Unity (GPU)**. Stack Edge performs as a cloud storage gateway, enabling data transfers to Azure while retaining local access to files on the appliance itself. There are two additional versions: Pro R, providing a GPU, and Mini R, with a **Vision Processing Unit (VPU)** for edge processing.
- **AWS Outposts:** Outposts is an AWS-managed appliance that can be hosted on-premises. It is delivered and installed either as a 1U or 2U rack server or as an AWS Outposts 42U rack. Customers can run native AWS services on-premises with AWS Outposts, including Edge Manager and IoT Greengrass.

Let's recap the challenges in managing IoT: it's about massive numbers of devices that we need to connect to a centralized platform for, among others, monitoring and data analysis. IoT devices are extremely distributed and typically only run with a small battery as a power source. Next, we need these devices to send and receive data in real time, at high speed, and across quite some distances. This requires different connection types and communication protocols. We will explore these in the next section.

Connecting IoT with IPv6, LoRa, and 5G

In the previous section, we discussed how we can connect IoT ecosystems to public cloud environments through gateways and edge computing. It's important to understand that IoT devices use different protocols to communicate with each other and eventually with the cloud. The reason for this is the massive number of devices that must be connected and with that the amount of data that is transferred from these devices to the cloud. The infrastructure must be capable of absorbing this, but it requires different means of communication.

We need connectivity that is able to connect machines with machines and continuously transport real-time, small chunks of data between the devices and cloud environments.

There are several emerging IoT standards, including the following:

- **IPv6 over Low-Power Wireless Personal Area Networks (6LoWPAN)**: The **Internet Engineering Task Force** (IETF) defines this as an open standard for connecting and communicating between low-power IoT devices and the internet. The 6LoWPAN standard includes 804.15.4, **Bluetooth Low Energy** (BLE), and Z-Wave for home automation.
- **Zigbee**: This is likely the best-known standard for IoT, mainly used in industrial settings as a low-power, low-data rate wireless network. Zigbee is based on the common 802.15.4 standard of the **Institute of Electrical and Electronics Engineers** (IEEE).
- **OneM2M**: This is a global standard for M2M communication, embedded in software and hardware for devices. Another standard for real-time M2M communication is **Data Distribution Service** (DDS), developed by the Object Management Group.
- **Advanced Message Queuing Protocol** (AMQP): This open-source standard is used for asynchronous, encrypted messaging by wire across various applications. This protocol is widely used in IoT device management and supported by all IoT solutions that we have discussed in this chapter.

- **Constrained Application Protocol (CoAP)**: This protocol was designed by the IETF to define how constrained, low-powered devices such as wireless sensors can communicate in networks. It uses a very simple syntax for messaging. The smallest CoAP message can be just 4 bytes.
- **Long-Range Wide Area Network (LoRaWAN)**: This is designed to support huge networks with IoT devices, such as smart cities.

One technology that we must mention here is **5G/LTE (Long-Term Evolution)**. The LoRaWAN network that we mentioned in the list is suitable for devices that occasionally need to be connected online. For devices that must be continuously connected and require a higher network speed and bandwidth, LTE-M and L4G M2M are better solutions.

The introduction and roll-out of 5G networks is the next phase with a huge potential to accelerate IoT. The biggest advantage of 5G is the ultra-low latency on the network, allowing for extremely fast response times, making it suitable for communication between services on the internet—the cloud—and fast-moving vehicles.

Some cloud providers have already developed services that use 5G. Azure Network Function Manager is a service that allows you to deploy specialized network functions—such as mobile packet core to enable a private LTE/5G solution on Azure Stack Edge.

Another example is AWS Wavelength, which allows for deploying ultra-low-latency applications. With Wavelength, parts of an application can be transferred to a Wavelength Zone, which is an extension of a virtual private cloud in AWS. IoT devices connected to 5G—think of moving vehicles—can connect to these applications in the Wavelength Zone without leaving the mobile network of a telecommunications provider. Typically, the device would communicate over different hops over the internet to reach the destination in the cloud, causing latency. Wavelength offers a solution for seamless communication.

More information on Azure Network Function Manager can be found at `https://azure.microsoft.com/en-us/products/azure-network-function-manager/`. For more details on AWS Wavelength, refer to `https://aws.amazon.com/wavelength/`.

Summary

In this chapter, we discussed the basic architecture principles for IoT ecosystems. It's fair to say that IoT architecture is extremely complex, exposing a lot of challenges. The first challenge is the scale of an IoT ecosystem that can easily hold hundreds to thousands of devices. All these devices must be secured and monitored. Next, these devices send data that must be analyzed. This requires a centralized platform.

We explored the IoT solutions of Azure, AWS, GCP, and Alibaba and what they have to offer. The services vary from the update management of devices to security monitoring to detect and respond to potential vulnerabilities in IoT ecosystems. An important element of IoT architecture is connectivity. In the final sections, we studied connectivity to the cloud using IoT gateways and edge appliances and the various protocols to enable low-powered devices to connect and communicate with each other. Last, we explored the possibilities of 5G and how this will impact the growth of IoT.

This concludes the section about setting up foundational environments for several use cases such as data analytics and IoT. The next part of this book will be about keeping financial control over cloud assets, using the principles of FinOps.

Questions

1. What AWS solution would you use to build, configure, and deploy applications to IoT devices?
2. Both Azure and AWS offer appliances that can be used for edge computing. Name the two solutions.
3. What does LoRaWAN stand for?

Further reading

- *IoT and Edge Computing for Architects*, by Perry Lee, Packt Publishing, 2020

10

Managing Costs with FinOps

Building and running a data center in public clouds can be more cost-effective than going the traditional way. However, running a data center in public clouds still costs money. Hence, as ever, you still need ways to control costs. **financial operations (FinOps)**—is all about cost control.

This chapter focuses on the starting point for managing FinOps in multi-cloud environments: the provisioning of resources and the costs that come with the deployment of resources. We will learn how to keep track of these costs in the public clouds of Azure, AWS, GCP, Alibaba Cloud, and **Oracle Cloud Infrastructure** (**OCI**). However, we start with a brief introduction to the principles of FinOps.

In this chapter, we're going to cover the following topics:

- Understanding the principles of FinOps
- Defining guidelines for provisioning cloud resources
- Defining cost policies for provisioning
- Understanding account hierarchy
- Understanding license agreements
- Defining tagging standards
- Validating and managing billing

Understanding the principles of FinOps

For starters, FinOps is not about saving money in the cloud, but about making money with the cloud. Hence, it should be tightly integrated into business planning and your architecture. FinOps has proven to be a must, with the advent of the cloud and DevOps, for staying in financial control.

Why would an enterprise need FinOps for its transition to the cloud? The simple answer to that question is: you need to have visibility into cloud spending. Achieving that sounds a lot easier than it actually is. In the cloud, engineers have the power to deploy resources "at will." There's no need to order new equipment, since the "equipment" is there and ready to use. While a finance department would need to sign off investments for hardware, they hardly know what's going on in cloud environments. Above all, with DevOps and multiple releases per month, week, or even day, cloud consumption is very dynamic. That makes it hard to keep track of spending. Moreover, it makes it hard to define whether the spending is contributing to the business goals—which it should. Enterprises need a methodology to track investments and enable them to relate investments to business goals, validating the business case. The FinOps Foundation, part of the Linux Foundation, has designed such a methodology. What FinOps tries to achieve is this:

- Inform
- Identify optimization targets
- Operationalize changes to realize the optimization targets

This is an iterative process, meaning that once a company has gained insights as to their cloud spending, they can identify possible targets where consumption can be optimized. The next step is to think about solutions to optimize the usage of cloud resources. The intent is not to cut costs per se, but to get more benefit out of cloud usage. In FinOps, this is an iterative approach, meaning it is not assumed that a company will do everything in one go. Teams start with a **minimal viable product** (**MVP**), build capabilities as they learn more, and then improve. The FinOps Foundation calls this lifecycle approach "**crawl, walk, run**." Teams will improve as they go through this lifecycle.

The most important principles of FinOps can be summarized as follows:

- Define principles and guidelines for the consumption of resources
- Define principles and guidelines for the provisioning of resources
- Define asset tagging standards
- Validate and manage billing
- Use chargeback models
- Ensure efficient purchasing options
- Carry out license management
- Automate

In the next sections, we will discuss these principles in more detail and explore examples of how they work with the major cloud providers.

Define guidelines for the provisioning of cloud resources

Before we dive into cost control in the provisioning of resources, we need to understand how resource provisioning works in the public cloud. There are lots of different ways to do this, but for this chapter, we will stick with the *native* provisioning tools that cloud providers offer.

There are basically two types of provisioning:

- Self-provisioning
- Dynamic

Typically, we start with self-provisioning through the portal or web interface of a cloud provider. The customer chooses the resources that are needed in the portal. After confirmation that these resources may be deployed in the cloud environment, the resources are spun up and made available for usage by the provider.

The resources are billed by hour or minute unless there is a contract for **reserved instances**. Reserved instances are contracted for a longer period—1, 3, or 5 years. The customer is guaranteed availability, capacity, and usage of the pre-purchased reserved instances. A benefit of reserved instances can be that cloud providers offer discounts on these resources. Over a longer period, this may be very cost-effective. It's a good way to set budget control: a company will know exactly what the costs will be for that period. However, it's less flexible than a pay-as-you-go model and, even more important, it requires up-front payments or investments, though most providers also offer the option to choose monthly payments at no extra cost.

Dynamic provisioning is more of an automated process. An example is a web server that experiences a spike in load. When we allow automatic scale-out or scale-up of this web server, the cloud provider will automatically deploy more resources to that web server or pool. These extra resources will also be billed at a pay-as-you-go rate.

Let's look at how these resources are deployed.

Deploying resources in Azure using ARM

Azure works with **Azure Resource Manager** (**ARM**), Azure's solution for **Infrastructure as Code** (**IaC**). ARM is a service that handles requests from users and makes sure that the requests are fulfilled. That request is sent to ARM. Next, ARM executes all the actions to actually deploy a VM, for example. What it does is assign memory, the processor, and disks to a VM and make it available to the user. ARM can do this with all types of resources in Azure: VMs, storage accounts, web apps, databases, virtual network resource groups, subscriptions, management groups, and tags.

ARM can be directly accessed from the portal. However, most developers will be working with PowerShell, the **Azure Command-Line Interface** (**CLI**), or REST APIs. In that case, the request goes to a **software development kit** (**SDK**). SDKs are libraries with scripts and code templates that can be called through a command in PowerShell or the CLI. From the SDK, the resources are deployed in ARM. The following diagram shows the high-level conceptualization for ARM:

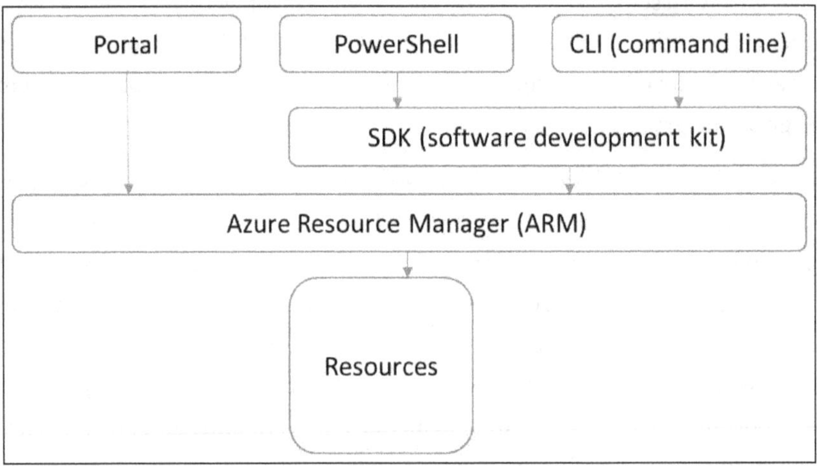

Figure 10.1: High-level concepts for ARM

In the next section, we will learn how to deploy resources in AWS.

Deploying resources in AWS using CloudFormation and OpsWorks

With AWS, we have to discuss two possible solutions to provision workloads. The first one is CloudFormation. As with ARM in Azure, CloudFormation is IaC, using JSON or YAML templates in a central repository, typically an S3 bucket in AWS. The code can be accessed through the browser console or CLI, allowing the architect or engineer to provision infrastructure from these templates in the designated AWS environments.

Another solution is AWS OpsWorks for the automatic deployment and configuration of resources. It works with a cookbook repository—a term that AWS borrowed from the automation tool Chef. That makes sense since, under the hood, OpsWorks works with Chef and creates managed instances of Chef and Puppet.

AWS OpsWorks Stacks allows for the automatic deployment of an entire stack, including infrastructure, load balancing, database, and the application itself. The stack itself is the core component for any deployment in AWS; it's the construct that holds the different resources, such as EC2, VMs, and Amazon **Relational Database Service** (**RDS**) database instances. OpsWorks makes sure that these resources are grouped together in a logical way and deployed as that logical group—we call this the cookbook or recipe.

It's important to remember that Stacks works in layers. The first layer is the **Elastic Load Balancing** (**ELB**) layer, which holds the load balancer. The next layer hosts the VMs, the actual servers that are deployed from EC2. The third layer is the database layer. If the stack is deployed, you can add the application from a different repository. OpsWorks can do this automatically as soon as the servers and databases are deployed, or it can be done manually.

Figure 10.2 shows the conceptualization of an OpsWorks stack:

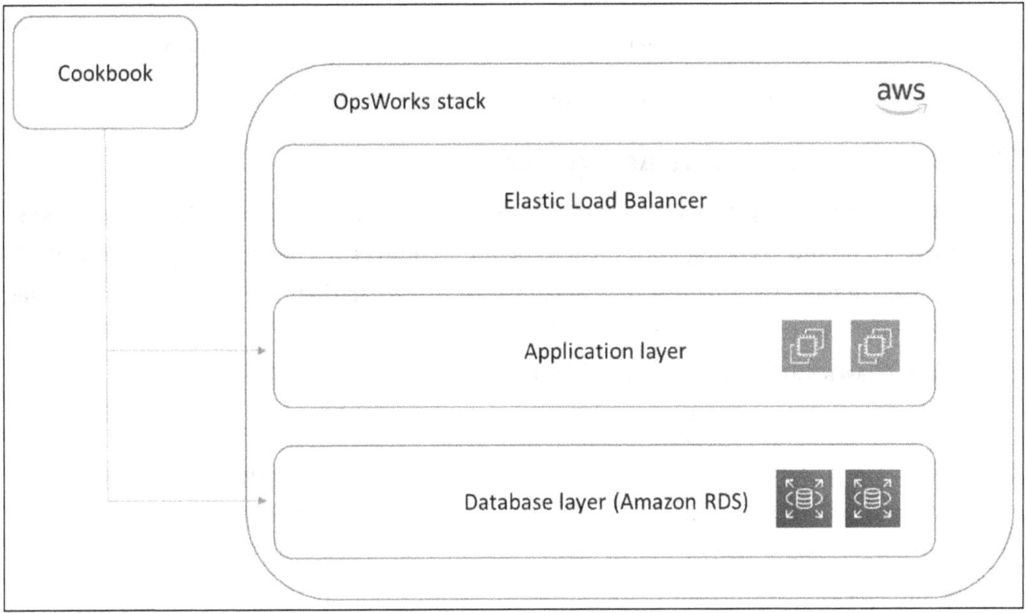

Figure 10.2: High-level overview of an OpsWorks stack in AWS

Next, we will look at Google's Deployment Manager.

Deploying resources in GCP using Deployment Manager

In GCP, the native programmatic deployment mechanism is Deployment Manager. We can create resources and group them logically together in a deployment. For instance, we can create VMs and a database and have them as one code file in a deployment. However, it does take some programming skills to work with Deployment Manager. To start, you will need to have the `gcloud` command-line tool installed. Next, create or select a GCP project. Lastly, resources are defined in a deployment coded in **Yet Another Markup Language** (**YAML**), which is commonly used. When the deployment is ready, we can actually deploy it to our project in GCP using `gcloud deployment -manager`.

As mentioned, it does take some programming skills. Deployment Manager works with YAML files in which we specify the resource:

- **Machine type**: A set of predefined VM resources from the GCP gallery
- **Image family**: The operating system
- **Zone**: The zone in GCP where the resource will be deployed
- **Root persistent disk**: Specifies the boot sequence of the resource
- **IP address:** The internet address the resource can be reached at

This information is stored in a `vm.yaml` file, which is deployed by Deployment Manager.

The final two clouds that we will discuss are Alibaba and Oracle's OCI.

Deploying to Alibaba using Terraform

Terraform is a tool to deploy IaC and can be used on every platform that we have discussed so far. Alibaba recommends using Terraform to deploy resources to their platform. Before we do that, we need to have Terraform on our machine. Next, we can define what resources we need in Alibaba, which starts with deploying the **Virtual Private Cloud** (**VPC**) and then the compute engine from **Elastic Cloud Service** (**ECS**)—indeed, the same terminology that AWS uses. We also need to configure networking, using a gateway, and assign IP addresses.

The following step is to check the configurations in the Terraform scripts. Alibaba has demo configurations in Git that contain best practices. The files can be downloaded `using git clone`: `https://code.aliyun.com/best -practice/017-int.git`. Take note of the fact that the original name of Alibaba Cloud is Aliyun.

From the demo files, you can check the configurations before the resources are deployed to Alibaba, taking the subsequent Terraform steps `init`, `plan`, and `apply`. In the ECS console, we can check if the resources have indeed been deployed.

A good explanation of how to work with Terraform on Alibaba Cloud is provided at `https://www.alibabacloud.com/blog/deploy-alibaba-cloud-resources-by-using-terraform_596412`.

Deploying resources to Oracle Cloud

Like Alibaba, OCI Resource Manager also uses Terraform as a deployment tool. Resource Manager can be launched from the main console, where it appears as one of the "quick actions," as shown in *Figure 10.3*.

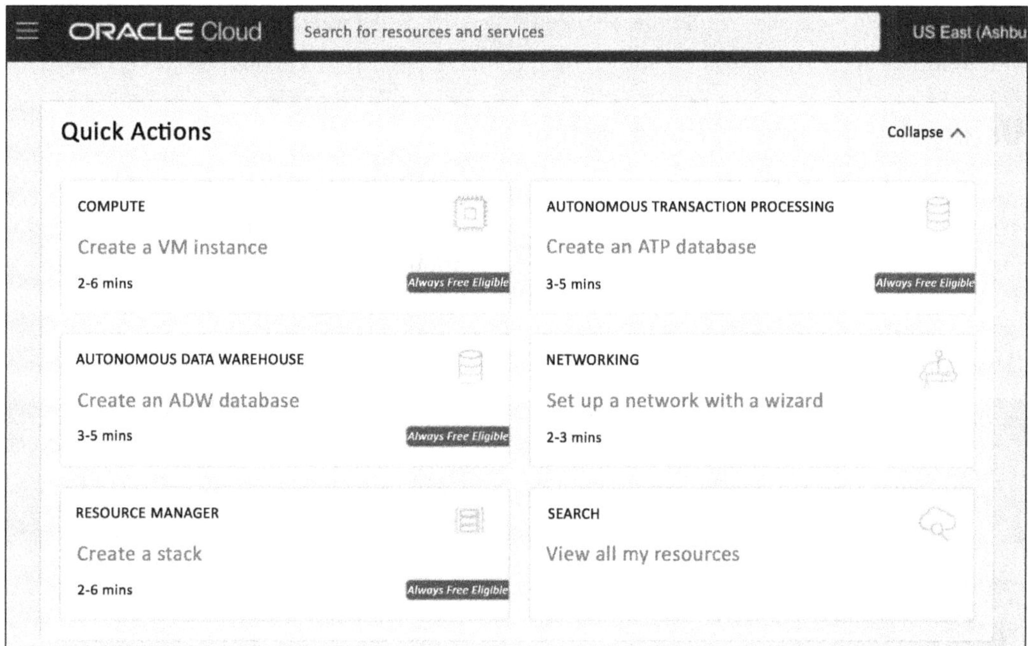

Figure 10.3: Quick access to Resource Manager in Oracle Cloud

Engineers can use various Terraform sources inside Resource Manager, such as GitHub, Git, and Bitbucket. Next, the engineer or developer creates a stack. For this stack, we can use pre-built Terraform configuration templates for Oracle. The Terraform configuration will automatically prepopulate variables to create the stack:

- Tenancy **Oracle Cloud Identifier** (OCID)
- Compartment OCID
- Region
- The OCID of the current user

These values are used when the Terraform commands `plan`, `apply`, and `destroy` are triggered. Remember that the Terraform code for the stacks is loaded in OCI Resource Manager.

Define cost policies for provisioning

In the previous section, we learned how to provision resources to clouds. This chapter is about keeping control of costs while provisioning resources. In *Chapter 11, Maturing FinOps*, we will learn about controlling methodologies and services, but we have to start with estimating our costs before we can set controls.

Let's start by saying that the sky is the limit in these clouds, but unfortunately, most companies do have limits to their budgets. So, we will need to set principles and guidelines for what divisions or developers are allowed to consume in cloud environments, to avoid budgets being overrun.

Using the Azure pricing calculator

It's easy to get an overview of what a resource, for example a VM, would cost us in Azure: the pricing overview on `https://azure.microsoft.com/en-us/pricing/calculator/` is a very handy tool for this and is, like all the other calculators and estimation tools that we will explore, completely free of charge to use.

If we open the page, we can see the **Virtual Machines** button, as shown in *Figure 10.4*:

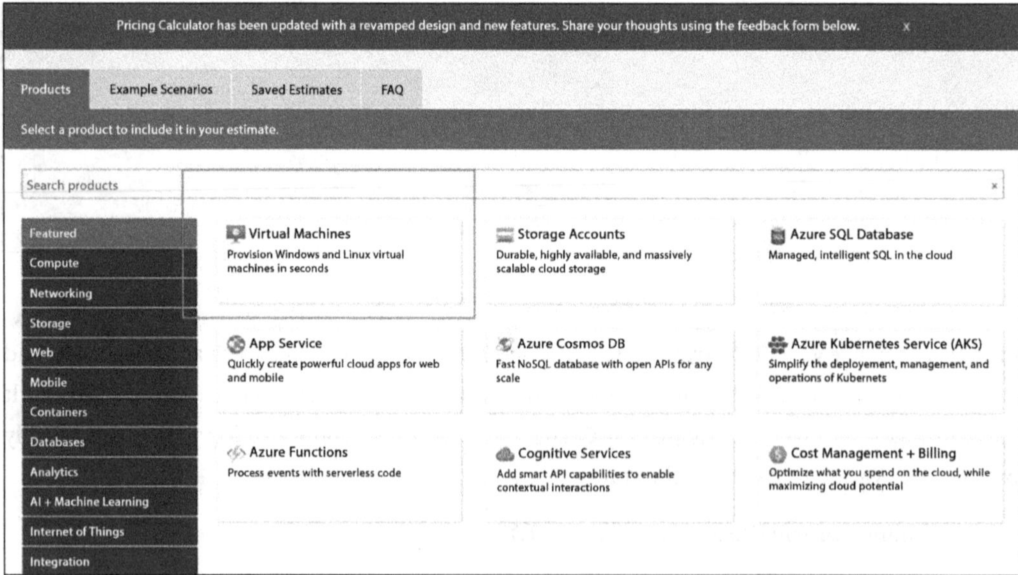

Figure 10.4: The Virtual Machines option in the Azure pricing calculator

The portal will display all the possible choices that are offered in terms of VMs, as shown in *Figure 10.5*:

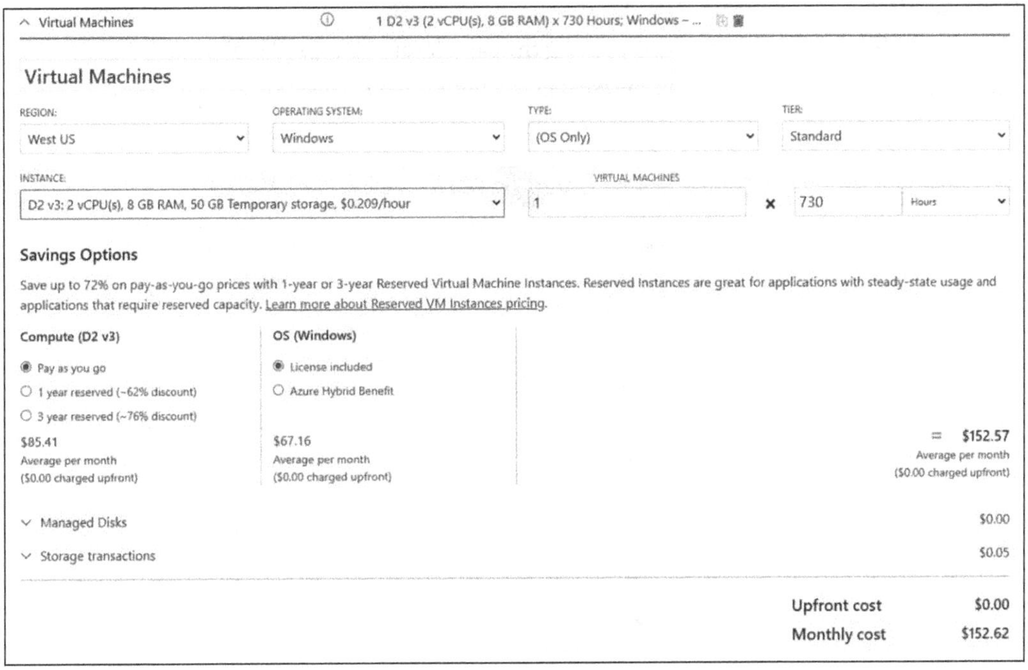

Figure 10.5: Tab details for VMs in the Azure pricing calculator

In the screenshot, the **D2 v3** VM has been selected. This is a standard VM with two virtual CPUs and 8 GB of memory. It also comes with 50 GB of ephemeral storage: this is temporary storage that exists as long as the VM runs. We can also see that it will be deployed in the **West US** Azure region, running **Windows** as the operating system. We purchase it for 1 month, or **730** hours, but under the condition of **Pay as you go**—so we will only be charged for the time that this VM is up and running. For the full month, this VM will cost us US$152.62. Note that prices will vary over time.

We could also buy the machine as a reserved instance, for 1 or 3 years. In that case, the VM cost would be reduced by 62% for 1 year and 76% with a 3-year commitment. The reason Azure does this is that reserved instances mean guaranteed revenue for a longer period.

The D2 v3 is a general-purpose machine, but the drop-down list contains well over 130 different types of VMs, grouped into various series. The D-series are for common use. The drop-down list starts with the A-series, which are basic VMs mainly meant for development and testing. To run a heavy workload such as an SAP HANA in-memory database, an E-series VM would be more appropriate. The E64s v4 has 64 vCPUs and 500 GB of memory, which would cost around US$5,000/

month. It makes sense to have this type of VM as a reserved instance.

Using the AWS calculator

The same exercise can be done in AWS, using the calculator at `https://calculator.aws/#/`. By clicking **Create an estimate**, the following page is shown. Next, select **Amazon EC2** as the service for creating VMs:

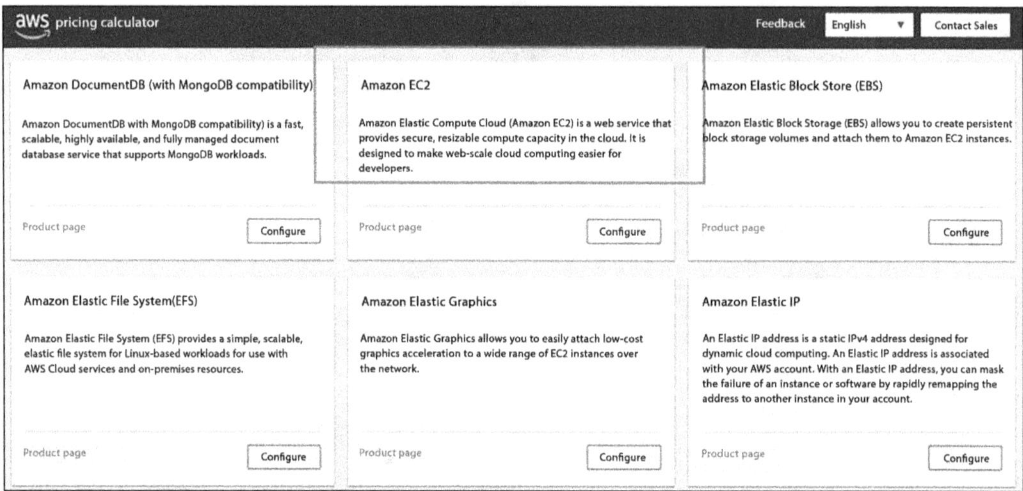

Figure 10.6: Option for EC2 VMs in the AWS pricing calculator

By clicking on that tab, a similar screen is displayed as in Azure. There is, however, one major difference. In Azure, the VM machine type is also taken into account.

It can be done in AWS, too; the requirements of the machine can be specified by indicating how many vCPUs and how much memory a machine should have. AWS will next decide what type of VM fits the requirements. The following example shows the requirements for a machine with two vCPUs and 8 GB of memory. AWS has defined `t3a.large`—**t3a** being a specific instance size—as a suitable machine:

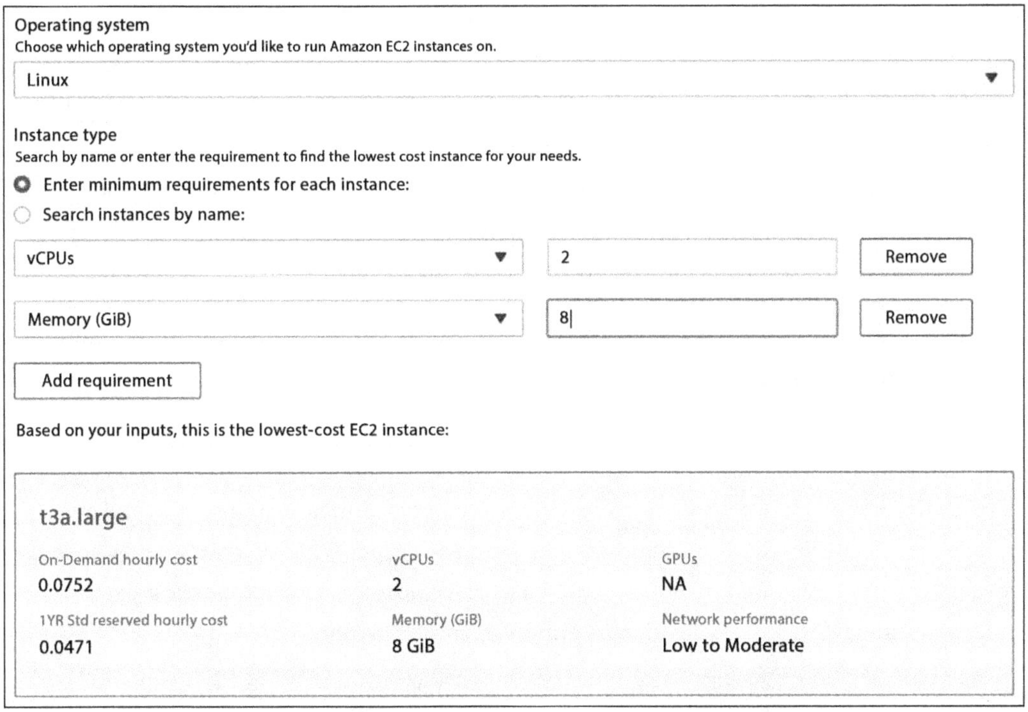

Operating system
Choose which operating system you'd like to run Amazon EC2 instances on.
Linux

Instance type
Search by name or enter the requirement to find the lowest cost instance for your needs.
Enter minimum requirements for each instance:
Search instances by name:
vCPUs 2 Remove
Memory (GiB) 8 Remove
Add requirement
Based on your inputs, this is the lowest-cost EC2 instance:

t3a.large

On-Demand hourly cost	vCPUs	GPUs
0.0752	2	NA
1YR Std reserved hourly cost	Memory (GiB)	Network performance
0.0471	8 GiB	Low to Moderate

Figure 10.7: Defining specifications for a VM in the AWS pricing calculator

The next decision to make is regarding the cloud strategy. AWS offers on-demand and reserved instances for 1 and 3 years, with the possibility of no, partial, or full up-front payments. With a relatively small VM such as our example, payment wouldn't become an issue, but also, AWS offers some huge instances of up to 64 vCPUs and up to 1 TB of memory, which would cost some serious money.

That's the reason for having guidelines and principles for provisioning. We don't want a developer to be able to *deploy* a very heavy machine with high costs without knowing it or having validated a business case for using this machine, especially since the VM is only one of the components: storage and networking also need to be taken into consideration. Costs could easily rise to high levels.

It starts with the business and the use case. What will be the purpose of the environment? The purpose is defined by the business case. For example: if the business needs a tool to study maps in a geographic information system, then software that views and works with maps would be needed. To host the maps and enable processing, the use case will define the need for machines with strong graphical power. Systems with **Graphics Processing Units (GPUs)** will fit best. In Azure, that would be the N-series; these machines have GPUs and are designed for that task. The equivalent in AWS is the G- and P-series, and in GCP and Azure, we can add NVIDIA Tesla GPUs to Compute engine instances.

Using the GCP instance pricing calculator

GCP doesn't really differ from Azure and AWS. GCP has a full catalog of predefined instances that can be deployed to a GCP project. The E2 instances are the standard machines, while the M-series is specially designed for heavy workloads with in-memory features, running up to and over 1 TB of memory. Details on the GCP catalog can be found at `https://cloud.google.com/compute/vm-instance-pricing`. And obviously, GCP has a calculator too. In the following figure, we've ordered a standard E2 instance with a free Linux operating system:

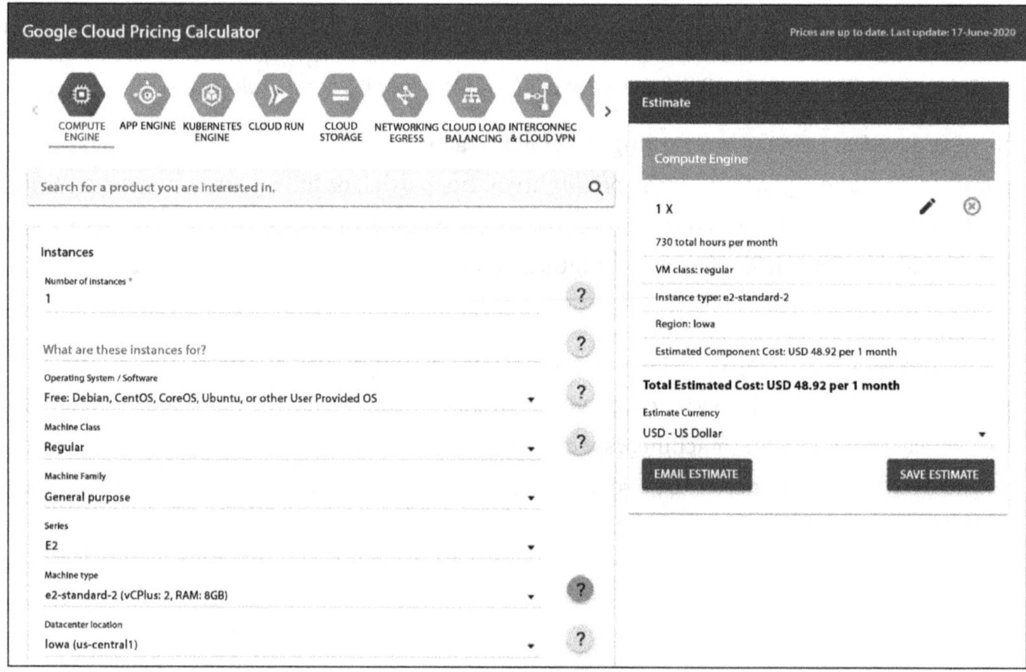

Figure 10.8: Defining specifications for a VM in the GCP pricing calculator

The pricing calculator for GCP can be accessed at `https://cloud.google.com/products/calculator`.

Understanding pricing in Alibaba Cloud

The pricing calculator for Alibaba Cloud can be found at `https://www.alibabacloud.com/pricing-calculator#/commodity/vm_intl`. The first thing that users will notice is that the calculator has three tabs, comparable to the Azure calculator.

The first tab contains the pricing list. The user picks a region at the top and the service for which they want the price tag at the left-hand side of the screen. You can also already choose options for a resource, for instance, pay-as-you-go or reserved instances.

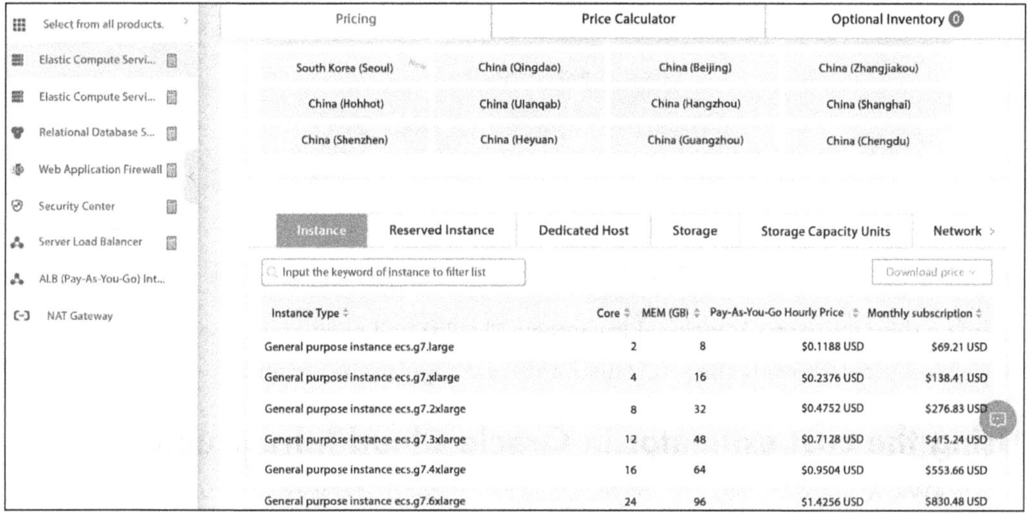

Figure 10.9: The price list for Alibaba Cloud

The second tab is the pricing calculator. It will calculate the resources against the list prices. The user must log in to the console to get the discounted prices. Refer to the message at the top of the screen in *Figure 10.10*:

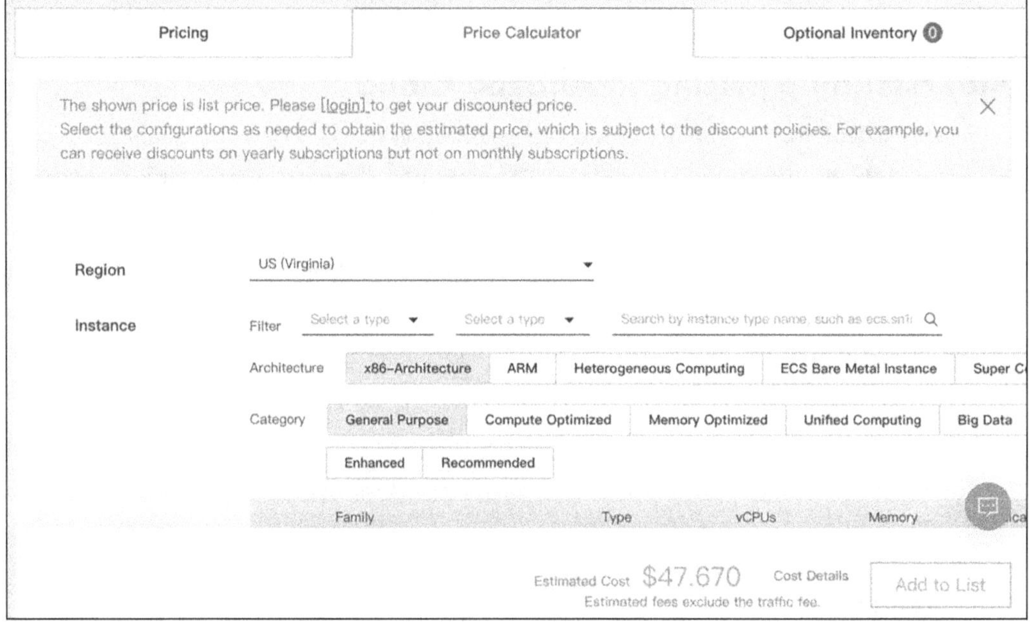

Figure 10.10: The pricing calculator for Alibaba Cloud

There's a third tab, named **Optional Inventory**. This list will populate as soon as you have purchased resources, showing the actual discounts that apply to the account.

Using the cost estimator in Oracle Cloud Infrastructure

Where AWS, Azure, GCP, and Alibaba have one calculator that covers all services for all regions, Oracle offers a cost estimator per region. As an example, you can go to `https://www.oracle.com/uk/cloud/costestimator.html` to find the cost calculator for the UK. It's shown in *Figure 10.11*:

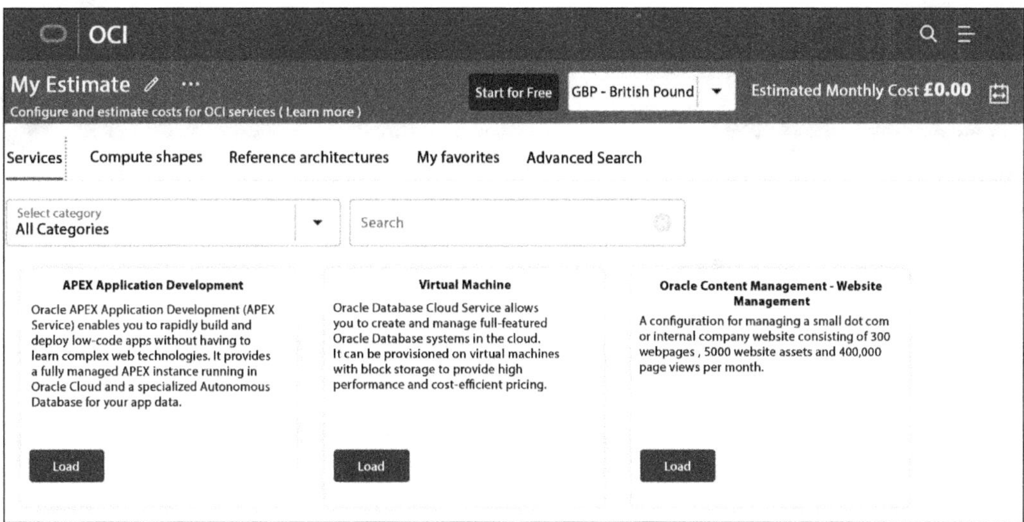

Figure 10.11: The pricing calculator for OCI

Under the tab named **Compute shapes**, you will find the various resources and corresponding prices. Reference architectures will help you in setting up specific stacks in OCI, for instance, a small stack to do data analytics. In that case, we can load **Oracle Analytics Cloud (Small)**, as shown in *Figure 10.12* below:

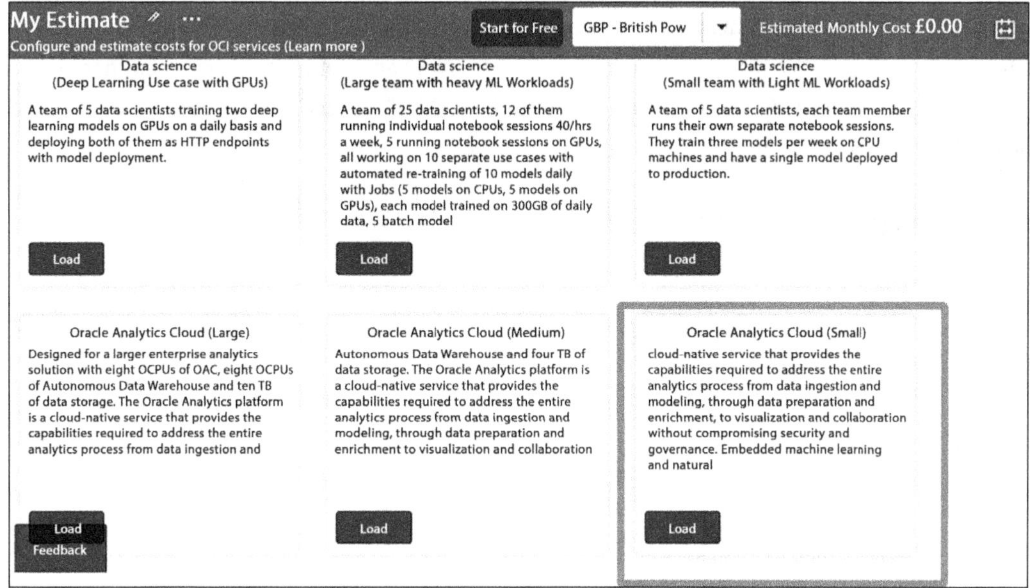

Figure 10.12: Loading a calculation from the OCI cost estimator

Now we can view the pricing details, presented in British pounds, shown in the next image:

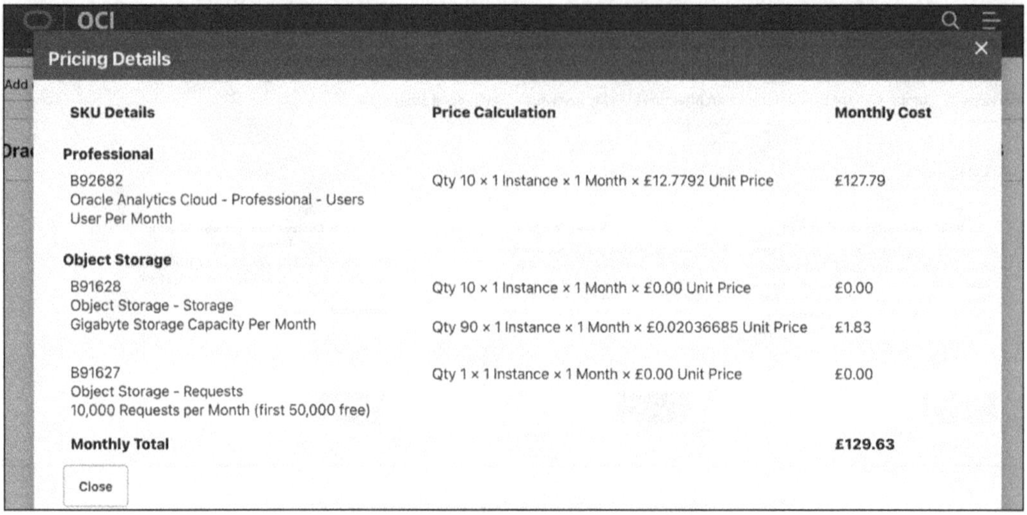

OCI

Pricing Details

SKU Details	Price Calculation	Monthly Cost
Professional		
B92682 Oracle Analytics Cloud - Professional - Users User Per Month	Qty 10 × 1 Instance × 1 Month × £12.7792 Unit Price	£127.79
Object Storage		
B91628 Object Storage - Storage Gigabyte Storage Capacity Per Month	Qty 10 × 1 Instance × 1 Month × £0.00 Unit Price	£0.00
	Qty 90 × 1 Instance × 1 Month × £0.02036685 Unit Price	£1.83
B91627 Object Storage - Requests 10,000 Requests per Month (first 50,000 free)	Qty 1 × 1 Instance × 1 Month × £0.00 Unit Price	£0.00
Monthly Total		**£129.63**

Close

Figure 10.13: Example of pricing details in OCI

Be aware that this only shows the estimated costs for a specific solution or a stack. It does not enroll the configuration to OCI from the estimator.

So far, we have looked at the major cloud platforms and how to purchase and provision VMs to the cloud environments. The next step is to track the costs of resources that we have deployed. One of the first questions that we must answer is from what level we want to track these costs and how we can identify these costs. It's the topic of the next sections, where we will discuss account hierarchy and tagging as one of the key elements of cost control.

Understanding account hierarchy

It's important to understand from what level enterprise management wants to see costs. Enterprises usually want a full overview of the total spend; hence, we need to make sure that they can view that total spend from the top level all the way down to subscriptions that are owned by specific business divisions or even DevOps teams. These divisions or teams might have a full mandate to run their own subscriptions, but at the top level, the enterprise will want to see the costs that these units are accruing at the end of the day.

This starts with the setup of the tenants, the subscriptions, and the accounts on public cloud platforms. This has to be set up following a specific hierarchy. The good news for financial controllers is that these structures in the public cloud closely follow the rules of the **Chart of Accounts (COA)** hierarchy, which is used for financial reporting.

This hierarchy has one top level. There can be many accounts underneath, but at the end of the day, they are all accountable to that top level. There's no difference when setting up an account hierarchy in the public cloud.

Let's look at a few examples in Azure, AWS, and GCP.

Enterprise enrolment in Azure

In Azure, we work with enterprise enrollment, the top level where we can manage our enterprise administrators and view all usage across all accounts in our Azure environment. The next level is the departments. Beneath the departments, we can create accounts. Both the top level—enterprise enrollment—and the departments are created through the Enterprise Agreement portal at `https://ea.azure.com`.

Be aware that you need an enterprise account in Azure before you can enter the portal. More information on enterprise enrollment in Azure can be found at `https://learn.microsoft.com/en-us/azure/cloud-adoption-framework/ready/landing-zone/design-area/azure-billing-enterprise-agreement`.

Now we can create accounts. These will be the account owners, who can view all of their subscriptions. The account owners will have the rights to create subscriptions and appoint service administrators that can manage the subscriptions. The following diagram shows the account hierarchy in Azure:

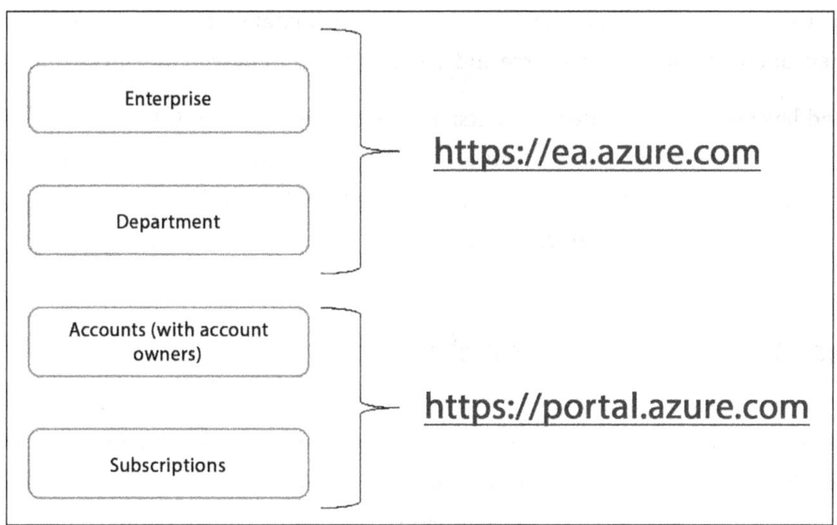

Figure 10.14: A high-level overview of enterprise enrollment in Azure

Organizations in AWS

In AWS, we can also enroll multiple accounts and centrally manage them. For this, AWS offers a service called AWS Organizations, where we can provision accounts using AWS CloudFormation and group them into organizational units that we can manage. Organizations also allows us to have a centralized cost management platform in AWS.

A service that needs mentioning here is AWS Control Tower, which allows the central management of multi-accounts in AWS, including AWS Organizations, AWS Identity and Access Management, AWS Config, AWS CloudTrail, and AWS Service Catalog. What Control Tower does is launch a landing zone in AWS with automated workflows and blueprints, adding configurations compliant with defined guardrails and setting up a structure to manage the workloads in this landing zone, with the underlying accounts. In short, we could specify Control Tower as a central governance and management dashboard to facilitate and automate enrollments.

To start enterprise enrollment, AWS advises contacting a sales representative directly. This is indeed strongly advised, since AWS has some interesting enterprise volume-driven discount programs such as the **Enterprise Discount Program** (**EDP**), which would be part of an Enterprise Agreement with AWS.

Organizations in GCP

The setup in **GCP** is very similar to Azure. In GCP, we also have a top level, the organization resource. This resource requires an organization node. We create the organization node through Cloud Identity. The node can match the corporate internet domain. Beneath the organization node, we can view and manage every resource and account that is deployed under the organization.

The second level in Azure was departments; in GCP, these are called **folders**. The final layer in the GCP hierarchy is the projects. Projects are functionally similar to subscriptions in Azure. Everything in GCP is created and managed through the Google Cloud console or the `gcloud` tool. In the console, we create an organization ID. Whoever creates this ID is automatically assigned as the super-administrator.

Account hierarchy in other clouds

The principles of account hierarchy in Alibaba and OCI do not differ from the examples that we discussed for Azure, AWS, and GCP. OCI works with compartments to organize and isolate resources. The highest level in OCI is the Administrators group. From this group, engineers have access to compartments. The structure for account hierarchy in Alibaba Cloud looks very similar to Azure, where we will have an enterprise management account at the top with a root folder.

From this account, the departments, as Alibaba refers to isolated environments, can be viewed and managed. The principle is shown in the following diagram.

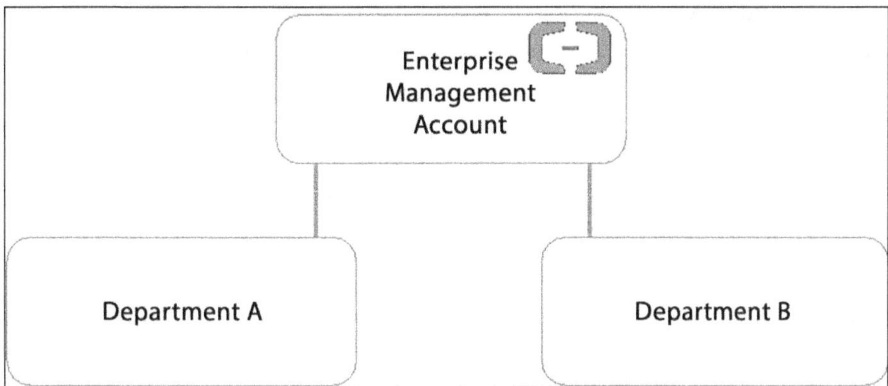

Figure 10.15: Simplified root account hierarchy in Alibaba Cloud

Consolidated billing can be executed from the top, the root enterprise management account.

This concludes the section about understanding the hierarchy of enterprise organizations in the major public clouds. The main conclusion is that the models resemble one another very closely, for a very good reason: at the top level, it is best for an enterprise administrator to have one single view of everything that is deployed in a cloud environment.

Understanding license agreements

License agreements are complicated, but in essence, there are three types of agreements to start using services in the public cloud:

- **Consumption-based**: This is often referred to as the pay-as-you-go model. The enterprise only pays for the actual usage in the public cloud, without any up-front commitment. Cloud providers issue a monthly invoice with the actual consumption of resources. These resources—for example, VMs, database instances, and storage units—are charged against the rates that are published on the public portals of the providers.
- **Commitment-based**: For most enterprises, this is the preferred model. In this case, the enterprise commits to the usage of a specific amount of resources in the cloud for a longer period of time, typically 1, 3, or 5 years. Now, public clouds such as Azure, AWS, and GCP were invented to enable maximum flexibility and agility.

If we allow enterprises to commit resources for a longer period, then this will have an impact on the resources that a public cloud can offer to other customers. For that reason, public cloud providers want to be certain that enterprises do really commit to the consumption of these reserved resources. Typically, an enterprise will need to pay upfront for these resources, whether they use them or not. Cloud consumption has become a formal contract that entitles an enterprise to have these resources available at all times.

- **Limited agreements**: These are agreements that are limited by time to an amount of resources that a customer can use. Typically, these are the type of agreements that are used for trial periods where resources are not charged for a specific period. Not all services will be part of these agreements—such as heavy instances with a lot of memory and terabytes of storage—and after 1, 2, or 3 months, the environment will be suspended by the cloud provider. A limited agreement can also hold a certain number of credits that can be used for a given time. If the credit amount is used, the trial period ends.

We need licenses to get started. Next, we can start provisioning resources. How do we identify and track these resources? Tags can be of great help. We will discuss this in the next section.

Define tagging standards

The major benefit of cloud provisioning is that an organization doesn't need to make large investments in on-premises infrastructure. In the public cloud, it can deploy and scale resources whenever needed and pay for these resources as long as the organization uses them. If it doesn't use the resources, it will not receive an invoice—unless a company has contracted reserved instances.

Another advantage of cloud provisioning is the agility and speed of deployment. Developers can easily deploy resources within a few minutes. But that's a budget risk at the same time. With on-premises investments, a company knows exactly what the costs will be over a certain period: the investment itself and the depreciation are a given. The cloud works differently, but an organization needs to be able to forecast the costs and control them.

A way to do this is by tagging resources. Tags allow a company to organize the resources in its cloud environment in a logical way, so they can easily be identified. By grouping resources using tags, it's also easy to see what costs are related to these resources and to which department or company division these costs should be transferred.

Tags are likely the most important attribute in terms of cost management and control in cloud environments. Naming conventions are much more focused on the identification of resources and are also crucial to the automation of cloud management. Tags are metadata that allow additional information on resources that can't be stored in a name. Tagging helps in understanding cost allocation, since we can use tags to categorize cloud resources.

All cloud providers offer extensive ways to apply tags to our resources. However, standard tags can be utilized across these different clouds. It's recommended to have tags for at least the following attributes:

- **Application**: Typically, a resource is part of an application or an application stack. To categorize resources—meaning VMs, storage, databases, and network components—that belong to one application or application stack, a tag should be added to identify to what stack or application a resource belongs.
- **Billing**: Especially large enterprises will have divisions, business units, or brands. These entities might have budgets or might be separate cost centers. Tags will ensure that resources are billed to specific budgets, or the accounting cost centers.
- **Service class**: Tags can indicate what service level is applicable to resources. Are they managed 24/7, what is the patch schedule, and what is the backup scheme? Often, enterprises have a tiered categorization for resources, such as gold, silver, and bronze. Gold is the highest level for production systems and may have disaster recovery solutions and uptime of 99.999%; silver and bronze would be for single systems with a much lower service level. A tag indicating gold, silver, or bronze will make clear what the service class of that particular resource is.
- **Compliance**: These tags indicate whether compliance rules apply. These can be industry compliance regulations, such as for healthcare or financial institutions, as well as internal compliance rules. These can be important in, for example, granting access to specific resources or the way data is securely stored.

Tags are a must to identify resources and the costs these resources generate. They will help us in validating invoices, which is the topic of the next section.

Validate and manage billing

It's very likely that a multi-cloud strategy will place several migrated systems into multiple different public clouds. With that, we are generating costs for pay-per-use instances and services, reserved instances for which companies have longer-term obligations, and licenses. Invoices will arrive from different providers. How do we keep track of all that?

Let's have a look first at billing in the major cloud platforms being discussed in this book: Azure, AWS, GCP, Alibaba Cloud, and OCI. These platforms share the same billing approach: as soon as services are consumed on the platform, charges will begin to accrue to which the CSPs can send invoices. Typically, this is referred to as the billing account. We will be using the cost or billing dashboards of the clouds to view costs and invoices.

Using cost management and billing in Azure

Azure billing has three types of billing:

- **Microsoft Online Services program**: Every user in Azure starts in this program. As soon as you sign in to Azure through the portal, you will get a billing account. This is also the case when you sign up for a free account. It's also needed for all pay-as-you-go services and for a subscription to Visual Studio, the Microsoft tool for development in cloud environments.
- **Enterprise Agreement**: An organization can sign an **Enterprise Agreement** (**EA**) with Microsoft to use Azure, which is valid across a lot of other products and services of Microsoft. An EA is a monetary commitment. An organization is entitled to extensive support from Microsoft, but it comes with contractual obligations such as a minimum spend.
- **Microsoft Customer Agreement**: If an organization signs up for Azure, in most cases a billing account will be issued for a Microsoft Customer Agreement. In some Azure regions, a Microsoft Customer Agreement can be issued if a free account is upgraded.

When a billing account is activated, a billing profile will be attached to it. This profile enables managing invoices and payments. Azure creates monthly invoices at the beginning of each month. Depending on the billing profile, the person who owns the account will see all costs associated with subscriptions and services in those subscriptions that are purchased under that specific account.

For example, if the billing profile is set to enterprise level, the billing account lists all costs that a company generates in Azure, in all subscriptions within the enterprise tenant. It's advised to define more billing profiles with specific invoice sections. This is done in the Azure portal under the **Cost Management + Billing** option, as shown in the following screenshot:

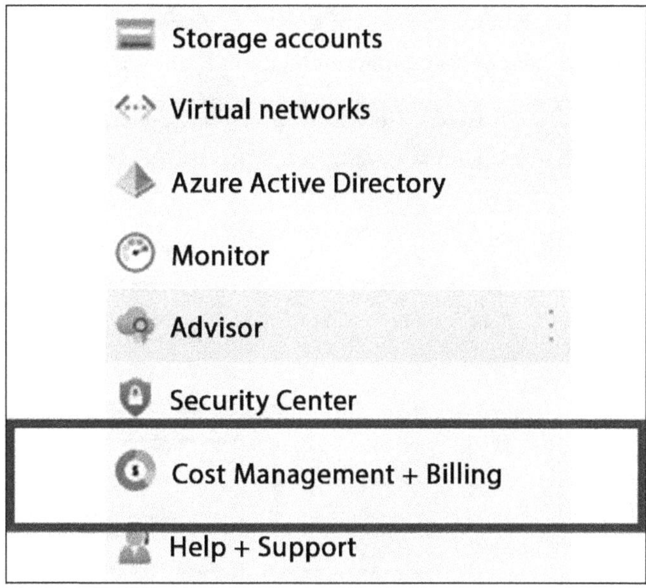

Figure 10.16: The Cost Management + Billing option in the Azure portal menu

It's true that the overall concept of cost management and billing is pretty much the same in the different clouds, but there are some differences in possible implementations for our own organization. Next, we will take a look at AWS and GCP.

Using AWS Cost Management for billing

In AWS, the free tier is the typical entry point and provides a lot of services for us to use. Organizations will typically enter into a customer agreement with AWS. Be aware that if you sign up on behalf of a company, AWS considers you the person with the legal authority to do so. Make sure that you are entitled to get into a commitment on behalf of your organization.

Similarly to the way Azure sets it up, cost and billing management for AWS is viewed through its portal. It's under the **AWS Cost Management** menu item, as shown in the following screenshot:

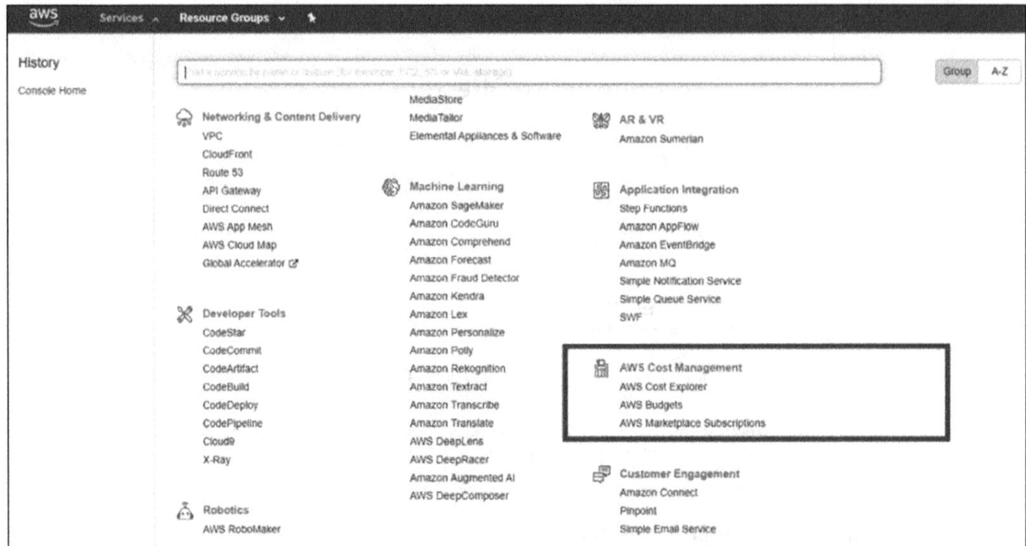

Figure 10.17: Cost and billing in the AWS console menu

It's common in both AWS and Azure for an organization to have separate divisions of accounts. With AWS Organizations, consolidating billing can be activated. There's one account for the whole organization or multiple accounts reflecting the organizational structure of the company. In the latter case, there will be multiple accounts. These accounts can be grouped under one, consolidated master account to have an overview of all AWS costs generated.

In AWS Cost Management, we can analyze costs with Cost Explorer, get usage reports, and manage our payments. Billing preferences can be set, such as receiving billing alerts and invoices being emailed as PDFs. Payment preferences, such as paying through a credit card or bank account, can also be set in Cost Management. In Europe, the **Single Euro Payments Area** (**SEPA**) is commonly used. In India, payments can be submitted through **Amazon Internet Services Private Limited** (**AISPL**).

Using billing options in GCP

As soon as billing is activated in GCP, the portal will prompt the user to set a billing account, as shown in the following screenshot:

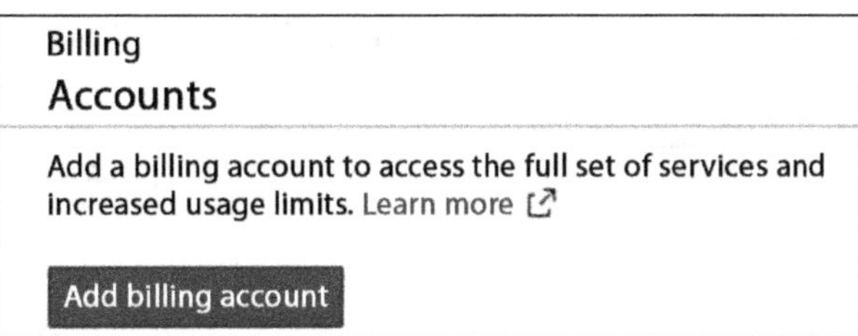

Figure 10.18: Adding a billing account in GCP

In GCP, cloud billing accounts are always associated with projects, which are the equivalent of subscriptions in Azure and accounts in AWS. Like in Azure, the billing account is coupled with a Google payments profile. There are two types of billing account roles:

- **Billing account admin**: This is typically someone in the finance department. This account can view all costs, set budgets and billing alerts, and link or unlink projects.
- **Billing account user**: The only thing the user can do is link a project to a billing account and see the costs associated with that project. The user can't unlink the project, unlike the admin.

The payment profile contains information about the legal entity that is responsible for the accounts. It also stores information on tax obligations such as VAT, bank accounts, payment methods, and transaction information such as outstanding invoices. Only the billing account admin role can view and alter this information.

In the Google Cloud console, we can enable interactive billing reports where the views and reports on billing information can be customized. For example, cost breakdowns can be added per project or per service used in GCP. In the Google Cloud console, this is all featured under **Billing** in the main menu, as shown in the following screenshot:

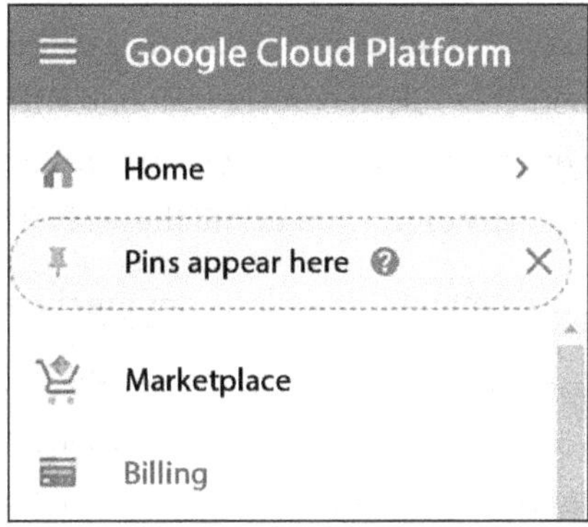

Figure 10.19: The Billing tab in Google Cloud's console

To view costs in Alibaba Cloud, we need to go to the Alibaba Cloud Management Console and browse to the **Expenses** tab:

Figure 10.20: The Expenses tab in the Alibaba Cloud Management Console

Before we do that, we must ensure that a role has specific access to the billing information—no different from the other clouds. In Alibaba, we need to grant access to the **Resource Access Management** (**RAM**) user. In fact, this RAM user must have an `AliyunBSSFullAccess` policy attached to the role. Only then will the user be able to view the billing information. On the billing page, we can then select **Spending Summary**, which will show the expenses per period or per subscription.

It's no different in OCI: to enable users to view and track costs in an OCI account, the user will need access provided through a policy. With the appropriate access, the user can manage payments and budgets, view cost and usage reports, and also do cost analysis. Costs can be filtered through dates, compartments (remember that isolated environments in OCI are called compartments), and/or tags.

In this section, we have explored the various billing options in Azure, AWS, GCP, Alibaba, and OCI using the billing dashboards in these clouds. In the next section, we will learn how we can validate invoices.

Validating invoices

Validating invoices has nothing to do with checking whether a cloud provider has charged us the correct amounts. Cloud providers have fully automated this process, so you may rest assured that if you or your company uses a resource in their cloud, it will show up on the bill. Validating invoices is about checking whether invoiced costs correspond with the forecasted usage of your company. Are you on budget or are you overspending? Are there resources on the bill that you aren't using anymore? And if so, why didn't you delete these resources?

Some key decisions will need to be made. These decisions are the same for all clouds covered:

- Will the organization use one or multiple billing accounts? If you want a project manager to be able to validate the costs for a specific project or in a particular environment, then they should be granted access to view these costs. As we have seen in the previous section, we can set these privileges granularly in roles and profiles that are attached to billing accounts.

- How will payments be processed? As discussed in the previous section, cloud providers offer various ways to process payments. Credit cards are popular, but most enterprises do their payments through invoiced billing and their respective bank accounts. The latter is strongly recommended for optimized cost control.

Next, we must define the validation process. It might sound overdone, but the truth is that organizations tend to have significant overspending in public clouds—simply because they lack insights and control into the billing process and because they don't have accurate cost management in place.

A recommended approach comes in three steps:

1. **Project control**: A project manager, product owner, or Scrum Master should be aware of what costs are generated from a project. If a team works in Agile Scrum and uses Sprints, it is advised to validate deployed resources after each Sprint. Are the designed resources deployed and what other services are related to that? These overviews should match the costs that are allocated to the project.
2. **Architecture control**: The role of the architect is to verify that only resources that are agreed-upon artifacts are included. A simple example: if it is agreed that only VMs of a certain series may be used for deployment to production, then the architect should check that this requirement is met. The deployment of other resources could inflict higher costs.
3. **Finance and accounting**: Based on the checks by project management and architecture control, the finance department can be sure that resource deployment is done correctly and that costs can be accounted for. Finance now has the task of checking invoices on terms of payment conditions and contractual agreements.

All cloud providers have dashboards that show exactly how much resource consumption will cost. That's a reactive approach, which can be fine. But if we want to force teams and developers to stay within budgets, we can set credit caps on subscriptions and have alerts raised as soon the cap is reached. All discussed cloud providers offer services to set budgets and alerts from the billing or expenses pages in their management consoles.

In multi-cloud, this would however imply that we need to view the dashboards of the various cloud providers. Enterprises would likely want to have a single pane of glass view: a central dashboard showing all the financial data of the total cloud consumption. One tool that can help with that is CloudHealth by VMware, which provides insights and advice for cloud financial management across multiple clouds. CloudHealth is a member of the FinOps Foundation, and hence follows the principles of the foundation.

Summary

This chapter started with a brief overview of the principles for FinOps: financial operations in the cloud or cloud financial management. We studied how we can provision resources for various clouds and then learned how we can track costs that are related to these resources. Before we can track resources, view the associated costs, and validate invoices, we must understand how cost management works in the cloud. We discussed the cost tools in Azure, AWS, GCP, Alibaba Cloud, and OCI. All these providers offer comprehensive toolsets to provision and identify resources from their respective management consoles. However, we must understand some general principles, such as license agreements and tagging.

In this chapter, we discussed the foundation of FinOps. In the next chapter, we will elaborate on how organizations can implement and develop cloud financial management, including the setup of a FinOps practice, using the FinOps maturity model.

Questions

1. If we want to run a trial period in a public cloud, what type of agreement would fit our needs?
2. Cloud providers use different technology to provision resources. What technology do both Alibaba Cloud and OCI use?
3. What is the discount program for large accounts in AWS called?
4. True or false: the pricing calculators of cloud providers are free to use.

Further reading

- *The Road to Azure Cost Governance*, by Paola E. Annis and Giuliano Caglio, 2022, Packt Publishing
- *AWS FinOps Simplified*, by Peter Chung, 2022, Packt Publishing

11

Maturing FinOps

In the previous chapter, we discussed the principles of cloud financial management, or FinOps. Now we know how we can provision workloads and track the costs of these workloads, it's time to take the next step and professionalize our FinOps practice. We will learn how to improve cloud financial management by studying various maturity models.

This chapter is about the transformation to managed FinOps in our organization, by setting up a FinOps team that has a major task in the adoption of the FinOps principles. Adoption starts with awareness. Hence, we will learn how to integrate FinOps principles in a cost-aware design process, making developers and engineers aware of the fact that every decision they take and every resource they implement has a financial consequence. That's OK, as long as the implemented solutions add to business value. That's the key theme of FinOps.

In this chapter, we're going to cover the following topics:

- Setting up a FinOps team
- Using maturity models for FinOps
- Introducing a cost-aware design
- Transformation to managed FinOps in multi-cloud
- Avoiding pitfalls in FinOps transformation

Setting up a FinOps team

For starters, FinOps is not about saving money in the cloud, but about making money with the cloud. Hence, it's tightly integrated into business planning and the forthcoming architecture. FinOps has proven to be a must to stay in control with the entrance of cloud and DevOps.

The challenge is evident: where engineers used to order physical equipment and had to go through a financial approval process to get that equipment delivered to a data center, now they can get that virtual equipment with a click of a button. However, although the equipment is virtual and basically just a piece of code deploying a resource in the cloud, it still costs money. The main difference is that companies shift from CAPEX to OPEX.

CAPEX—capital expenditure—concerns upfront investments; for example, in buying physical machines or software licenses. These are often one-off investments, of which the value is depreciated over an economic life cycle. **OPEX—operational expenditure**—is all about costs related to day-to-day operations and, for that reason, is much more granular, meaning that costs can be more detailed and defined per resource. Usually, OPEX is divided into smaller budgets that teams need to have to perform their daily tasks. In most cloud deployments, the client only pays for what it's using. If resources sit idle, they can be shut down and costs will stop. A single developer could—if mandated for this—decide to spin up an extra resource if required.

That's true for a **pay-as-you-go (PAYG)** deployment, but we will discover that a lot of enterprises have environments for which it's not feasible to run in full PAYG. You simply don't shut down instances of large, critical **Enterprise Resource Planning (ERP)** systems such as SAP, just as an example. So, for these systems, businesses will probably use more stateful resources, such as reserved instances that are fixed for a longer period. For cloud providers, this means a steady source of income for a longer time and therefore, they offer reserved instances against lower tariffs or apply discounts. The downside is that companies can be obliged to pay for these reserved resources upfront. Indeed, that's CAPEX. To cut a long story short: the cloud is not OPEX by default. The best advice here is to check the conditions with the cloud provider.

Hence, the journey to the cloud will require financial expertise. At least, that is strongly advised for businesses. It's not a matter of just pulling out a credit card to get started: businesses need a plan, just like they did when they still had data centers. Cloud is about doing investments with the right business justifications. That's the reason why every **Cloud Adoption Framework (CAF)** has governance and finance as one of the essential pillars.

We need the cloud, financial experts, and a plan. The FinOps Foundation calls this "laying the groundwork." This includes the collaboration between engineers, finance, and executive leadership. Technical and financial experts need to work together in moving environments to the cloud. But someone has to take the lead: this role is called the **Driver**.

In this book, we will use the terminology of the FinOps Foundation as the industry standard. More information can be found on the official website of the foundation: `finops.org`.

The Driver is the persona that glues the required disciplines together and drives the process of establishing the FinOps practice. This role is fulfilled by a FinOps practitioner. The main task is to advocate the principles of FinOps and develop the capabilities. This is not something that the FinOps practitioner can do in isolation. A good FinOps practice takes a FinOps team that works closely together with product or application teams, IT, finance, and the business.

The main tasks of the FinOps team are:

- Setting up control guidelines for cloud costs.
- Be aware that in multi-cloud environments, there can be different guidelines for different clouds. Hence, the team must have knowledge of the clouds that are used by the business. Refer to *Chapter 10, Managing Costs with FinOps* for more detail.
- Setting standards for tagging.
- Setting standards for cloud cost allocation.
- Businesses will have divisions, units, and teams who all can provision resources in the cloud and initiate costs. These costs must be allocated to the right team: the team that is responsible for the resources. This is where chargeback models are used. This is the topic for *Chapter 12, Cost and Value Modeling in Multi-Cloud Propositions.*
- Matching budgets with actual spends.
- Collaborating with application and development teams to improve efficiency in cloud resource provisioning, including purchasing strategy, and right-sizing of resources and automation.

The FinOps team must be knowledgeable about the cloud and cloud technologies, but also about buying and licensing strategies of the cloud providers. This team is typically the team that views the costs, validates the invoices, and advises application teams and businesses on improvements, leading to cost efficiency and more added business value. This is not an easy task.

A good FinOps practice is not set up in one day. Businesses will need to grow the FinOps practice. Maturity models will help with this. We will study the models in the next section.

Using maturity models for FinOps

In *Chapter 3, Starting the Multi-Cloud Journey*, we discussed the generic maturity model as guidelines for an assessment, defining how mature an organization is and what its ambition is. We used the **Capability Maturity Model** (**CMM**) as an example. It's shown once more in the following diagram:

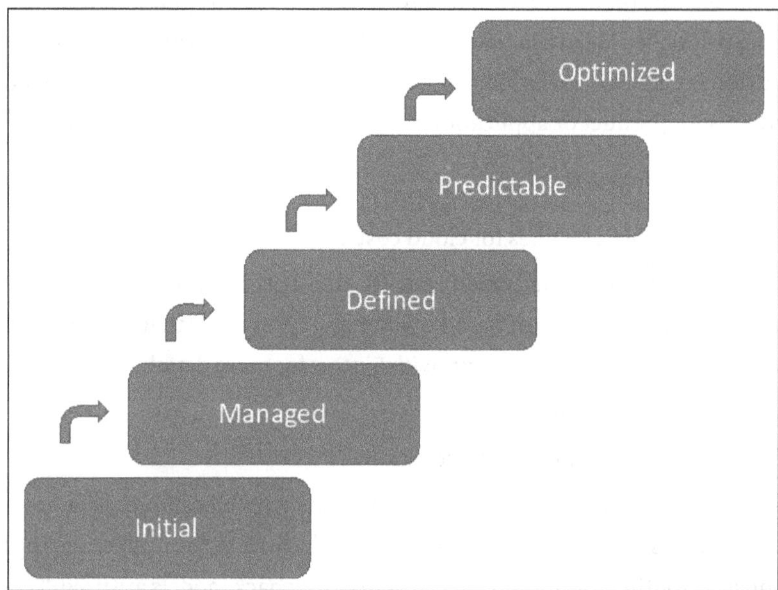

Figure 11.1: The Capability Maturity Model (CMM)

We can plot this model on cloud financial operations too. The principles are the same, where on the initial level processes are poorly controlled, and outcomes—in financial terms, this would refer to costs—are highly unpredictable. Level 5 is the highest level, where organizations can focus on improvements since projects and management are well-defined, processes are under control, and costs are predictable and measurable.

Let's look at another maturity model that will help in establishing financial management. This model is inspired by an originally developed model by the research company Gartner.

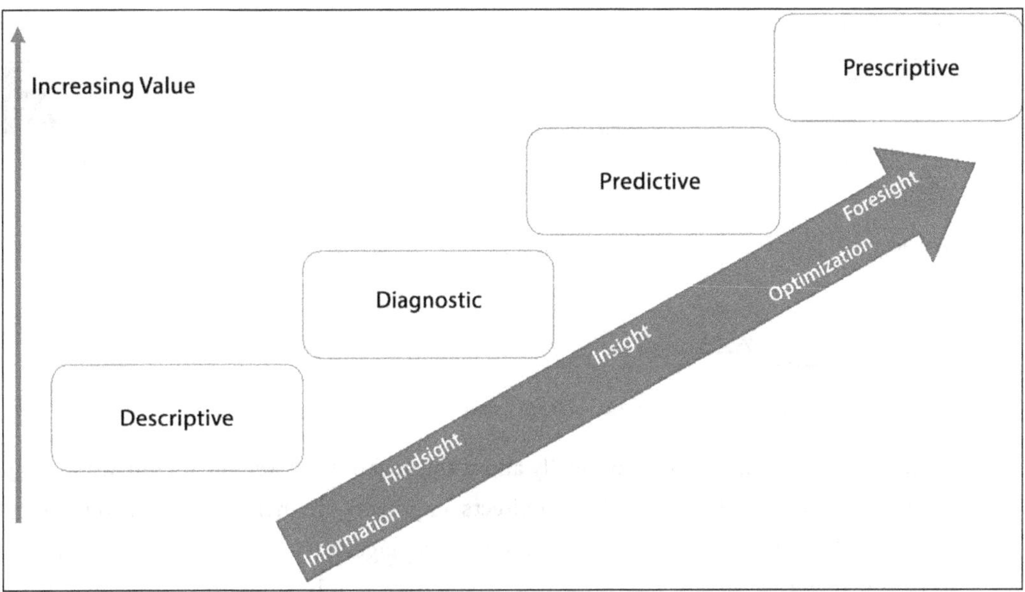

Figure 11.2: Maturity model showing stages from information to optimization

The main themes are hindsight, insight, and foresight, and that's exactly what FinOps aims to do: provide insight and enable foresight through financial analysis. It means that we need data and processes to analyze the data. We need metrics to validate the outcomes of the data and eventually predict future cost developments. Using automation, we could build in processes to automatically perform actions in our clouds to stay in control of costs.

These steps—hindsight, insight, and foresight—are integrated into the maturity model that was developed by the FinOps Foundation. That model has five lenses:

- Knowledge
- Process
- Metrics
- Adoption
- Automation

They are shown in the figure:

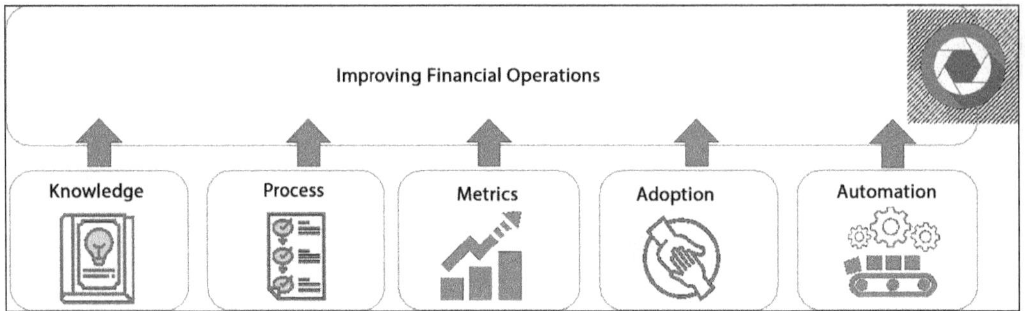

Figure 11.3: Maturity—The five lenses for improving FinOps

Knowledge, process, and metrics are basically about the vocabulary in financial operations. It's about having one common language that architects, engineers, financial experts, and management understand. If all stakeholders speak the same language, have the same information and data, and know how to interpret that information and data, combined with shared goals, then adoption will be easier.

Adoption requires a common understanding of data, a commonly shared vision of how business should evolve, and how IT can help enable this vision. The following step would then be to optimize; in the case of FinOps, this should be through automation of processes, for instance, using mechanisms in the cloud that automatically scale resources to align them with budgets and forecasts in usage of resources.

But how do we get there? The FinOps Foundation developed a methodology with three steps or stages: crawl, walk, and run. This is a very simple, yet comprehensive way to grow the maturity in implementing and applying FinOps. The best part of it is probably that it is completely agnostic to any cloud platform. The model simply tells a business how to grow in complexity and scope of the cloud services and manage the costs that are associated with these services, on any platform. An organization will typically go through three stages:

- **Crawl:** In this phase, the organization starts applying the basics of FinOps. There's very little reporting and only on basic metrics. Measurements can be made, and costs can partly be allocated to the business, by mapping resources to business processes or products. Reporting is only in hindsight, and forecasting versus actual cloud spend is not accurate, meaning that there are significant deviations in actual spend and forecasting.
- **Walk:** A much larger part of the cloud spend can be allocated to business functions such as sales and delivery processes of products. Collecting data for metrics is already largely automated.

- **Run**: Cost allocation is almost entirely automated, where cloud resources can be allocated to business functions and cloud spend is automatically charged back to that business function. Accuracy in forecasting versus actual cloud spend is high.

The big question now is: how do we take this into real-life practice? In the next section, we will try to provide some guidelines.

Introducing cost-aware design

This book, and even this specific chapter, started with the observation that implementing cloud solutions must add value to the business. It's the most important principle to keep in mind when we're discussing financial operations. FinOps is not about cutting costs, but about adding value using cloud technology. But inevitably, we will have to deal with costs. The cloud simply doesn't come for free.

As such, we need to know what the costs are when we design for the cloud. Next, we need to identify risks, since these will also inflict costs when risks materialize. These can be direct costs, but also costs that help us mitigate the risks. Lastly, we must define the expected benefits when we design and implement a solution. Costs, risks, and benefits will facilitate the calculation of the **Total Cost of Ownership** (**TCO**) and the **Return on Investment** (**ROI**), as shown in *Figure 11.3*.

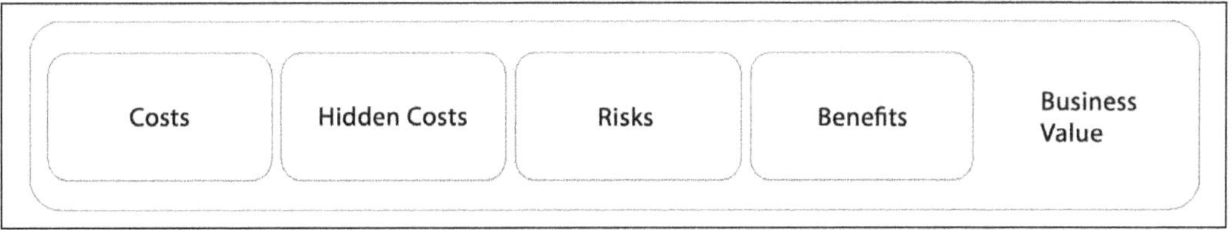

Figure 11.4: Enterprise TCO/ROI model

The benefits are all about the business outcome or the business value. The business value can, for instance, be increased revenue or faster time to market. If we look at the Cloud Adoption Frameworks of the cloud providers, we will notice that all frameworks work toward the business outcome as the justification to go to the cloud. But these frameworks also address the risks and the costs that a company will make to move resources to the cloud.

What are the costs? First, we have direct costs, sometimes referred to as out-of-pocket:

- Licenses
- Costs for resources such as virtual machines, firewalls, gateways, and other infrastructure services in the cloud

- Time that is spent by architects, engineers, and other professionals on designing, implementing, and operating cloud environments

There might also be hidden costs or costs that can't directly be associated with resources in the cloud. Think of the training of staff.

In terms of risks, we have to mention one particular tricky element: software licenses that we use in cloud environments.

When an enterprise uses software, it needs to purchase licenses for its usage. That goes for both proprietary and open-source software. The main difference between these two categories is that open-source software does allow modifications and changes in the software, as long as the changes are committed back to the development of that software. Proprietary software is typically closed source, where the source code may not be modified.

As with a lot of licenses, this can become quite complicated. If a product that a company uses is not sufficiently licensed, the company can be forced to pay fines along with the license fees it should have paid from the start. Using non-licensed software is illegal. That doesn't change when we're moving environments to the cloud.

How do you know when a software product is properly licensed? There are just a few types of software licenses—with a lot of variations—but stripped to their essence, they come down to these categories:

- **License on a user basis**: This is often the model that is used for end user licenses and it's probably the most straightforward way of licensing. For each user, there's a license that entitles the use of the software. A good example of this is Microsoft 365, for which a company can order a license per user, per month. In that case, there's a one-to-one relationship between the user and the use of the specific software product.
- **Licenses based on resources**: This is a more complicated licensing model. An example is software licenses based on the usage of a specific number of CPUs or the number of database instances. This is still a popular way of licensing proprietary software: the license fee is calculated according to how many CPUs or instances are being used. The issue in cloud infrastructure is that resources are quite often shared resources, virtualized on top of the real physical machine. Which CPU is licensed then? Typically, the license in cloud environments is based on the virtual CPU, in that case. Keep in mind that support agreements on software might change as soon as the software is deployed in cloud environments. Not all software is supported when used in Azure, AWS, or Google Cloud Platform.

- **Lump sum fee**: Software is purchased and paid in full upfront. This may even apply to major updates, whereas in subscription-based models, you will have access to major updates too. For software vendors, the lump sum is not a very attractive model, since they will only receive a one-time payment. But for companies, this is also not very attractive, since they will be confronted with high, upfront, cash-out investments.

How do you know when a software product is properly licensed? For that, specialized **Software Asset Management** (**SAM**) tools can be very convenient. SAM tools do a lot more than just make an inventory of all the software that a company uses: these solutions evaluate the whole life cycle of software, from purchase to deployment and, indeed, to utilization.

Other risks are an increase in costs or a lack of skilled staff. Both are becoming more and more a fact of life than a risk. A lot of companies already suffer from a lack of skilled personnel and are confronted with increasing costs. It's obvious that risks will influence the business value severely when they materialize.

To manage costs and risks we need a managed process. That process starts with a **cost-aware design**. The process is represented in the diagram.

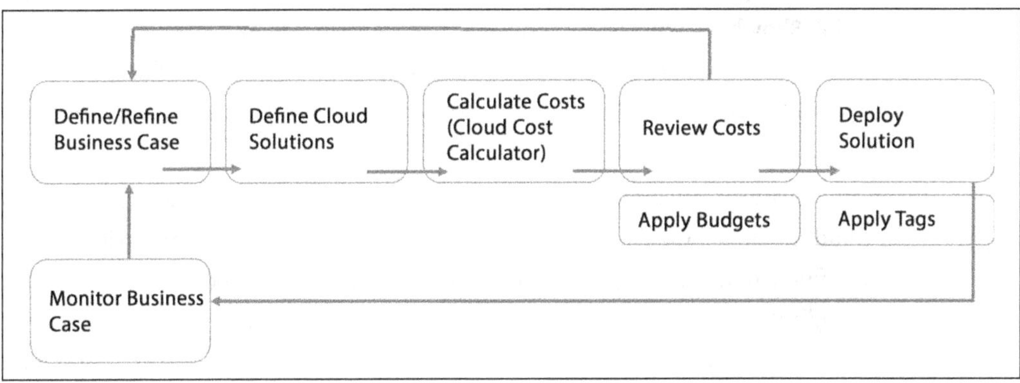

Figure 11.5: Cost-aware design process

The process begins with the business case and then helps the business to stay in control of costs and investments in the cloud by executing the following activities. Note that all activities loop back to the business case:

- **Mapping the data about cloud spend to the business**: An organization will have cost centers, business units, application owners, and, of course, applications and data. This is represented in an organizational breakdown. The cloud spend must be mapped to this organizational hierarchy.

This will take a lot of time, but it must be done. This is typically an activity to start with in the crawl phase, but it can become more granular and detailed in the next phases. It's essential that every dollar that is spent in the cloud is accurately allocated to a cost center, unit, or application owner. First, it's important to agree on how the organizational hierarchy is reflected in the design of the cloud platforms.

- In multi-cloud, we have to ensure that this is done identically in each of the platforms that we use. The process of setting up account hierarchy must be the same across the platforms, otherwise, the recharging model won't work. Costs will not be accurately allocated or, worse, completely missed in the models and left unallocated.
- **Define tagging strategy**: This must be aligned and applied throughout the entire organization. Agree on what type of tags are mandatory in every cloud platform. Make sure that tags allow for accurate reporting, meaning that resources are uniquely identifiable through tags and show up as unique resources in financial reporting. But tags should also allow fast, fault-tolerant allocation of the specific resource.
- **Define a clear showback and chargeback model**: When we have an organizational structure in place and agreement on the tagging policies, we can define a showback and chargeback model. Showback will clearly show what costs are made by a specific application, unit, or cost center; in other words, who is accountable for cloud spend. Chargeback will actually push the costs to that organizational unit, charging these costs to the budgets of the unit. In the next chapter, we will discuss this in more detail.
- **Define budgets and forecasts**: In a later stage, organizations will be able to use data about cloud spend to do accurate forecasting and set budgets for the organizational units that we defined in the first step of our cost-aware design. This can be automated too, for instance, by automated sizing of resources or suspending resources when these are not used for a certain time. By setting budgets, we can get alerts when resources in the cloud reach thresholds. All this information can next be used again for adjusting forecasts.
- Forecasts must be done for every organizational unit and must take every aspect or artifact in the cloud into consideration. Hence not only the use of virtual machines but also containers and serverless functions. Containers and functions are workloads too and cost money.

Now we can start building a cost model. This will be discussed in detail in *Chapter 12, Cost and Value Modeling in Multi-Cloud Propositions*. First, we will learn how to get operational excellence by implementing FinOps.

Transformation to managed FinOps in multi-cloud

In terms of maturity, we have reached the level of understanding financial management in the cloud and how we can set up a cost-aware design process. But our challenge is multi-cloud. We will likely have to deal with more than one cloud platform, which makes cloud financial management even more complicated. Still, we can use the FinOps principles to get to managed multi-cloud financial operations.

The FinOps Foundation provides a comprehensive model for managed operations, using three stages: inform, optimize, and operate. We will recognize this from the model that we showed in *Figure 11.2*.

This is a continuous cycle and part of the entire cloud operations. In multi-cloud, this implies that this cycle must be fulfilled for every cloud platform that we are utilizing. In multi-cloud, we need a single pane of glass, covering all the platforms in our landscape.

First, let's explore the three phases in the cycle:

- **Inform**: This phase is about visibility and getting the right data. This is actually harder than it sounds in the cloud. Cloud is by default dynamic. We have discussed this in the first two chapters of this book: companies migrate to the cloud as part of their digital transformation, making the business more agile. The development of new products can be accelerated because companies can respond faster to new market demands.
    - Companies can lift and shift their traditional IT into the cloud without making use of the real advantages of cloud and cloud-native; in that case, the environment in cloud platforms will remain rather static and data about the environment will not change much. But if cloud-native services are used with, for instance, automatic scaling, then the environment will be very dynamic. This has to be constantly verified against the business case: does the cloud deliver the expected business value that we defined during our cost-aware design? Continuously gathering the right data is crucial to enable optimization in the cloud.
- **Optimize**: This is the phase where companies can optimize their cloud environments. This is not only about enhancing automation, but also about planning and forecasting. That's why the inform phase is so important. If companies use specific workloads for a longer period, it might be worthwhile to consider a different deployment schedule for these workloads. Cloud providers offer reserved instances or committed use discounts (as Google refers to reserved instances) that have financial benefits in cloud consumption.

A workload on a reserved instance in the cloud can save up to 70 percent in costs. But it means that the provider can't use the reserved compute and storage capacity for other companies anymore, since it's now reserved and committed.

- The provider will likely ask for a sort of guarantee that the customer will consume the resource for one, three, or five years. The provider might even require upfront investments to cover this decrease in capacity that now can't be freely allocated. Of course, providers will have mechanisms in place to still enable shifting with capacity and reserve part of the capacity for one customer as reserved or committed use discount. There's also an advantage for the providers since they are assured that a customer is committed to using the cloud for a guaranteed period.
- Optimization can also be resizing and even redesigning workloads. Resizing can be done automatically. If a workload is hosted on an instance that is only using half of the compute and storage capacity in normal operation, automated scripts or the autoscaling capabilities provided by the cloud can reduce the compute and storage capacity. When usage peaks, the capacity can be scaled out again, also automatically.
- The redesign will require an evaluation of the architecture. Using cloud-native services, such as serverless or migrating applications to containers instead of full-size VMs, can lead to a decrease in costs and, more importantly, lead to more flexibility, increased efficiency, and better performance. As an end result, this can lead to better business results—the main goal of FinOps. However, this is often not an easy task.

- **Operate**: The data that we gather in the inform phase, leading to optimization, must be continuously evaluated against the business case and with that, the business goals. In the end, it's all about knowing what's going on, delivering the right resources to achieve business goals, and continuous improvement by constant optimization, and by doing all of this, increasing value to the customers. In a nutshell, this is operational excellence and FinOps can have a great attribution to achieving that excellence. FinOps can help in monitoring metrics that indicate how business objectives are met. It is important to have clearly defined metrics, formulated in a **SMART** way: specific, measurable, achievable, relevant, and timely. The FinOps team that we discussed in the first paragraph of this chapter can make this happen, working together with all other stakeholders, including the business.

The next step is to implement that single pane of glass since we are dealing with multi-cloud. We need tools for that. VMware's CloudHealth and Apptio's Cloudability are the leading tools in the space of financial management in multi-cloud. They offer real multi-cloud visibility and with that, the starting point for optimization and business-case-driven operational excellence:

- **CloudHealth by VMware: CloudHealth** will help organizations in managing clouds across the leading platforms AWS, Azure, GCP, Oracle, and, of course, **VMWare**. It monitors cloud spending and provides advice in optimizing the cloud resources with right-sizing and the purchase of reserved instances. It follows the FinOps maturity model with visibility, optimization, and finally, business integration in operations. More information about CloudHealth can be found at `https://cloudhealth.vmware.com/`. Be aware that the branding recently has been changed to VMWare Aria.
- **Apptio Cloudability: Cloudability** by **Apptio** works cross-platform too and provides extensive dashboards on cloud spend, budgets, and forecasting. It extracts billing and consumption data from the cloud platforms, while allowing business mapping. This can be very finely granular, up to the team level, which makes the tool suitable for use in DevOps streams. A strong asset in this tool is anomaly detection with alerts when resources overrun budgets and forecasting without direct reasons. More information on Apptio can be found at `https://www.apptio.com/solutions/cloud/cost-management/`.

We can have all the best-in-class tools and have our organization set up according to the best practices of FinOps, but there are still a few pitfalls that we should avoid. This is the topic of the next section.

Avoiding pitfalls in FinOps transformation

Let's start by repeating this message: FinOps is not about cutting costs in the cloud. It's about adding value to the business through cost management in the cloud. That is a major difference. The mindset of cost-cutting can lead to unwanted situations that can cost a company a lot of money. Think of this example: development environments that are not used during the night might be switched off during the nighttime and brought back online in the morning when developers get to work again. This will save costs, but the question is if it brings value to the company when developers can't work on environments outside of office hours.

Note: This is not advocating that developers should work 24/7, but there might be occasions where it's OK to allow people to work at times that are convenient for them. It raises productivity.

In international companies, shutting down instances in a specific timeframe might even be completely impossible. Think of European or American companies that have their development outsourced to India or other offshore countries.

One other major pitfall is buying a tool for cloud management without having the processes and people in place. There are first steps that must be taken: getting the right people on board, setting the objectives, and defining the metrics.

The most important lesson in implementing FinOps is probably that it should not be done following a top-down approach. It will not lead to the adoption of FinOps practices, and we won't reach our objectives: making teams that work in cloud platforms aware of financial consequences. Teams should have a great responsibility in using the right resources and services in the cloud, but also must be cost-aware. Every design decision that they take has a financial consequence. That is awareness and it's the start of adoption.

We have learned how to grow the maturity in FinOps. In the next chapter, we will start building cost models in the cloud, using all the different deployment and service models in multi-cloud environments.

Summary

In this chapter, we discussed how we can professionalize our FinOps practice by using maturity models. The key learning was that teams that work in the cloud must become cost-aware. We studied the principles of cost-aware design and how the maturity lenses of the FinOps Foundation can help us in adopting this way of working and, eventually, achieve operational excellence in managed FinOps.

We learned that getting the right data on resources and cloud spend is crucial and that we should map this to business processes and functions using showback and chargeback. From there, we can start optimizing our cloud resources, for instance, through right-sizing and purchasing reserved capacity. We have stressed that all these activities must lead to increased business value, rather than cutting costs in the cloud. In the final paragraph, we discussed a few major pitfalls.

We are now ready to go one step deeper in financial management in multi-cloud, and that's developing cost models. This will be the topic of the next chapter.

Questions

1. Name the five lenses in the FinOps Maturity Model.
2. What are reserved or committed use discount instances?
3. True or false? FinOps is mainly about cutting costs in cloud usage.

Further reading

- Information about and certification for FinOps practitioners can be found on the website of the FinOps Foundation: `finops.org`.

12

Cost Modeling in the Cloud

In the previous chapter, we looked extensively at the FinOps processes and how we can implement these in an organization. We also looked at how costs are generated in the various cloud platforms by evaluating the provisioning of resources. Now we have to make sure that costs are allocated in the right way and that they are booked at the right level in the organization.

In this chapter, we will learn how to develop and implement a cost model that allows us to identify cloud costs (showback) and allocate (chargeback) costs to the budgets of teams or units. Before we do that, we must understand the principle of cost coverage, the types of costs in the cloud, and how rates are set by providers.

In this chapter, we're going to cover the following topics:

- Evaluating the types of cloud costs
- Building a cost model
- Working principles of showback and chargeback

Evaluating the types of cloud costs

Before we get to define a cost model, we must understand what types of costs we will be faced with in the cloud. It's important to have teams understand how costs are being allocated and to have a centralized, controlled, and consistent cost allocation strategy.

Cost coverage

A typical cloud deployment model is **pay-as-you-go** (**PAYG**) and that's OK when the usage of the cloud is limited. Most enterprises will have larger landscapes in the cloud, including critical systems for their business. The company would want to have the guarantee that workloads will always be available, and that capacity is always available. For this, there is the principle of reserving capacity. Reserving capacity—or pre-committing—will bring companies benefits. The first benefit is that they're sure that capacity is reserved for them and thus available for a longer, contracted period.

The other benefit is that they can have discounts on reserved capacity. The reason for that is that the cloud provider is now ensuring that the customer is committed to using the capacity for a longer period, securing income for the cloud provider. However, there's also a bit of a downside. Cloud providers can't freely use the capacity for other customers now, limiting their flexibility in allocating resources in their data centers for those on the PAYG tariff. Hence, cloud providers may charge the customer upfront for reserved capacity. This means that the customer will have to make an investment to secure the capacity. We have to take that into account in our financial and cost models.

For cloud providers, it is all about capacity management. The reserved capacity allows predictability at the data centers. In the current context of global chip shortage affecting many industries, this is very important. But you have the option to pay a reservation monthly and even cancel it (at least in Azure) and apply for other resources/scope or ask for a refund. In case of refunds, additional conditions may apply.

In essence, we will have cloud usage that is covered by reservation and cloud usage that is not. Some usage might not be coverable through reservation. Think of spike usage. A customer could make a reservation for spike usage, but when the resources are removed after the spike has declined, then the reservation would only cost money, without usage. In FinOps this is labeled as **wasted usage**. Resource capacity that is not used but is provisioned will still be charged by the cloud provider. There will likely always be some underutilization of reservation, but then we must make sure that the costs of the underutilization are not exceeding the amount of the discounts. In that case, the reservation will cost the company more than without the discount.

We will try to get to covered usage as much as we can since that will provide the best insights and enable accurate forecasting. Covered usage is a resource that is covered by a reservation, resulting in a lower rate. Now, let's talk about the rates.

Cloud rates

Rates are a difficult thing in multi-cloud usage since cloud providers calculate rates in different manners. But there are some guiding principles in understanding how rates are presented to the customer.

We discussed how we can track costs in the cloud. Now, there might be a surprise when the company actually receives the invoice. There might be decreasing rates on the invoice. There might even be different rates calculated for the same type of resource. If a company uses more resources or resources for longer periods, the cloud provider might invoice different rates using, for instance, volume discounts. This can be confusing since the invoice can indeed show various rates for the same type of resources. In that case, FinOps talks about unblended rates.

If there are unblended rates, there will also be blended rates, where costs are evenly distributed across all used resources. Hence, every resource will have the same rate. This might look like a good approach, but it's not, since blended rates will not tell you exactly what the real costs per resource are.

The rates that eventually show up in the invoice will almost certainly differ from the rates that are shown in the costing portals of the different providers. The public pricing will show the list prices or on-demand rates. Companies hardly pay these full rates. There will be volume discounts, special programs, or otherwise contractually agreed discounts leading to a rate reduction. These reductions can lead to cost savings. The amount of savings will be visible in the invoice and then we can also calculate the savings potential by applying the commercial agreements to our existing and planned cloud environments.

It's worthwhile to study the various discount programs of cloud providers. In Azure the main ones are:

- **Azure Hybrid Benefit**: This program allows for optimal use of licenses for Windows, SQL Server, Red Hat, and SUSE Linux on virtual machines in Azure (`https://azure.microsoft.com/en-us/pricing/hybrid-benefit/`).
- **Azure Migration and Modernization Program**: This is a program wherein Microsoft offers expert help and best practices to migrate workloads to Azure. Cost optimization and incentives are included to reduce migration costs (`https://azure.microsoft.com/en-us/solutions/migration/migration-modernization-program/#benefits`).

In AWS, an interesting program is **Enterprise Discount Program** (**EDP**). AWS offers enterprise agreements that allow for tailored solutions in AWS. Enterprise discounts for large environments might be applicable. Enterprises that are eligible for this program typically get their own AWS account team that will work closely with the enterprise.

Other clouds have discount programs for big customers too. In most cases, customers will work together with account representatives of the specific cloud provider. In the case of Oracle Cloud, for instance, there are significant benefits when a customer runs Oracle software on top of OCI with software support discounts.

Licenses and agreements influencing rates are a complicated area in the cloud: the advice is really to work with the provider. Rest assured that the provider will try its best to get the best deal for a company that wants to use its cloud. There's a simple reason for that: without customers, their platforms wouldn't exist.

Amortized and fully loaded costs

We discussed reservations, pre-committing, committing, and discounts. How does that show in costs? We recognize two different sorts of costs here: amortized and fully loaded.

As we have seen in the previous section, reserving capacity might come with upfront investments. In the cost model, we must take these initial upfront costs into account and divide these costs by the actual usage. This is referred to as **amortized** costs.

The **fully loaded** cost is the amortized cost and the actual costs, including the discount rates, plus the shared costs divided by the actual usage. To get to fully loaded costs, we must do a complete breakdown of all cloud components, which we show in the next section.

Building a cost model

To build a cost model, we have to follow three basic steps:

1. **Identify cost drivers**: This is something that we discussed in the first two chapters of the book. Cost drivers are closely related to business processes. Think of the number of orders that customers of a company place in a defined time. Identifying the cost drivers will help in setting up the environment. In this particular case, we must think about scaling capacity when ordering intake spikes. But we can also think of event-driven architecture, where the placement of an order triggers a number of events, including invoicing and payment, fetching the order in a warehouse, and distributing the product for delivery. This will likely involve various systems and applications that must communicate with each other. An important question would be what could be automated and how to create the optimal solution since this will have an impact on the costs.

    - One thing that we haven't touched on yet is the fact that costs will largely define the price that a company can ask for a product or a service. The price of that product is basically cost and margin. A company that simply covers the cost will not make a profit. But the company can also not put any price on their product: it needs to "make sense" to a customer. Anyway, having very detailed, accurate insights into the costs for a company and how it can influence these costs is essential.

- **Define a full breakdown of all cloud components**: We simply need to know all elements that are used in our clouds. But at the same time, this must be presented in a comprehensible manner. That's what a cost model does. All expenses must be identified: infrastructure, software, development, security, and so on.
- **Quantify the components**: We must also answer the question of how much of the specific components we will use. As we have seen in the previous section, is not as simple as **P*Q (price times quantity)** or, in this case, **UC*Q (unit cost times quantity)**. We have to take amortized costs into account and various discounts to get to the right levels of cost. In this phase, we should also consider what shared cloud services we could use. Shared services will lower costs, but be aware that there might be a trade-off in compliance.

Let's start with the breakdown. A best practice is to choose the perspective of the application, assuming that every application will have an application owner. From the application, we can start defining what resources are needed to run that application. There will be unique resources and shared resources. In fact, there will be costs for landing zones that have to be managed. We can have a basic split in costs:

- **Consumption-based**: Everything that responds to the usage of the application.
- **Fixed cost**: Costs that will be made even when there are no applications hosted on the infrastructure. This must be shared among all applications.

Next, we have to think about managed and unmanaged services:

- **Managed**: Services that are completely managed by the provider. There's no need to add extra costs for activities that a customer must execute themselves to operate the service.
- **Unmanaged**: The cloud provides the service, but the customer has to manage the service itself. This means that there will be additional costing that must be considered. Think of labor, but there might be other costs involved, such as licenses.

This will result in a breakdown per service. In the table below, we list all components of IaaS that will result in a cost. First, the components of the virtual machine. In this example, we used EC2, the service from AWS, but it works exactly the same for virtual machines in any other cloud.

- EC2
    - General
        - Operating Systems
        - Licenses
        - Extensions
        - RDP Settings
        - Serial Console
        - Availability
        - Virtual Machine SLAs
        - VM Types
        - Auto-Shutdown
        - Just-in-Time Access
        - Boot Diagnostic
        - Virtual Server Templates and Infrastructure as Code
        - Virtual Machine Sizes and Tiers
        - Virtual Machine Naming Convention
    - Virtual Machine Deployment and Image Management
        - Golden Image
        - Post-Deployment Configuration
    - Monitoring
    - AWS EBS
        - Disk Caching
        - Drive Type
        - Drive Performance
        - EC2 Instance Additional Storage
        - Managed Drives Reserved Instance
        - Disk Bursting
        - Ephemeral OS Disks
        - Shared Disks
        - Network Connectivity

• Disk Naming Convention
• Disk Encryption

Table 12.1: Example of a list of components for IaaS

This is only the VM for the workload. We also need connectivity and storage for that workload. Last but not least, we have to make sure that the workload is secure. The list of components regarding these topics is shown in *Table 12.2*.

• Network
• Private IP Addresses
• Public IP Addresses
• Accelerated Network
• Proximity placement group
• Security
• Virtual Machine Encryption
• Virus Scanning and Anti-Malware
• Operating System Update Management
• Software Inventory
• Host Firewall
• Operating Systems Editions
• OS Hardening
• Storage Accounts
• Storage Accounts Security
• Authorization
• Encryption
• Recovery
• Tracking
• Backup and Recovery
• Cloud-Native Backup
• AWS Backups
• Windows and Linux Server Backup
• RDS Backup
• File-Based Restore
• Backup Monitoring

Table 12.2: Example of a list of components for securing a workload

All these components have a cost associated with them. All these costs must be tracked and managed. Hence, we also need to specify cost management settings. *Table 12.3* shows the activities for cost management.

• Cost Management
• Cost Visibility
• Cost Reporting and Analysis
• Management of Invoices and Payments
• Setting and Management of Budgets
• Control Usage via Policy
• Cost Optimization
• Reserved Instance
• Spot Instances
• Serverless
• Trusted Advisor

Table 12.3: Components for cost management

This is an extensive exercise but is needed to create a cost-aware design. You've guessed it: this must be done for every artifact in the cloud landscape, for every cloud that we use.

The second step is defining the quantities to start calculating the **total cost of ownership** (TCO). In Azure, we can use the TCO calculator for this. There are two ways of working with this calculator. We can start defining our workloads one by one. In that case, we visit the calculator at `https://azure.microsoft.com/en-us/pricing/tco/calculator/`. This is the screen that will be presented:

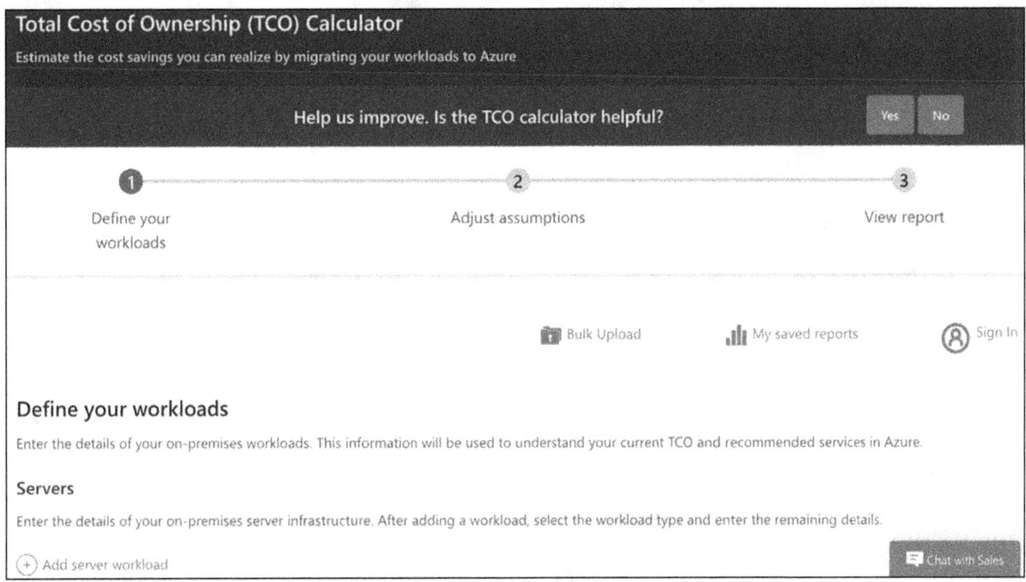

Figure 12.1: Starting screen of the Azure TCO calculator

At the bottom of the screen, we see that we can add a server workload. If we scroll further down, the calculator will present the basic services that we need to host workloads in Azure: databases, storage, and networking. At the top of the screen, we will find the option to do a "bulk upload," saving a lot of time. Next, we can adjust assumptions and view the report, providing us with the expected TCO when we move workloads to Azure.

Of course, other cloud providers have similar services. Next to pricing calculators, AWS offers Migration Evaluator (`https://aws.amazon.com/migration-evaluator/`). However, this has a different approach. This is an assessment that needs to be requested, as shown in the screenshot:

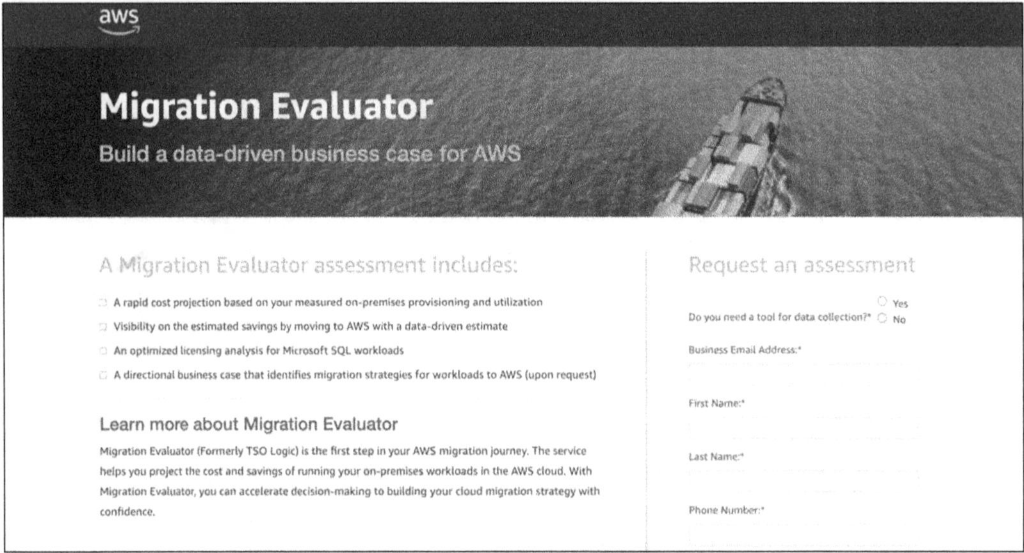

Figure 12.2: Starting the request for Migration Evaluator in AWS

Next, AWS will ask us to upload data, either through, for instance, spreadsheets or automatically collected by installing an agentless collector on existing systems that gathers the data. Migration Evaluator next analyzes the data, estimating the costs and savings in AWS. So, this tool calculates the business case when workloads are migrated to AWS. In Azure, this can be compared to Azure Migrate.

Last, Google Cloud, like AWS, offers an assessment of cloud migration and modernization costs. This assessment can be requested through `https://inthecloud.withgoogle.com/tco-assessment-19/form.html`.

We've built the cost model and also assessed the total cost of ownership. The final step is allocation. **Allocation** will help to determine if teams, business units, or departments that are accountable for a product or a service are profitable or not. Cost allocation will reveal cost utilization and help management make decisions. Hence, we must define a logical way to implement cost allocation, which is not easy in the cloud and especially not in multi-cloud. We must map the utilized cloud services to business artifacts or functions to get a clear view of what services are used by who, for what, and what the spend is. A way to do this is by taking the perspective of the application or application group. We will elaborate on this in the next section.

We have now prepared a cost model and we can start monitoring our cloud spend. These costs must be allocated to groups that we have defined. This can be done by showback and chargeback, the topic of the final section of this chapter.

Working principles of showback and chargeback

In the previous sections, we learned all about the type of costs, and the various discount programs that cloud providers offer, and we built a cost model. The latter is an extensive exercise, where we list all components that we use in our clouds.

In multi-cloud, an application could span services over various clouds. By choosing the perspective of the application, all services will be captured. If we have done cost modeling right, we will know exactly what resources are connected to the application and, with that, what running the applications costs. Then there's one last thing to do and that's coupling the application to the application owner, and with that, the team, department, or business unit.

Before we can start the process of chargeback, we need visibility on the application chains, the resources and services that are used in the cloud, and the respective costs that are associated with these resources and services. This process is referred to as **showback** in FinOps: analysis of all data about usage in the cloud, enabling reporting to relevant stakeholders. This data must cover all aspects that we discussed in the previous sections: usage, cost elements, rates, and discounts applied to rates. This data must be identifiable to an application or an application chain and, with that, to an application owner, team, or business unit.

The following step is **chargeback**, where the costs are sent to the accountable part of the organization and charged to the budgets of a team or a unit. The challenges here are the shared costs and the way discounts are distributed over the various resources, including the shared resources. This will require financial engineering, one of the key tasks of the FinOps team that we set up in the previous chapter. The first step is to be able to allocate the costs to the right budget. In a truly mature organization, the chargeback is automated and integrated into the financial system, and shared costs are allocated based on the actual usage of the shared resources.

This concludes the part about financial operations in the cloud. The next part of this book will be about security and, specifically, implementing security using the DevSecOps principles.

Summary

In this chapter, we learned how to build a cost model that allows us to track costs in the cloud and how these costs can be allocated to specific budgets. The key is the understanding that we must cover all resources and services that an organization uses in the cloud, including all shared resources.

We also saw that to get the right costing level, we must consider that we're dealing with various types of rates. The rates that are shown in the pricing calculators of cloud providers are often not applicable: discount programs might be applied, influencing the true costs of cloud environments. We learned about blended and unblended rates, managed and unmanaged services, direct and shared costs, and consumption-based and fixed costs. All of this must be reflected in the cost model to get to an accurate showback and eventually chargeback.

This was the final chapter about FinOps. The next part of this book will be about security.

Questions

1. What is the main benefit of reserved instances for companies?
2. On invoices, various rates for the same type of resources might appear. What do we call these types of rates?
3. Microsoft offers a program that allows for the optimal use of licenses for Windows Server, SQL Server, and various Linux distributions in Azure. What is this program called?

Further reading

- The FinOps Foundation releases the *State of FinOps* yearly, showing the trends in cloud financials. The *State of FinOps* can be found at `https://data.finops.org/`.

13

Implementing DevSecOps

The typical reason why most enterprises adopt the cloud is to accelerate application development. Applications are constantly evaluated and changed to add new features. Since everything is codified in the cloud, these new features need to be tested on the infrastructure of the target cloud. The final step in the development process is the actual deployment of applications to the cloud and the handover to operations so that developers have their hands free to develop new features again based on business requirements and feedback from the customer and operations.

To speed up this process, organizations work in DevOps cycles, with continuous development and the possibility to test, debug, and deploy code multiple times per week or even per day so that these applications are constantly improved. Consistency is crucial: the source code needs to be under strict version control. That is what **CI/CD** pipelines are for: **continuous integration and continuous delivery and deployment.** But we also need to make sure that code, pipelines, applications, and infrastructure remain secure. Hence, we must embed security into DevOps and aim for DevSecOps.

We will study the principles of DevSecOps, how CI/CD pipelines work, and how they are designed so that they fit multi-cloud environments and comply with security policies.

In this chapter, we're going to cover the following topics:

- Understanding the need for DevSecOps
- Starting with implementing a DevSecOps culture
- Setting up CI/CD
- Working with CI/CD in multi-cloud
- Exploring tools for CI/CD

- Following the principles of Security by Design
- Securing development and operations using automation

Understanding the need for DevSecOps

Before we dive into the layers of DevSecOps, it's good to understand why DevSecOps is important in multi-cloud. First, we must understand the layers of securing the cloud. There are four layers to be considered:

1. Organizational level, or the overarching governance
2. Enterprise level, ensuring the security across accounts, auditing centralized compliance through monitoring and logging, and promoting automation
3. Subscription level, using **Role-Based Access Control** (**RBAC**), threat detection, and in-depth defense
4. Solution level, using CI/CD with validated templates, blueprints, and images

We must define security at all levels. The following diagram shows all levels of defense in the cloud.

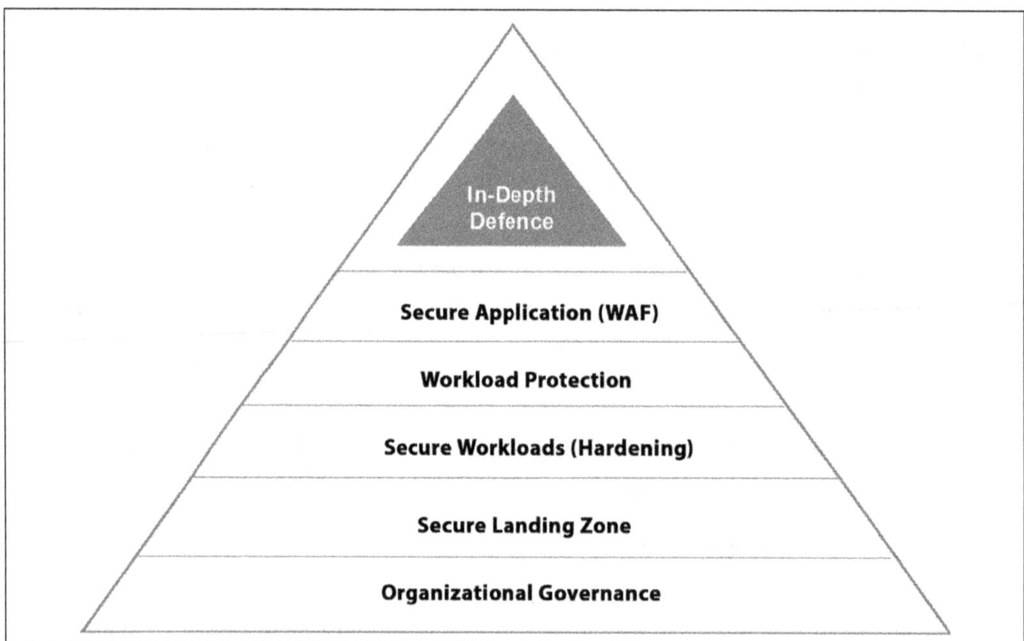

Figure 13.1: Levels of security in application stacks

The top of the stack is formed by the application payloads. In multi-cloud, enterprises will likely use containers and CI/CD pipelines. With multi-cloud, enterprises acquire the capabilities to:

- Manage and provision workloads across multiple cloud platforms in various regions
- Scale workloads for the best performance, leveraging the benefits of each specific platform
- Create high-availability solutions using the resilience of multiple cloud platforms
- Simultaneously operate applications across multiple clouds

How do we do that in multi-cloud? The best answer to this is: with containers that we can deploy on these various cloud platforms. Hence, we first must understand how containers and container orchestration work in multi-cloud. Containers allow the software to run reliably, regardless of the environment or cloud. Developers only have to take care of what's in the container, while the operations of the containers themselves are typically fully automated using container orchestration, which enables, for instance, unlimited scaling without manual intervention.

Containers work differently from VMs. With VMs, we have a hypervisor telling a server that it's multiple servers, able to run multiple applications on one of the virtual servers. However, each virtual server requires its own operating system, making a VM rather heavy. Containers don't require their own operating system; they simply use the operating system that's running on the host machine. This makes containers very light and easily transportable. The difference between containers and VMs is explained in *Figure 13.2*:

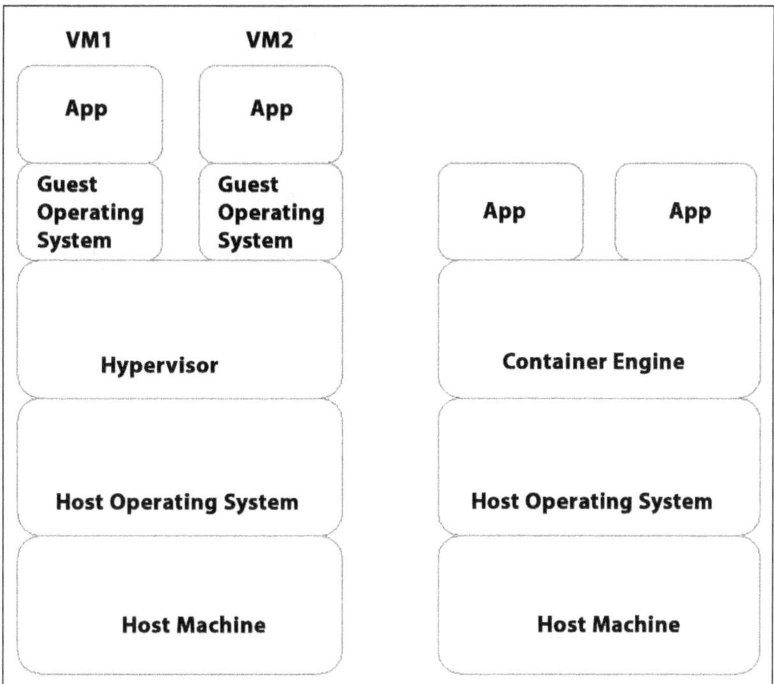

Figure 13.2: Difference between Virtual Machines (left) and containers (right)

So, containers can run on any platform: they are portable, meaning that a container can easily be deployed to different servers on different platforms. That is true, but we need to prepare our clouds to run containers. In practice, it means that we need a runtime environment for our containers, the container engine. The industry standard to orchestrate containers in any cloud is Kubernetes, originally invented by Google.

Kubernetes takes care of the deployment and management of containers. It handles all the heavy operational tasks, such as allocating networking, compute, and storage to the workloads that are hosted in the containers. It also enables automated scaling: if workloads need more capacity, Kubernetes will scale out the container infrastructure, allowing for more containers to run. If usage decreases, Kubernetes will scale down again. The best part: it can do that on all clouds. All cloud providers offer Kubernetes services:

- **Azure Kubernetes Services (AKS)**
- **Elastic Kubernetes Services (EKS)** in AWS
- **Google Kubernetes Engine (GKE)** in GCP
- **Alibaba Cloud Container Service for Kubernetes (ACK)**
- **Container Engine for Kubernetes in OCI (OKE)**

Now we have a runtime environment, and we have containers. Both must be secured. First, we must understand that Kubernetes is software that runs on servers, forming a Kubernetes cluster with multiple virtual machines or nodes. As with any server, we need to protect the servers and the software that runs on top of it, preventing unauthorized access, malicious injections, or other breaches. This is a shared responsibility: PaaS Kubernetes cloud offerings protect and secure the nodes in the cluster.

Our first concern is related to the hardening of the cluster. In the design and implementation, we have to think about:

- Enabling RBAC
- Enabling authentication
- Protection with firewalls
- The isolation of Kubernetes nodes
- Monitoring network traffic and audit logging

Then, we must secure our containers. For this, we can refer to the **National Institute of Standards and Technology (NIST)** Application Container Security Guide (`https://nvlpubs.nist.gov/nistpubs/SpecialPublications/NIST.SP.800-190.pdf`).

To secure containers, we must consider hardening the host operating system but also make sure that we use secure images to build our containers. Especially as the image is crucial in securing containers. The image is a file with executable code and includes everything that is required to run a container: the container engine (for instance, Docker), system libraries, and configuration settings. Components in images are typically reusable to prevent that we have to build new images over and over again.

These components and images are stored in controlled container registries. These registries are crucial in every enterprise environment using containers so that the CI/CD pipelines only use approved images from these registries. Containers are immutable by their nature. That's another reason why it's crucial to protect the source images. With that, we are stepping into DevSecOps.

In this chapter, we will look at DevSecOps in three layers. The first layer is about creating awareness in the teams. Every member in a DevOps team must be aware of the fact that every decision they make will have consequences, financially and in terms of security. Developers must learn the basics of writing secure code and understand the implications of writing code that leaves "holes" and introduces vulnerabilities. Engineers must be proficient in managing the environments in which developers work, keeping this secure. Think of the landing zones and the CI/CD pipelines that must be secured. Hence, education is important at all levels, in all roles.

The second layer is about applying the principles of Security by Design, making security an integrated part of everything that developers and engineers build and manage in the cloud. A lot of tasks can, however, be automated, limiting the fault tolerance of manual work. As such, the third layer is securing by automation.

In the next sections, we will explore these layers one by one and discover why these are important in multi-cloud, starting with education and creating awareness. We will then also discover how we can set up CI/CD pipelines and what tools we can use in the various clouds, but ultimately, we want a process and a pipeline that is able to address multi-cloud.

Starting with implementing a DevSecOps culture

In the previous section, we already mentioned the NIST guide to secure containers. That guide starts with something non-technical: a mindset. The first advice NIST gives is:

> *Tailor the organization's operational culture and technical processes to support the new way of developing, running, and supporting applications made possible by containers.*

Why would we need to change the culture in the way we do IT? Because with cloud, cloud-native, and containers, the way of doing software development changes drastically. Developers and operations might be less concerned with traditional IT processes such as patching and upgrading systems. We want to integrate security in to the builds of the applications, including the way applications utilize the underlying infrastructure without having to worry about the physical infrastructure or even the virtual machines. Let software take care of it, but then the software must be programmed, configured, and maintained in the most secure way.

There's one question that we must ask upfront: where does security start? The answer to that question is: at the very beginning, also in CI/CD pipelines. The following diagram shows a high-level representation of a pipeline. The triangles here indicate the security practices that must be followed as a minimum:

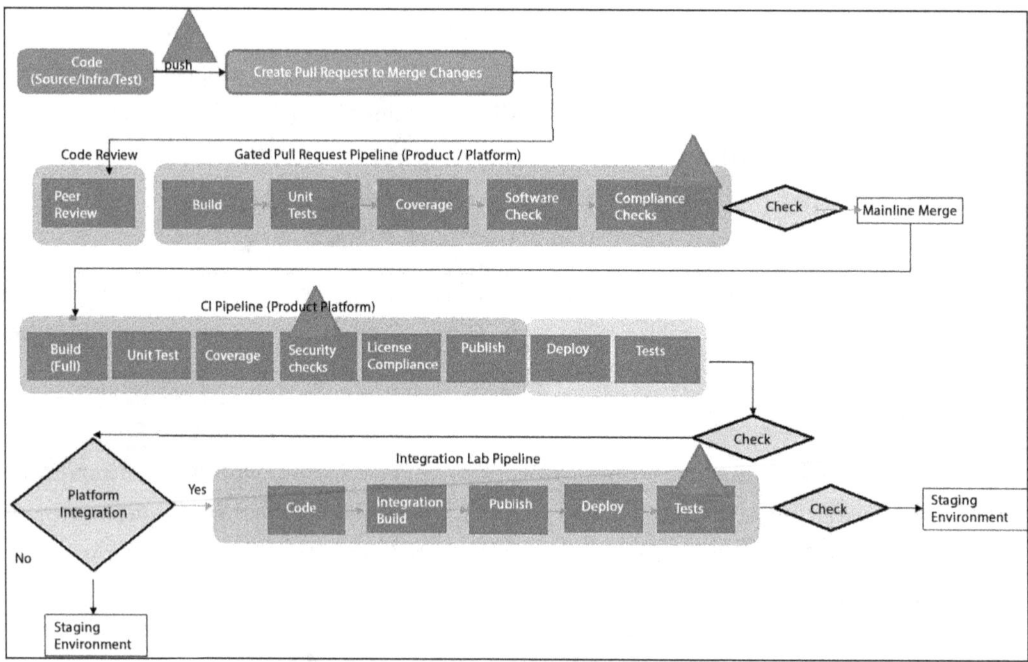

Figure 13.3: CI/CD pipeline with security indication

This is a good point to explain the principle of shift-left: the principle where we shift testing as early as possible in the lifecycle of application development, so indeed, already at the very beginning of the pipeline where the code is pulled or written. It's a crucial principle in DevSecOps. Another principle is the close cooperation between developers and operations, including security definition and management.

There are a lot of views on DevOps, but this book sticks to the definition and principles as defined by the **DevOps Agile Skills Association** (**DASA**). They define a DevOps framework based on six principles:

- **Customer-centric action**: Develop an application with the customer in mind—what do they need, and what does the customer expect in terms of functionality? This is also the goal of another concept, domain-driven design, which contains good practices for designing.
- **Create with the end in mind**: How will the application look when it's completely finished?
- **End-to-end responsibility**: Teams need to be motivated and enabled to take responsibility from the start to the finish of the application life cycle. This results in mottos such as "you build it, you run it, and you break it, you fix it". One more to add is "you destroy it, you rebuild it better".
- **Cross-functional autonomous teams**: Teams need to be able and allowed to make decisions themselves in the development process.
- **Continuous improvement**: This must be the goal—to constantly improve the application.
- **Automate as much as possible**: The only way to really gain speed in delivery and deployment is by automating as much as possible. Automation also limits the occurrence of failures, such as misconfigurations.

It's a new way of thinking about developing and operating IT systems based on the idea of a feedback loop. Since cloud platforms are code-based, engineers can apply changes to systems relatively easily. Systems are code, and code can be changed as long as changes are applied in a structured and highly controlled way. That's the purpose of CI/CD pipelines.

What do we need to do to build a DevSecOps culture? The answer: education and experimenting. Organizations should support experimenting in isolated environments so that developers and operations can learn. Instructor-led and self-paced training, peer reviews, and support in architecture, development, and security will accelerate the adoption of the DevSecOps mindset.

The following step is to integrate this mindset and the forthcoming principles in the setup of the CI/CD pipelines. We will discuss that in the next section.

Setting up CI/CD

Before we start building pipelines, it's good to have a definition of CI and CD.

Continuous Integration (**CI**) is built on the principle of a shared repository, where code is frequently updated and shared across teams that work in the cloud environments. CI allows developers to work together on the same code at the same time. The changes in the code are directly integrated and ready to be fully tested in different test environments.

Continuous Delivery and Deployment (**CD**) focuses on the automated transfer of software to test environments. The ultimate goal of CD is to bring software to production in a fully automated way. Various tests are performed automatically. After deployment, developers immediately receive feedback on the functionality of the code.

CI/CD enables the DevOps cycle. Combined with CI/CD, all responsibilities from planning to management lie with the team, and changes can reach the customer much faster through an automated and robust development process. The following diagram shows the DevOps cycle with CI/CD:

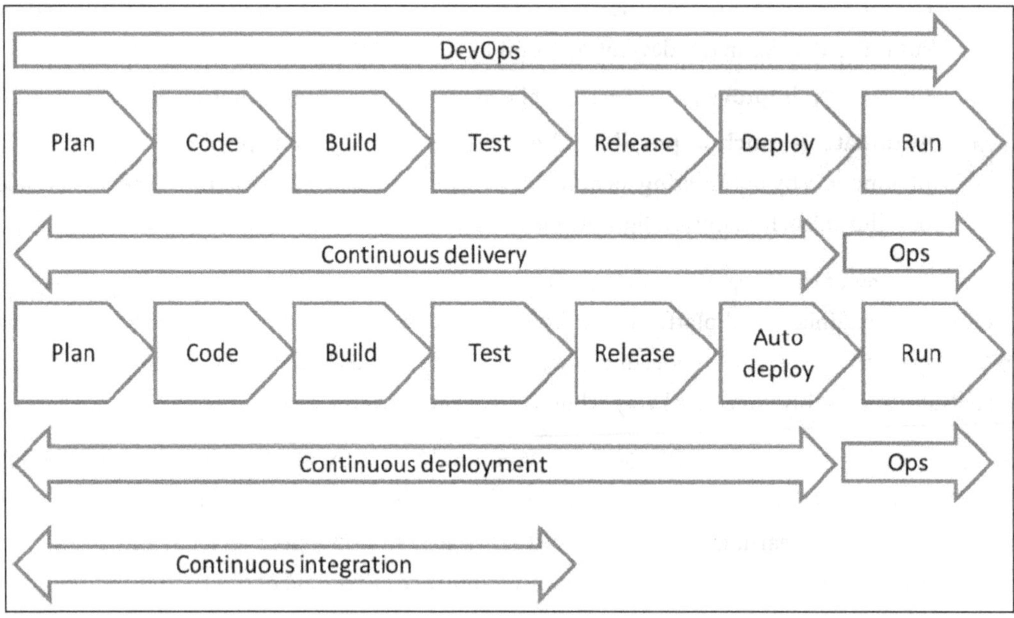

Figure 13.4: DevOps cycle with CI/CD

CI/CD is widely adopted by enterprises, but a lot of projects fail, typically because of a lack of consistency. With CI, development teams can change code as often as they want, leading to the continuous improvement of systems. Enterprises will have multiple development teams working in multi-cloud environments, which makes it necessary to have one way of working. Fully automated processes in CI/CD pipelines can help keep environments consistent. That's why CI/CD and DevOps are not about tools in the first place. They're about culture and sticking to processes.

To get to a successful implementation of DevOps, an organization is advised to follow these steps:

1. Implementing an effective CI/CD pipeline begins with all stakeholders implementing DevOps processes. One of the key principles in DevOps is autonomous teams that take end-to-end responsibility. It's imperative that the teams are given the authority to make decisions and act on them. Typically, DevOps teams are agile, working in short sprints of 2 to a maximum of 3 weeks. If that time is wasted on getting approval for every single detail in the development process, the team will never get to finish anything in time.
2. Choose the CI/CD system. The repository is really important, and it's one of the first steps. Then you can decide on the CI/CD system: one that includes a repository such as GitHub, Azure DevOps, or GitLab. There are a lot of tools on the market that facilitate CI/CD. GitHub Actions and Jenkins are popular, but a lot of companies that work in Azure choose to work in Azure DevOps. Involve the people who have to work daily with the system and enable them to take a test drive. Then, make a decision and ensure all teams work in that system. Again, it's about consistency.
3. It's advised to do a proof of concept. An important element of CI/CD is the automation of testing, so the first step is to create an automated process pipeline. Enterprises often already have quality and test plans, possibly laid down in a **Generic Test Agreement (GTA)**. This describes what and how tests must be executed before systems are pushed to production. This is a good starting point, but in DevOps, organizations work with a **Definition of Done (DoD)**.
    - The DoD describes the conditions and the acceptance criteria a system must meet before it's deployed to production. The DoD is the standard of quality for the end product, the application, or the IT system that needs to be delivered.
4. Automating as much as possible is one of the principles of DevOps. This means that enterprises will have to adopt working in code, including IaC. In CI/CD, teams work from one repository, and this means that the application code and the infrastructure code are in the same repository so that all teams can access it whenever they need to.

In DevOps, teams work with user stories. An example of a user story is: *as a responsible business owner for an online retail store, I want to have multiple payment methods so that more customers can buy our products online*. This sets requirements for the development of applications and systems. The DoD is met when the user story is fulfilled, meaning that unit testing is done, the code has been reviewed, the acceptance criteria are signed off, and all functional and technical tests have been passed.

The following diagram shows the concept of implementing a build and release pipeline with various test stages. The code is developed in the build pipeline and then sent to a release pipeline, where the code is configured and released for production. During the release stages, the full build is tested in a test or **Quality and Assurance Environment (Q&A)**. In Q&A, the build is accepted and released for deployment into production:

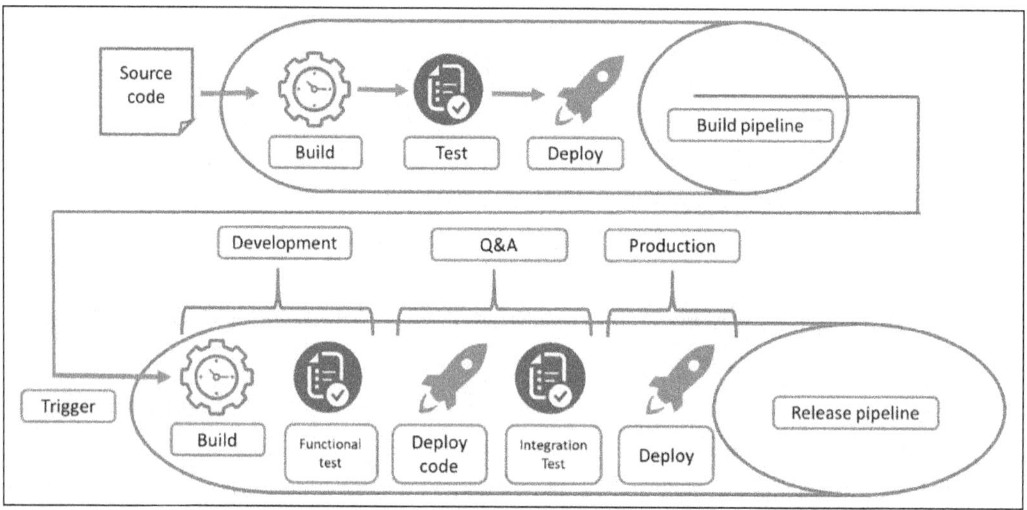

Figure 13.5: Conceptual diagram of a build and release pipeline

If all these steps are followed, an organization can start working in DevOps teams using CI/CD. But how will this work in multi-cloud? We'll discuss this in the next section.

Working with CI/CD in multi-cloud

The development of code for applications can be cloud-agnostic, meaning that it doesn't matter to which cloud the code is pushed: the functionality of the code remains the same. That's the reason why we use containers: to abstract the code from the underlying infrastructure. However, a lot of developers will discover that it does matter and that it's not that simple to develop in a truly multi-cloud fashion.

In multi-cloud, developers also work from one repository, but during deployment, platform-specific configuration might be added and tested, even when we utilize Kubernetes platforms in our clouds. You will find that Kubernetes is the same, but the various implementations of Kubernetes platforms in clouds might differ.

There are a few steps that developers have to take to make it successful. First, the DevOps way of working should be consistent, regardless of the platform where applications will eventually land. A company might want to run applications in Azure, AWS, GCP, or even on-premises, but the way application code is written in DevOps cycles should be the same. As said, each of these platforms will have specific features to run the code, but that's a matter of configuration. The process called *staging* is meant to find out if the application package, including the configuration of the underlying infrastructure, is ready for release to production.

We need to think in terms of layers: abstracting the application layer from the resources in the infrastructure and the configuration layer. That's the only way to get to a consistent form of application development with CI/CD. One other challenge that developers need to tackle in multi-cloud environments is the high rate of changes in the various cloud platforms. DevOps tools must be able to adapt to these changes in deployment, but without having to constantly change the application code itself.

Developers, however, need to have a good understanding of the target platforms and their specifics. It also makes sense to study the best practices in DevOps for these platforms and the recommended tooling: this will be discussed in more detail in the *Exploring tooling for CI/CD* section of this chapter about tooling for CI/CD. Most common tools are cloud-agnostic, meaning that they can work with different clouds, leveraging native APIs.

So, the ground rules for the successful implementation of DevOps are as follows:

- One repository
- One framework or way of working
- A unified toolset that can target multi-cloud environments

In terms of one framework, SAFe by Scaled Agile should be mentioned. **SAFe** stands for **Scaled Agile Framework**, and it's used by a lot of enterprises as the foundation of DevOps. One of the key assets of SAFe is the Agile Release Train.

The **Agile Release Train** is built around the principle of the standardization of application development and the release management of code. This is done by automating everything. The steps that must be included are:

- Build by automating the process of compiling the code
- Automated unit, acceptance, performance, and load tests
- Automated security tests

- Continuous integration, automated by running integration tests and releasing units that can be deployed to production
- Continuous delivery by automated deployments to different environments of the version-controlled code
- Additional automation tools for configuration, provisioning, security by design, code review, audit trail, logging, and management

This supports the application life cycle management, continuous improvement, and release of the application code. Details and courseware on SAFe can be found at `https://www.scaledagileframework.com/DevOps/`.

Exploring tools for CI/CD

The tooling landscape for CI/CD and DevOps is massive and changes almost every month. There's no right or wrong answer in choosing the toolset, as long as it fits the needs of the enterprise and people are trained in the usage of the tools. In this section, the native CI/CD tooling in the major clouds is discussed: Azure DevOps, AWS CodePipeline and CloudFormation, and Google Cloud Build. We will also look at tooling for Alibaba Cloud and OCI.

First, we will look at Kubernetes since that will likely be the platform that we'll use in multi-cloud. The key components of Kubernetes CI/CD are:

- Container management
- Cluster operations
- Configuration management for the underlying infrastructure
- Version control system
- Image registries for container images
- Security testing and audits
- Continuous monitoring and observability

To create a good pipeline, we should consider the following design principles:

- All-in-one CI/CD tool instead of cloud-specific solutions when we plan to have CI/CD multi-cloud
- Managed or self-managed CI/CD, including the deployment, configuration, and management of infrastructure
- Automated or manual code testing and validation
- Deployment methodologies: rolling or blue/green

Now, what do clouds have to offer? We will look at the various solutions.

Azure DevOps

Azure DevOps enables teams to build and deploy applications; it caters to the full development cycle. Azure DevOps contains the following:

- **Boards**: This is the planning tool in Azure DevOps and supports scheduling with Kanban and Scrum. Kanban works with cards, moving tasks through stages, while Scrum works with short sprints to accomplish tasks.
- **Repos**: This is the repository in Azure DevOps for version control based on Git or **Team Foundation Version Control** (**TFVC**). The term Team Foundation still refers to the original name of Azure DevOps: Visual Studio **Team Foundation Server** (**TFS**).
- **Pipelines**: This is the CI/CD functionality in Azure DevOps, which supports the build and release of code to cloud environments. It integrates with repos and can execute scheduled tasks from Boards.
- **Test Plans**: This allows teams to configure test scripts manually and automatically.
- **Artifacts**: This feature allows developers to share code packages for various sources and integrate these into pipelines or other CI/CD tooling. Artifacts support Maven, **Node Packet Manager (NPM)** (for Node.js and JSON), NuGet packages, and Python package feeds.

The following screenshot shows the main menu of Azure DevOps with a project defined as Scrum that divides the work items into backlog items and sprints:

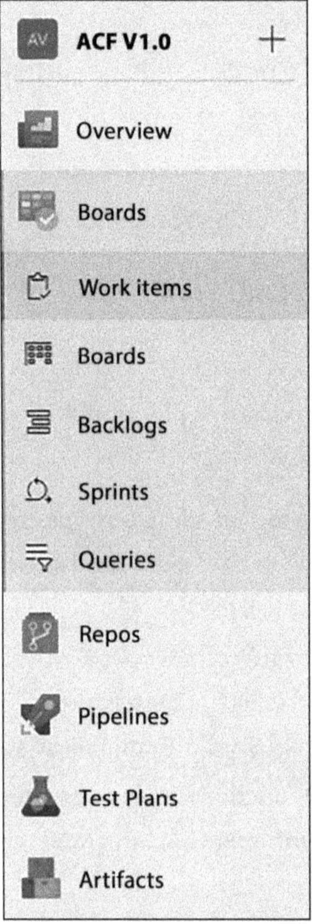

Figure 13.6: Main menu of Azure DevOps

When a team starts in Azure DevOps, the first thing to do is to define a project and assign project members. Next, the project manager or product owner defines the development methodology. Depending on that choice, DevOps presents the possibility to define work items, such as features, backlog items, and tasks, and a board to plan the execution of these items. A work item or product can be a piece of code that may be deployed; in DevOps, this can be automated with Azure Pipelines, which after review, deploys the code to the target cloud environment.

Developers can use Azure DevOps for AWS too using the AWS Toolkit for Azure DevOps. The toolkit even accepts the use of AWS CloudFormation templates to provision and update AWS resources within Azure DevOps. In this chapter's *Further reading* section, a link to the documentation has been added.

AWS CodePipeline

AWS CodePipeline is the CI/CD tool for AWS and offers development teams a tool to deploy applications and infrastructure resources. CodePipeline provides the following:

- **Workflow modeling**: You could see this as the planning tool in CodePipeline. Workflow modeling defines the different stages for the release of code: build, test, and deploy. Teams can create tasks that need to be executed in the different stages.
- **Integrations:** As with any CI/CD tool, CodePipeline works with version control for the source code. With Integrations, developers can use various sources, such as GitHub, but also the native AWS service CodeCommit (the default in the main menu of CodePipeline), Amazon **Elastic Container Registry** (**ECR**), and Amazon S3. Provisioning and updating code is done with AWS CloudFormation. AWS Integrations can do a lot more, such as continuous delivery to serverless applications with the **Serverless Application Model** (**SAM**) and automating triggers with AWS Lambda functions to test whether application code has been deployed successfully.
- **Plugins**: It looks like AWS mainly uses its own tools, but developers absolutely have freedom of tools. AWS Plugins allow the use of GitHub for version control and Jenkins for deployment, for example.

The following screenshot shows the main menu of CodePipeline:

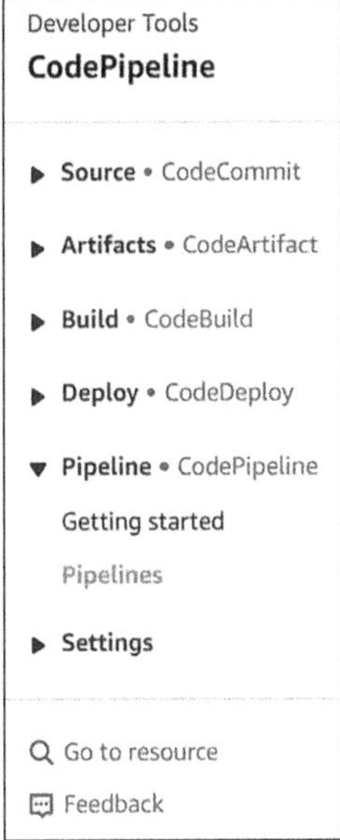

Figure 13.7: The main menu of AWS CodePipeline

As shown in the screenshot, creating a pipeline from the `CodePipeline` main menu will start by pulling code from a repository that sits in `CodeCommit`. The pipeline itself is built in `CodeBuild` and deployed in `CodeDeploy`.

Be aware that `Artifacts` is not the same as it is in Azure. In AWS, Artifacts uses an S3 artifacts bucket where CodePipeline stores the files to execute actions in the pipeline.

Google Cloud Build

The CI/CD tool in GCP is Cloud Build. The main functions in Cloud Build are as follows:

- **Cloud Source Repositories**: These are private Git repositories that are hosted on GCP. This is where the pipeline workflow starts: developers can store, access, and pull code from this repository using Cloud Build and Cloud Pub/Sub. Creating a repository can be done through the GCP UI portal or Google Cloud Shell with the `gcloud source repos create` command. After the creation of the repository, developers can start pushing code to it with the `git add`, `git commit`, and `git push` commands from the `gcloud` console.
- **Artifact Registry**: This is basically the same service as Artifacts in Azure DevOps. It allows the creation and management of repositories that hold Maven and NPM packages. In GCP, Artifact Registry is also used to create repositories for Docker container images.
- **Cloud Build**: This is the engine of the CI/CD functionality in GCP. In Cloud Build, developers define the pipelines. It imports source code from Cloud Source Repositories but can also pull code from other sources, such as GitHub. Cloud Build tests and deploys the code to the targeted GCP infrastructure. Cloud Build integrates with a lot of different solutions—for example, with Jenkins and the open source tool Spinnaker for automated testing and continuous delivery. These solutions can also be used to work with GKE to enable CI/CD on container platforms running Kubernetes.

Figure 13.8 shows the menu of Cloud Build:

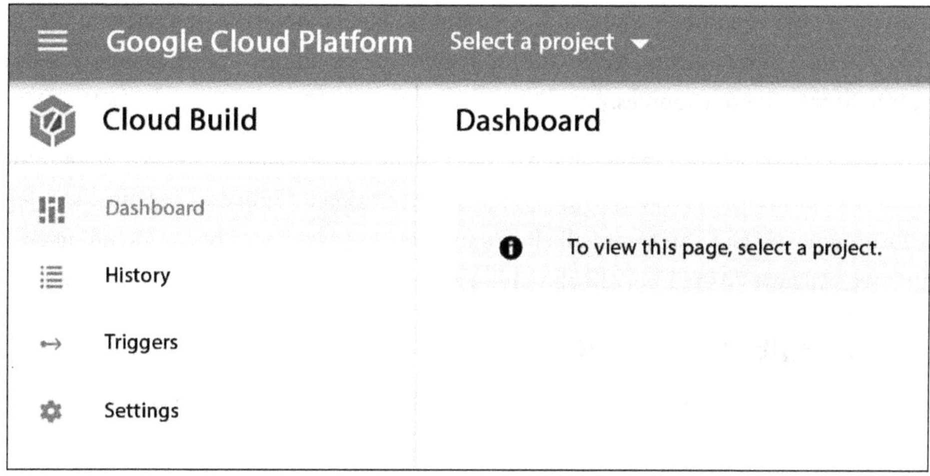

Figure 13.8: Introduction screen to start with Cloud Build in the GCP console

The main menu is very lean, as shown in the preceding screenshot. Only when a project is defined and started can developers start using Cloud Build to create the code repository. That service is available from the console, as shown in *Figure 13.9*:

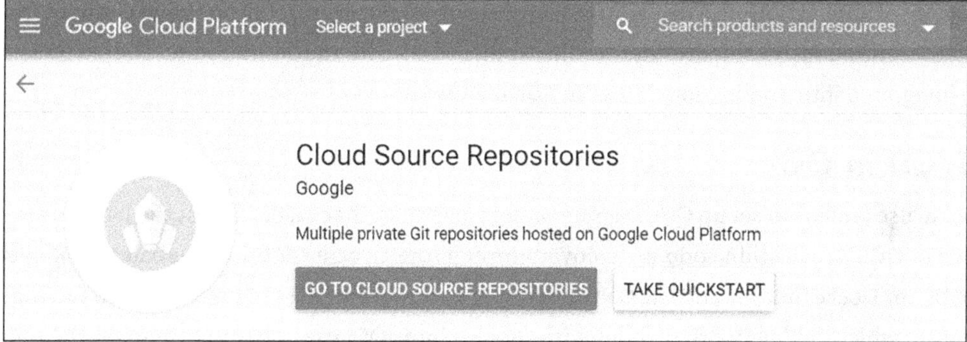

Figure 13.9: Starting a repository in GCP's Cloud Source Repositories

So far, we have discussed the CI/CD tools that are available in the major public clouds. It must be noted that these tools are provider-centric, meaning that AWS CodeLine, for instance, is very suitable to build in AWS. There are extensions and toolkits available to run pipelines from Azure DevOps to AWS and GCP. In that case, the toolkit will pick up the actions from Azure DevOps Pipelines and transport these to AWS CloudFormation, which will deploy the resources and code using CloudFormation Stack.

Alternatives are to use agnostic tools such as GitHub and Jenkins. GitHub is a relevant CI/CD tool. It allows you to deploy to Azure, AWS, GCP, and any other cloud provider that supports OpenID Connect to access cloud resources.

Over the years, Jenkins has also evolved to become an industry-leading standard for automated CI/CD since it offers different tools, languages, and automation tasks in creating pipelines. Jenkins is platform-agnostic and can target all clouds that we discuss in this book, including the final two that we still have to explore: Alibaba Cloud and OCI.

CI/CD in Alibaba Cloud

Key in the Alibaba Cloud solution for CI/CD are the **Kubernetes Container Service** and the **Alibaba Cloud Image Service**. The latter is used for image replication and image signature. The service enables fast replication across clusters and regions in Alibaba Cloud. With image signature, we can use Key Management Service to add signatures to Kubernetes images and security policies.

Like other public clouds, Alibaba Cloud utilizes commonly known and industry-standards technologies and tools in the CI/CD chain. We can, for instance, use ArgoCD to deploy applications to clusters. **ArgoCD** is a Kubernetes controller that automates the deployment and lifecycle management of applications in a declarative way through application definitions and configurations. ArgoCD continuously compares the running applications with the desired state. The technologies are integrated into Application Center in Alibaba Cloud.

CI/CD in OCI

We can use Jenkins to set up CI/CD architectures in OCI. In that case, we host the Jenkins master node in OCI. Application code is deployed through the **Oracle Cloud Infrastructure Registry** (**OCIR**) for Docker images and the **Container Engine for Kubernetes** (**Oracle Kubernetes Engine — OKE**), which deploys worker nodes with Oracle Linux. OCI uses GitHub to manage the source code and Terraform to automate the infrastructure.

A cool feature is that we can immediately deploy this solution to OCI from the website `https://docs.oracle.com/en/solutions/cicd-pipeline/index.html#GUID-D741A2F7-7E15-44ED-8B57-CB7197FE5D07` by clicking **Deploy to Oracle Cloud** or get the Terraform code in Github. It's shown in the image below:

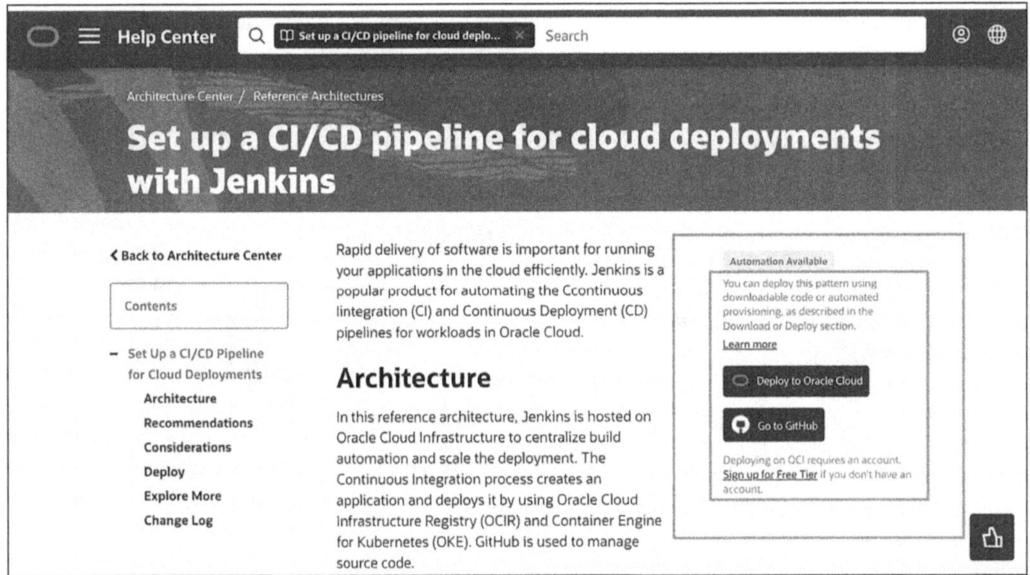

Figure 13.10: Option to deploy CI/CD with Jenkins to OCI

It is worth mentioning that OCI has recently leased a new service with OCI DevOps, which is a complete CI/CD platform. More information about this can be found at `https://www.oracle.com/devops/devops-service/`.

Tools for multi-cloud container orchestration and application development

The tools and technologies that we discussed so far are mostly cloud-native and hosted by a specific cloud provider. In multi-cloud, we might have Kubernetes clusters and containers in various clouds. Some of the tools have APIs to use in other clouds, such as Azure DevOps, which can be used for AWS too, using the AWS Toolkit as an extension.

Two other popular options for multi-cloud container orchestration are VMware's Tanzu Mission Control and Red Hat OpenShift. Tanzu allows for managing Kubernetes clusters in multiple clouds through one console, including centralized policy management. Tanzu Mission Control supports the deployment, scaling, and management of clusters on AWS, Azure, GCP, OCI, and, of course, clouds that are built on vSphere.

Red Hat OpenShift is also an agnostic environment to deploy, scale, and manage Kubernetes clusters using the Red Hat suite with Red Hat Enterprise Linux and Ansible for configuration management of clusters. OpenShift has editions for Azure, AWS, and IBM Cloud. A fully managed RedHat OpenShift Dedicated edition is available for AWS and Google Cloud.

At the beginning of this section, it was mentioned that there are a lot of tools available to developers for developing, storing, testing, and deploying code. The main principles of CI/CD are the same, but there are differences in the way that these tools deploy code and especially how they test and validate code in an automated way. The tools discussed cover a lot of functionality already, but typically, additional tools are required. Just to name a few of the most used and well-known ones:

- GitHub and CI/CD with GitHub Actions to manage repositories
- **GitLab**: To manage code, but also a tool that makes deployment of Kubernetes clusters easy with Auto Build, executing Docker builds, and managing Docker files
- **GitOps** and **Spinnaker**: For cloud-agnostic continuous deployment

We have established a DevOps culture and built CI/CD pipelines. The next step is securing the pipelines and the code that we are developing. This is the topic of the next section.

Following the principles of Security by Design

In the previous sections, we designed our CI/CD pipelines. But as we have concluded, security starts at the very beginning of DevOps and should be integrated throughout the entire process, from the moment the code is pulled or new code is written up until deployment to production. We need to apply security by design. This is the second layer of DevSecOps and includes the following activities:

- **Securing pipelines**: A best practice is to apply zero-trust principles to the pipeline. Pipelines should only be accessed through least privilege policies. Also, continuous testing must be integrated into the pipeline. This includes **Static Application Security Testing (SAST)** and **Dynamic Application Security Testing (DAST)**, but also penetration testing to find any backdoors in the pipelines or any other vulnerabilities.
- **Clean code practice**: This one is subject to multiple interpretations, but the key is the principle of KISS: keep it simple, stupid. Try to keep the code as simple and short as possible. Document it well, so other developers know what the code is about. Most important of all: be consistent.

- **Application security design principles**: We can use the **Open Web Application Security Project (OWASP)** security design principles for this. These principles are based on CIA: confidentiality, integrity, and availability. In essence: allow only access to data to which the user is authorized, make sure that data is not altered in an unidentified way or by unauthorized users, and that data is available to authorized users when they request it. Following this, OWASP lists security principles, amongst others:
    - Limit the attack surface as much as possible.
    - Applications are secured by default, meaning that users have to 'prove' they are eligible to use the application and the associated data.
    - Least privilege.
    - Defense in depth using layered security controls.
    - Separation of duties, ensuring that users can only perform actions that they are entitled to without having control over the lifespan of an entire transaction in an application. Simply explained: someone initiating a request can't authorize the request by themselves.
- **Microservices with containers and sidecars**: Microservices are loosely coupled services that together form the functions of an application. They are considered to be more secure than monolith architectures where all functions are integrated into one big application. Because microservices are decoupled and communicate with each other using APIs, breaches that exploit vulnerabilities are often limited to one particular microservice instead of breaching the entire application.
- Keep in mind that the security of a system depends on a multitude of factors, and the architecture is just one aspect of it.
- Microservices with containers and sidecars can offer security benefits, such as increased isolation and modularity, which can make it easier to manage security vulnerabilities in individual components. However, the complexity of managing multiple containers and services can also introduce new security challenges, such as securing the communication between components and ensuring the proper configuration of security settings.
- Monolith architectures, on the other hand, may have fewer components to secure, but a single vulnerability in the monolithic codebase can have a larger impact. The size and complexity of a monolithic codebase can also make it difficult to identify and remediate security vulnerabilities.

- However, building a microservices architecture is complex, and developers would still want to be able to reuse artifacts instead of programming every microservice over and over again. This will likely be the case with functions that apply to every application. These functions can be coupled using sidecars. In sidecars, components of an application are deployed as a separate process using containers, providing isolation and encapsulation. The sidecar is, like a real sidecar to a motorcycle, attached to a parent application and provides supporting features for the application. We can use sidecars to deploy monitoring, logging, and configuration files. *Figure 13.11* shows the pattern for sidecars:

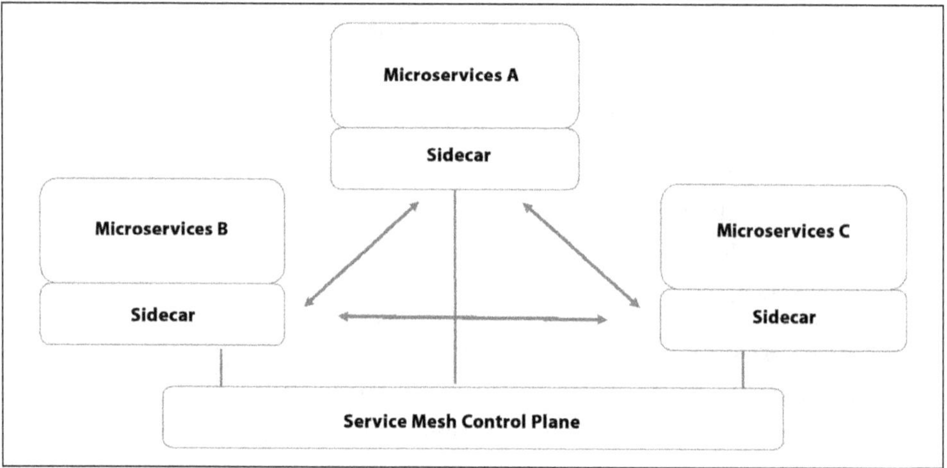

Figure 13.11: Architecture pattern for sidecars

- To integrate the sidecars with the other components of the applications, we need a service mesh, which enables service-to-service-communication between microservices. An industry-leading technology for using service mesh with sidecars is the open source Istio service mesh. This is supported by all public clouds using Kubernetes as a container orchestration platform.
- **Conduct threat modeling**: Organizations should adopt the process to identify potential threats and evaluate these against the security policies and implemented safeguards in the application and data stacks. Threat modeling will reveal vulnerabilities and indicate what mitigating actions must be taken to prevent these vulnerabilities from being exploited. A common way to perform threat modeling is using **Tactics, Techniques, and Procedures (TTP)**, amongst others used by the Mitre Att@ck framework. Tactics describe the behavior of an attacker and what and how techniques are used to breach a system.

Testing applications and infrastructure is a critical component of software development lifecycles. This includes testing to verify that vulnerabilities and potential risks have been addressed. With that, it's best practice to consider the full stack:

- Code
- CI/CD pipelines
- Registries and repositories holding container images and services
- IaC used to deploy resources in the cloud
- Hosts in the cloud (virtual machines)
- Container runtime environments (Kubernetes)
- Containers

Something that we haven't touched upon yet is serverless functions to run containers in the cloud. Examples are Azure Container Instances, AWS Fargate, and Google Cloud Run; both offering compute as a service. Fargate is a serverless, completely managed service that builds infrastructure and is compatible with EKS in AWS. Google Cloud Run is not quite the same but it does run applications without you having to worry about the underlying infrastructure. Hence, there's no need to deploy your own virtual machines and configure clusters. It's a high level of automation that we can use in the cloud. In the next section, we will discuss automation in more detail.

Securing development and operations using automation

Automation is a must in DevSecOps. At a minimum, we should consider automating the following processes:

- Version Control
- Continuous Integration
- Continuous Testing
- Configuration Management and Deployment
- Continuous Monitoring

In the previous section, we talked about containers, sidecars, and orchestration. Clouds also offer tools to automate this as a completely managed service. Containerization with automated orchestration offers great benefits to companies. Orchestration platforms such as Kubernetes and the various managed Kubernetes services in public clouds take care of the installation, scaling, and management of containerized workloads and services, including debugging and deployment of new versions of applications.

Containers are isolated by default. One failing container will not impact other containers. Individual containers can be updated without causing downtime for other containers. This also means that containers are protected very well from malicious code spreading through entire environments. However, it does mean that we have to test intensively. These tests can be automated too to a large extent:

- Application and application runtime security testing
- **Software composition analysis (SCA)**
- IaC validation and testing
- Container image scanning
- Dynamic threat analysis
- Network scanning
- Vulnerability scanning and management

One item that we must discuss is the method of deployment from a DevSecOps perspective. There are two common ways to do deployments: rolling and blue/green. Deciding between one of the two is a decision that is key in DevSecOps thinking. In **rolling** deployments, we replace the old versions of the application and the underlying infrastructure with new versions. In containerized environments, we will replace all containers that run the previous versions of the application one by one. This is generally the fastest way of deployment.

However, there will be no isolation between the old and the new containers. A more secure way of deployment is **blue/green**. The blue environment is the old environment and the new one is green. Both will run in production at a certain time, where traffic is gradually transferred from blue to green. The two environments are strictly separated. Only when the traffic on blue is completely transferred and the green line tested can the old environment can be decommissioned. The blue line will be a fallback along the entire process.

This way of deployment is safer but also more costly since multiple resources will run at the same time. Container pods are copied and used to set up the green line. Tools such as Kubernetes and Jenkins support the blue/green deployment very well using the principle of Declarative Deployment, which allows for automated, fast copying of container pods.

This concludes the chapter about setting up DevSecOps and CI/CD. But we're not done yet with security. In the next chapter, we will look at the security policies themselves and how to apply them to cloud environments.

Summary

After completing this chapter, you should have a good understanding of the DevOps way of working and the use of CI/CD pipelines in cloud environments. Everything is coded in the cloud, from the application to the infrastructure and the configuration. Code needs to be stored in a central repository and brought under version control. That's where the CI/CD pipeline starts. Next, the DevOps team defines the phases of the pipeline, typically build, test, and deploy. Actions in these phases are automated as much as possible.

We learned that we must integrate security from the moment developers start writing or pulling code. Access to the code, the code itself, the pipelines, and the target clouds that we use must be secured and protected from breaches by identifying vulnerabilities and taking appropriate mitigating actions. We studied the various technologies and tools that provide CI/CD functionality and explored how we can protect our applications when we work with CI/CD by applying the principles of security by design.

Prevention, protection, and the defense of our applications and data are defined in policies that developers and operations need to comply with. In the next chapter, we will discuss how we can define security policies in cloud environments.

Questions

1. All cloud providers offer solutions to run and manage Kubernetes clusters. Name three of these solutions.
2. What is the function of GitHub?
3. Microservices will likely share functions such as monitoring and logging. We can use sidecars to attach these functions to microservices. We need to "merge" these functions with our microservices. What do we call this process of integrating services using sidecars?
4. We mentioned two ways of deployment: rolling and blue/green line. Which one is considered to be more secure?

Further Reading

- **DevOps Agile Skills Association** (**DASA**): `https://www.DevOpsagileskills.org/`
- Documentation on Azure DevOps: `https://docs.microsoft.com/en-us/azure/DevOps/get-started/?view=azure-DevOps`
- Information on integrating Azure DevOps with AWS: `https://aws.amazon.com/vsts/`

- Documentation on AWS CodePipeline: `https://aws.amazon.com/codepipeline/`
- Documentation on Google Cloud Build: `https://cloud.google.com/docs/ci-cd/`
- Documentation on CI/CD for Alibaba Cloud: `https://www.alibabacloud.com/solutions/devops/CI-CD`
- Documentation on CI/CD for OCI: `https://docs.oracle.com/en/solutions/cicd-pipeline/index.html`
- *Hands-On DevOps for Architects*, by Bob Aiello, Packt Publishing

Join us on Discord!

Read this book alongside other users, cloud experts, authors, and like-minded professionals. Ask questions, provide solutions to other readers, chat with the authors via. Ask Me Anything sessions and much more.

Scan the QR code or visit the link to join the community now.

`https://packt.link/cloudanddevops`

14

Defining Security Policies

Whatever we do in the cloud needs to be secure. Cloud providers only provide tools. You need to define how to use these tools. In order to determine what these tools should do, you need to think about what type of assets you want to protect and how you need to protect them. There are quite a number of security baselines—for example, the baseline as defined by the **Center for Internet Security** (**CIS**), which provides guidelines.

In this chapter, we will learn what a security framework is and why it's important as a starting point for security policies. We will discover what we need to protect in our cloud environments. Next, we will look at the globally adopted CIS benchmark for Azure, AWS, GCP, Alibaba Cloud, and OCI and learn how to implement CIS using the security suites of these platforms. We will then learn what the difference is between security governance and management, and lastly, study **Cloud Security Posture Management** (**CSPM**) to control our cloud configurations and ensure they are secure and compliant.

In this chapter, we're going to cover the following main topics:

- Managing security policies
- Understanding security policies
- Understanding security frameworks
- Understanding the dynamics of security and compliance
- Defining the baseline for security policies
- Implementing security policies
- Implementing security policies in Azure, AWS, GCP, Alibaba Cloud, and OCI
- Manage risks with Cloud Security Posture Management

Understanding security policies

Let's start with our traditional on-premises data center—a building traditionally used to host physical equipment that runs applications and stores data. The building is very likely secured by a fence and heavy, locked doors that can only be opened by authorized personnel. Access to the computer floors is also secured. There may be guards in the building or CCTV systems watching over equipment 24 hours a day. The next layer of defense is access to the systems and data. Access to systems is strictly regulated; only authorized and certified engineers may access the systems. It's all common sense when it comes to running systems in a physical data center.

You will be surprised to see what happens when companies move these systems to cloud environments with IaaS, PaaS, and SaaS solutions. For some reason, companies tend to think that by moving systems to the cloud, those systems are secured intrinsically, by default. That is not the case.

Platforms such as Azure, AWS, GCP, Alibaba Cloud, and OCI are probably the best-secured platforms in the world. They have to be since they are hosting thousands of customers globally on them. But this doesn't mean that a company will not have to think about its own security policies anymore. The platforms provide a huge toolbox that enables the securing of workloads in the cloud, but what and how to protect these workloads is still completely up to the companies to implement themselves. We will need to establish and enforce our security policies in the cloud, think them through very carefully, and stick to them. That is what this chapter is about.

As with physical data centers, access needs to be regulated first by defining which identities are authorized to enter systems, and next, by determining what these entities are allowed to do in these systems. This is all part of identity and access management, a topic that we will cover in full in *Chapter 15, Implementing Identity and Access Management.*

The foundation for security policies is the CIA principle:

- **Confidentiality**: Assets in the IT environment must be protected from unauthorized access.
- **Integrity**: Assets in the IT environment must be protected from changes that are not specified and unauthorized.
- **Availability**: Assets in the IT environment must be accessible to authorized users.

The security policy itself has nothing to do with technology. The policy merely defines the security principles and how these are safeguarded in the organization. The policy does not define what ports must be opened in a firewall or what type of firewall is required. The policy describes the requirement that assets belonging to a certain function in the enterprise must be protected at a certain level.

For example, a business-critical functionality that relies on a specific stack of applications needs to be available at all times and the data loss must be zero. That will lead to an architectural design using mirrored systems, continuous backups, disaster recovery options, and a very strict authorization and authentication matrix for people who must be able to access these systems.

Understanding security frameworks

Security policies and forthcoming principles do not stand on their own. Typically, they are defined by industry or public frameworks to which a company must adhere. There are two types of frameworks: mandatory industry frameworks and best practices.

Examples of industry frameworks are the **Health Insurance Portability and Accountability Act** (**HIPAA**) for health care and the **Payment Card Industry** (**PCI**) data security standard for financial institutions. These frameworks were created to protect consumers by setting standards to avoid personal data—health status or bank accounts—being compromised. The cloud architect must have a deep understanding of these frameworks since they define how systems must be designed.

Next to these industry frameworks, there are some overall security standards that come from best practices. Examples are the standards of the **International Organization of Standardization** (**ISO**) and the U.S. **National Institute of Standards and Technology** (**NIST**). Specific to the cloud, we have the framework of the CIS.

Cloud providers have adopted the CIS framework as a benchmark for their platforms, as it is the internationally accepted standard for cybersecurity. The reason is that CIS maps to the most important industry and overall security frameworks such as the ISO, NIST, PCI, and HIPAA. The controls of CIS take the principles from these frameworks into account, but it doesn't mean that by implementing CIS controls, a company is automatically compliant with the PCI or HIPAA. CIS controls need to be evaluated per company and sometimes per environment.

Basically, there are two levels of CIS controls:

1. Essential basic security requirements that will not impact the functionality of the workloads or services.
2. Recommended settings for environments that require greater security but may impact the workloads or services through reduced functionality.

In summary, CIS provides a security framework based on best practices. These are translated into benchmarks that can be adopted for specific platforms and systems: Azure, AWS, and GCP, and the instances in those clouds using operating systems such as Windows Server or various Linux distributions. These benchmarks lead to settings in the hardening of servers.

CIS offers recommendations for the following:

- Identity and access management
- Storage accounts
- Database services
- Logging and monitoring
- Networking
- Virtual machines
- Application services

Adhering to CIS or any other framework doesn't necessarily mean that our cloud environments are compliant by default. We will elaborate on this in the next section.

Understanding the dynamics of security and compliance

Security and compliance are two completely different things, yet they are closely related to each other. Security policies are required to achieve compliance. We will get into the relationship a bit further in this section.

First, let's get a good definition of security. **Security** involves all activity to protect the assets of a company and the users of their systems. This activity can be defined in security controls: physical, technical, and administrative. Typically, we don't have to worry about physical controls in the cloud. Microsoft, Amazon, Google, Alibaba, and Oracle will make sure that their data centers and all the hardware that's in them are well protected. However, we do have to worry about technical and administrative controls. We need to take action to, for instance, implement antivirus and antimalware software on workloads that we host in the cloud—these are technical controls. Administrative controls are procedures, protocols, and processes. They need to be in place too.

If we have taken actions to implement technical and administrative controls, then we can achieve compliance. However, the controls must be aligned with cloud usage standards and laws, which might be specific to industries, regions, or countries. These laws and rules are documented in frameworks.

In short, cloud security and compliance are two essential aspects of any organization's IT infrastructure. Ensuring that your organization's data and systems are secure and compliant with relevant regulations is essential for protecting a business and its customers. The question is, does the cloud help to make the IT infrastructure more secure? By moving to the cloud, organizations outsource the management and maintenance of infrastructure to a third-party provider, the cloud platform. This can free up valuable time and resources for your organization, but it also means that you need to rely on the security measures put in place by the cloud provider.

To ensure the security of your data and systems, it is essential to choose a cloud provider that has strong security measures in place. This includes physical security measures, such as data center security and employee background checks, as well as technical measures, such as encryption and access controls. As we already concluded, the major cloud providers have these physical controls well in place.

Next, we come to compliance. Different industries and countries have their own specific regulations and requirements when it comes to the storage and handling of sensitive data. For example, the **General Data Protection Regulation** (**GDPR**) in the European Union sets strict guidelines for the protection of personal data, while the HIPAA in the United States regulates the handling of medical information.

To ensure compliance with these regulations, it is essential to choose a cloud provider that has demonstrated its commitment to compliance. This may involve obtaining certifications, such as SOC 2 for security or the PCI DSS for handling credit card information, or undergoing regular audits to ensure that their practices align with relevant regulations.

It is also important for organizations to have their own processes and controls in place to ensure compliance with relevant regulations. This may include implementing access controls to ensure that only authorized individuals have access to sensitive data, regularly reviewing and updating policies and procedures, and providing training to employees on how to handle sensitive data.

The relationship between cloud security and compliance is dynamic and critical at the same time. Ensuring that your organization's data and systems are secure and compliant with the relevant regulations is essential to protect your business and its customers. By carefully evaluating the security measures and compliance commitments of your cloud provider and implementing your own controls, you can ensure that your organization's IT infrastructure is secure and compliant.

It all starts with setting the baselines for our security. In the next section, we will learn how to define this baseline.

Defining the baseline for security policies

It just takes a few mouse clicks to get a server up and running on any cloud platform. But in an enterprise that's migrating or creating systems in the cloud, there's a lot for an architect to think about—securing environments being the top priority. It is likely that IaaS, PaaS, and SaaS solutions will be used to build our environment. It could grow in complexity where a lack of visibility could lead to vulnerabilities. So, with every service enrolled in the cloud environment, we really need to consider how best to secure each service. Every service needs to be compliant with the security baseline and the policies defined in that baseline.

What are the steps to create policies and the baseline?

1. **Check regulations**: Every company is subject to regulations. These can be legal regulations such as privacy laws or industry compliance standards. Make sure the regulations and compliance frameworks your company needs to adhere to are understood. Be sure to involve internal legal departments and auditors. This is the starting point in all cases.
2. Also, check which security frameworks cloud providers have adopted. The major platforms—Azure, AWS, GCP, Alibaba, and OCI—are compliant with most of the leading compliance and security frameworks, but this may not be the case for smaller providers, for instance, specific SaaS solutions. Be aware that with SaaS, the provider controls the full stack: operating systems, hardware, network infrastructure, application upgrades, and patches. You have to be sure that this is done in a compliant way for your company.
3. **Restrict access**: This is what is often referred to as **zero trust**, although the term is even more related to network segmentation. But zero trust is also tightly connected to access management. We will have to design a clear **Role-Based Access Control (RBAC)** model. Users have specific roles granting authorization to execute certain actions in cloud environments. If they don't have the appropriate role or the right authorization, they will not be able to execute actions other than the ones that have been explicitly assigned to that particular role. One term that is important in this context is **least privilege**—users only get the role and associated authorizations to perform the minimum number of actions that are really required for the daily job and nothing more.
4. **Secure connections**: Cloud environments will be connected to the **wide area network (WAN)** of a company and the outside world, the internet. The network is the route into cloud environments and must be very well secured. What connections are allowed, how are they monitored, what protocols are allowed, and are these connections encrypted? But also, how are environments in the cloud tenant segmented, and how do systems in the tenant communicate with each other? Are direct connections between workloads in the cloud tenant allowed or does all traffic need to go through a centralized hub?

5. The security baseline should contain strict policies for all connectivity: direct connections, VPNs, in-transit encryption, traffic scanning, and network component monitoring. Again, we should think about the zero-trust principle; network segmentation is crucial. The architecture must be designed in such a way that users can't simply hop from one segment of the environment to another. Segments must be contained and workloads inside the segments must be protected. A zero-trust architecture typically has zones defined, for instance, a private zone where only inbound traffic is allowed or a public zone that has connections to the outside world. These zones are strictly separated from each other by means of a variety of security elements, firewalls, security groups, or access control lists.
6. **Protect the perimeter**: This is about protecting the outside of the cloud environment, the boundary. Typically, the boundary is where the connections terminate in the cloud environment. This can be a hub, and that's where the gateways, proxy servers, and firewalls will be hosted. Typically, it also hosts the bastion host or jump server as a single point of entry, where a user is allowed to gain access to the workloads in the environment.
7. **Protect the inside**: There will be workloads in our cloud: servers, applications, containers, and functions. Although there is boundary protection with gateways and firewalls, we must also protect our workloads, especially, but not limited to, the critical ones. These workloads must be hardened, reducing the vulnerability of systems with mandatorily applied security settings, such as removing software components or disabling services that are not required to run on a system.
8. **Perform frequent audits**: This is a step that falls within managing security policies, which will be covered in the last section of this chapter. Security policies need to be constantly assessed. Hackers don't sit on their hands and will constantly think of ways to look for vulnerabilities. Therefore, it's necessary to continuously assess and audit policies and evaluate identified vulnerabilities. How critical are those vulnerabilities and what are the odds that our environments will get breached? Are we protected well enough? But also, how fast can action be taken if a vulnerability gets exploited and we need to mitigate the consequences? This is not something that should be discussed once a month but instead should be at the forefront of our minds at all times, for everyone developing or managing cloud environments.

We will need to define the scope of our security policies. One way to do that is by thinking in layers, derived from defense-in-depth as a common methodology to design security architectures. Each layer comprises protective measures against specific threats. These layers are as follows:

- **Network layer:** As already stated in the previous section, the network is the entrance into our cloud environment. Networks need to be protected from unauthorized people getting in. Technologies to protect a network from threats are firewalls, **Intrusion Detection Systems (IDSes)** and **Intrusion Prevention Systems (IPSes)**, **Public Key Infrastructure (PKI)**, and network segmentation, preferably adhering to zero-trust principles.
- **Platform layer:** Typically, this is the layer of the operating system. Systems should be fully patched with the latest fixes for (possible) vulnerabilities and hardened. Also, pay attention to ports that are opened on a system. Any port that is not required should be disabled.
- **Application layer:** This layer is not only about an application but also about middleware and APIs communicating directly with the application. Application code must be secured. Static code analysis can be very helpful and is strongly advised. Static program analysis is performed without actually executing software, validating the integrity of source code so that any attempt to change code or software parameters is detected.
- **Data layer:** This is the holy grail for hackers, the ultimate target of almost every hacker. If a hacker succeeds to get through the first three layers—network, platform, and application—then the next layer is the data itself. We will extensively discuss data security in *Chapter 16, Defining Security Policies for Storing Data*. All critical data should be encrypted, in transit and at rest.
- **Response layer:** This is the layer for all security monitoring, typically the layer for **Security Information and Event Management (SIEM)** and **Security Orchestration, Automation, and Response (SOAR)** systems. This is the layer where all suspicious activity is captured, analyzed, and translated into triggers to execute mitigating actions.

Security policies must be defined and applied at each layer. Now, let's look at some best practices for security policies:

- **Access:** Only use named accounts to allow access to systems, including just-in-time access. Be extremely selective when granting global admin rights, implement RBAC, and use multi-factor authentication. In the next chapter, we will go into this subject in more detail.

- **Perimeter or boundary protection**: Implement firewalls or use the native firewalls from the cloud platforms. A recommended practice is to have the firewall set to "block all" as the default and then open up ports as per the requirements of a certain workload or functionality. Only have ports open when there's a valid reason.
- **PKI**: Public and private keys are used to verify the identity of a user before data is transferred. Breached passwords are still the number one root cause for compromised systems and data leaks. Therefore, it's recommended not to use passwords but instead keys, securely stored in a key vault. All major cloud providers offer PKI services and key vault solutions.
- **Logging and audit trails**: Be sure that you know what happens in your cloud environment, at all times. Even with the most rigid security policy, a company should never fully rely on security measures alone. Monitoring and an audit trail are highly recommended (or required, even) best practices.

Now it's time to discover how these policies should be implemented using the native security suites in Azure, AWS, and GCP.

Implementing security policies

We have studied the compliance and security frameworks, and we've defined our security baseline. Now we need to implement it in our cloud environments. In this section, we will explore implementations in the major clouds, using native security platforms. Since CIS is widely and globally adopted as the baseline for security policies, all sections will explore specific settings that CIS benchmarks recommend for the different platforms. Links to the benchmarks are provided in the *Further reading* section of this chapter. CIS not only provides recommendations but also documents how policies should be implemented.

For example, in GCP, there is a recommendation to "*ensure Cloud Audit Logging is configured properly across all services and all users from a project.*" CIS benchmarks also guide users to find where a setting needs to be configured and how—in this example, by going to audit logs at `https://console.cloud.google.com/iam-admin/audit` or by configuring it from the command line:

```
gcloud organizations get-iam-policy ORGANIZATION_ID
gcloud resource-manager folders get-iam-policy FOLDER_ID
gcloud projects get-iam-policy PROJECT_ID
```

The format in the CIS benchmarks is always the same, for all cloud platforms.

Implementing security policies in Microsoft Defender for Cloud

For starters, we have to understand what defines a policy, and this is basically the principle that applies to other clouds as well. A policy defines:

- **Mode**: Stating what resource types will be evaluated against the policy.
- **Parameters**: This sets the action a policy will/should have, for example, allow or deny.
- **Policy rule**: Typically a rule appears as an if-then statement. If a specific condition is met, then a specific action must be taken.
- **Effect**: This specifies the expected outcome when a policy is applied.

Microsoft Defender for Cloud is a native service of Azure where we can specify security policies. Microsoft Defender for Cloud automatically, at no cost, enables any of your Azure subscriptions not previously onboarded by you or another subscription user. Within minutes of launching Defender for Cloud for the first time, you might see:

- Recommendations for ways to improve the security of your connected resources.
- An inventory of your resources that are now being assessed by Defender for Cloud, along with the security posture of each.

Then, you can enable enhanced security features for unified security management and threat protection across your hybrid cloud workloads.

We don't need to install or configure anything; from the Azure console, Defender for Cloud can be accessed immediately by simply enabling it. It then starts monitoring workloads that you have deployed in Azure: virtual machines, databases, storage accounts, networking components, and other Azure services.

By default, every Azure subscription has the Microsoft cloud security benchmark assigned. This is the successor of **Azure Security Benchmark (ASB)**, which was rebranded in October 2022. However, additional policies will need to be configured in Defender for Cloud. You have dozens of regulatory standards you can apply (PCI, SOC, CIS, NIST, or ISO). As Microsoft Defender for Cloud is a multi-cloud product, you have different options for the cloud you are protecting (Azure, AWS, or GCP). You can even create your own custom security initiatives.

CIS lists some recommendations specific to Azure Security Center. The most important one is to activate the standard pricing tier in Security Center—this enables threat detection for all networks and VMs in the Azure tenant. Every CIS recommendation to implement a policy comes with an explanation.

Enabling the standard pricing tier and adjusting settings is done through the **Defender for Cloud** blade in the portal at `https://portal.azure.com/#home`, as shown in the following screenshot:

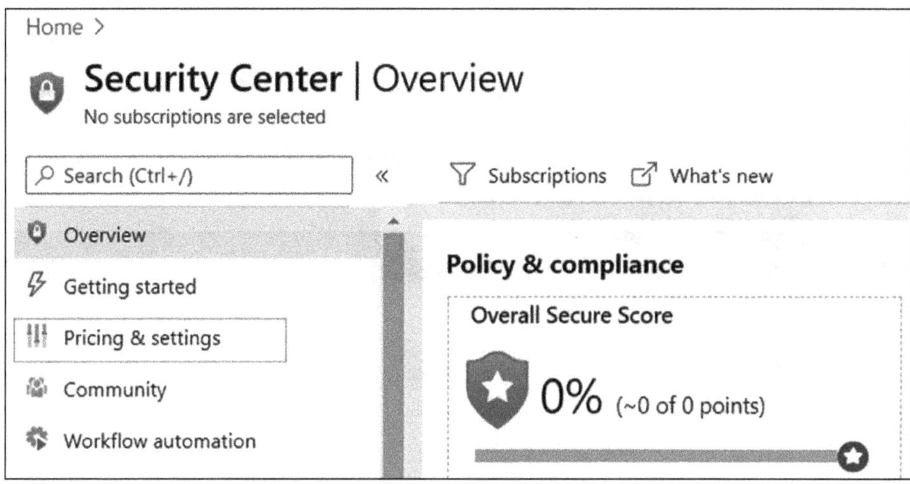

Figure 14.1: Overview of the Security Center blade in the Azure portal

Getting started with Defender for Cloud is best done by visiting `https://learn.microsoft.com/en-us/azure/defender-for-cloud/enable-enhanced-security`.

Azure has more than just Azure Security Center—Azure Sentinel, a native SIEM and SOAR solution. Sentinel is an intelligent defense-in-depth solution, and it is especially useful when activating its MITRE ATT&CK® security framework. ATT&CK is a knowledge base that is constantly updated with the latest threats and known attack strategies. A group of developers under the name of BlueTeamLabs have published templates and code to implement ATT&CK in Sentinel. It's worthwhile taking a look at this at `https://github.com/BlueTeamLabs/sentinel-attack`.

Implementing security policies in AWS Security Hub

AWS offers a single security dashboard with AWS Security Hub. The solution aggregates monitoring alerts from various security solutions, such as CloudWatch and CloudTrail, but also collects findings from Amazon GuardDuty, Amazon Inspector, Amazon Macie, AWS IAM Access Analyzer, and AWS Firewall Manager. CloudTrail, however, is the key element in Security Hub. CloudTrail constantly monitors the compliance of accounts that are used in the AWS environment. It also performs operational auditing and risk auditing, meaning it keeps track of all activity that is started from the console in your environment, enables analysis of changes to resources, and detects unusual activity. It's fair to say that CloudTrail is the engine underneath Security Hub.

Security Hub makes it easy to start monitoring all activity in your AWS environment. It's accessible from the AWS console, as shown in *Figure 14.2*:

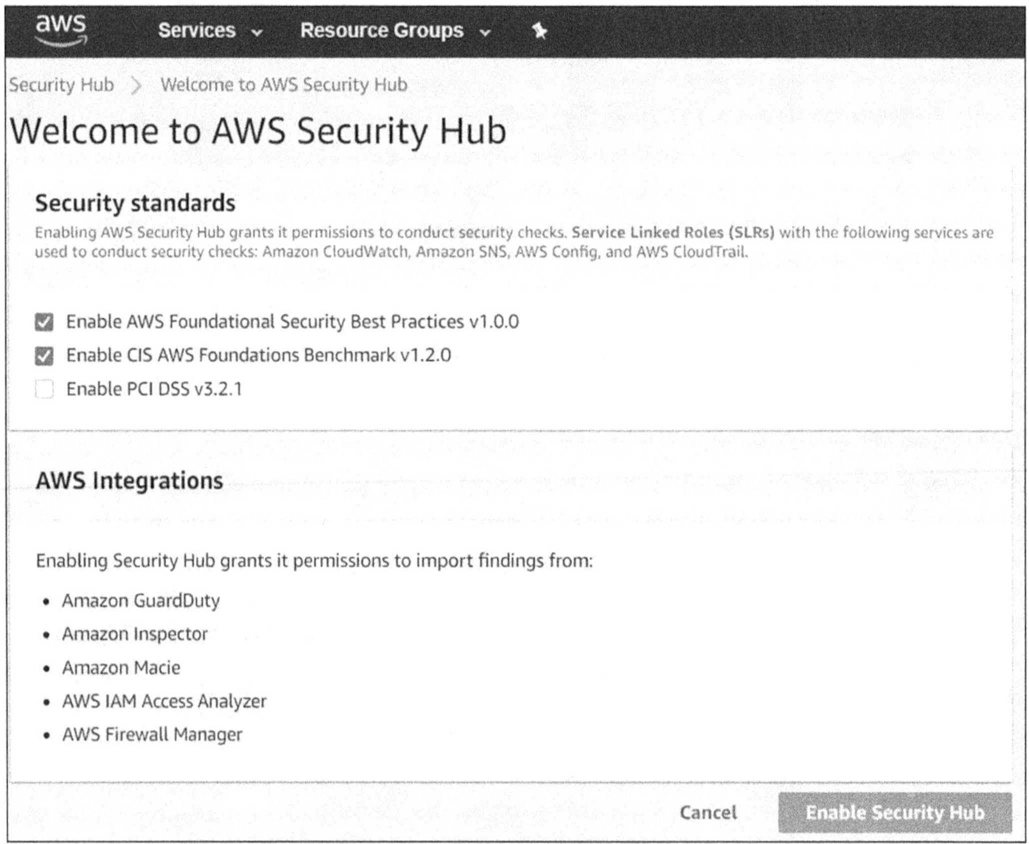

Figure 14.2: Accessing Security Hub in the AWS console

There are a couple of things that need explaining in the preceding screenshot. The top part of the screen shows the security baselines that can be selected—**Enable AWS Foundational Security Best Practices v1.0.0** and **Enable CIS AWS Foundations Benchmark v1.2.0** have been ticked by default. The third one is the PCI DSS framework. **PCI DSS** stands for **Payment Card Industry Data Security Standard** and is specific to financial institutions.

In the lower part of the screen, we see all the integrations that Security Hub offers:

- **GuardDuty**: Amazon's solution for threat detection.
- **Inspector**: This tool assesses applications for exposure, vulnerabilities, and deviations from best practices valid for these applications.
- **Macie**: This solution monitors the data security and data privacy of your data stored in Amazon S3 storage.
- **IAM Access Analyzer**: This tool keeps track of accounts accessing environments in AWS and whether these accounts are still compliant with security policies.
- **Firewall Manager**: This tool enables centralized management of all firewalls in the AWS environment.

By clicking the **Enable Security Hub** button, the aforementioned baselines with the named integrations will be enrolled.

The CIS baseline should definitively be implemented as the worldwide accepted standard for securing online environments. Specific to AWS, CIS includes the following recommendations for settings to control security policies:

- Ensure that CloudTrail is enabled in all regions
- Ensure that CloudTrail log file validation is enabled
- Ensure that an S3 (storage) bucket used to store CloudTrail logs is not publicly accessible
- Ensure that CloudTrail logs are integrated with CloudWatch logs
- Ensure that AWS Config is enabled in all regions
- Ensure that S3 bucket access logging is enabled on a CloudTrail S3 bucket
- Ensure that CloudTrail logs are encrypted at rest using **Key Management Services—Customer Master Keys (KMS CMKs)**
- Ensure that rotation for customer-created CMKs is enabled
- Ensure **Virtual Private Cloud** (**VPC**) flow logging in all VPCs

Obviously, these are not all the settings; these are the most important settings to get the logging and monitoring of security policies right. In the *Further reading* section, we include links to the various CIS benchmarks for the major clouds.

Implementing security policies in GCP Security Command Center

In GCP, we will have to work with Security Command Center. You can manage all security settings in Security Command Center and view the compliancy status from one dashboard. The concept is the same as AWS Security Hub—Security Command Center in GCP comprises a lot of different tools to manage security in GCP environments. In the GCP cloud console, we'll see **Security** in the main menu. Hovering over the **Security** subheading will pop up the products and tools that are addressed in **Security Command Center**, as shown in the following screenshot:

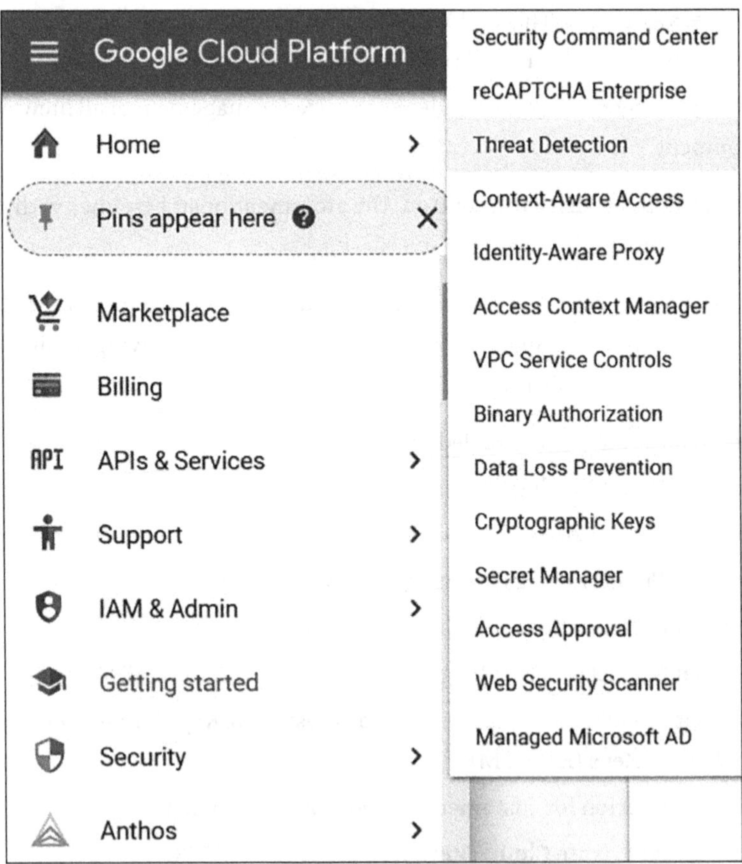

Figure 14.3: Launching Security Command Center in the cloud console of GCP

Security Command Center does an inventory and discovery of all assets in the GCP environments and, next, starts monitoring them in terms of threat detection and prevention. One special feature that needs to be discussed here is Google Cloud Armor. Cloud Armor started as a defense layer to protect environments in GCP from **Distributed Denial-of-Service** (**DDoS**) and targeted web attacks. Cloud Armor has since been developed into a full security suite in GCP to protect applications, using the functionality of **Web Application Firewalls** (**WAFs**).

Cloud Armor can be launched from the GCP console at `https://console.cloud.google.com/`. You won't find it under **Security Command Center** but under **Network Security**, as shown in *Figure 14.4*:

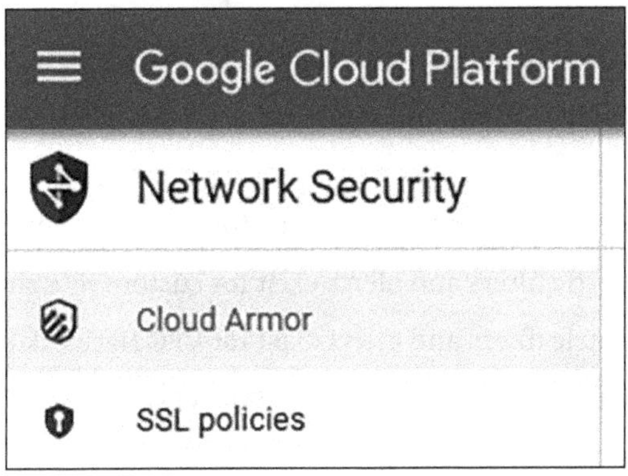

Figure 14.4: Menu of Cloud Armor in GCP

We can specify security policies in Cloud Armor, but GCP already includes a list of policies that can be evaluated. These preconfigured policies are based on the **OWASP** CRS—the **Open Web Application Security Project**, a community that strives to find methodologies and practices to constantly improve the protection of online applications. **CRS** stands for **Core Rule Set**. Cloud Armor includes the top 10 OWASP threats in rule sets. The number one threat is the injection of hostile code in order to breach an application and get access to data. In *Chapter 16, Defining Security Policies for Storing Data*, we will explore OWASP in more detail since this is all about securing applications and data.

However, OWASP does overlap with CIS, but OWASP merely identifies the threats, whereas CIS makes recommendations for avoiding vulnerabilities and assesses the chances of threats really being exploited. Misconfigured security, for example, is number 6 in the top 10 of OWASP. Insufficient logging and monitoring conclude the top 10. Both are heavily addressed by CIS.

The CIS 2.2.0 benchmark for GCP was released in March 2020. Specifically, for logging and monitoring, CIS recommends the following settings to audit security policies:

- Ensure that Cloud Audit Logging is configured properly across all services and users in a project.
- Ensure that sinks are configured for all log entries.

A sink will export copies of all the log entries.

- Retention policies on log buckets must be configured using Bucket Lock.
- Ensure that log metric filters and alerts exist for project ownership assignments and changes.
- Ensure that log metric filters and alerts exist for audit configuration changes.
- Ensure that log metric filters and alerts exist for custom role changes.
- Ensure that log metric filters and alerts exist for VPC Network Firewall rule changes.
- Ensure that log metric filters and alerts exist for VPC Network Route changes.
- Ensure that log metric filters and alerts exist for VPC Network changes.
- Ensure that log metric filters and alerts exist for cloud storage IAM permission changes.
- Ensure that log metric filters and alerts exist for SQL instance configuration changes.

As with Azure and AWS, these are the settings to audit security policies against the CIS benchmark. In the *Further reading* section, we include links to the various CIS benchmarks for the major clouds.

Implementing security policies in Alibaba Cloud

The main product in Alibaba Cloud is Security Center. This is the place where we can define and store security policies for assets that we host in Alibaba Cloud. Security Center offers extensive services for:

- Defining and guarding the security baseline
- Asset control
- Compliancy check
- Ransomware alerts
- Mining alerts

- Tamper-proofing
- AccessKey pair leaks
- Attack-source tracing
- Cloud security configuration monitoring
- Cloud asset risk monitoring

A nice feature is that users can have a 7-day trial in Security Center. But this does not cover everything in terms of security. Alibaba offers a wide range of additional services, including "business security." Fraud detection offers real-time analysis and identification of risks. One more service that is worthwhile mentioning is **Content Moderation**—a service that enables the detection of pornography, violence, and terrorism in images, text, audio, and videos. It includes daily updates on compliance intelligence. All these services are listed under **Security** in the console, as shown in *Figure 14.5*:

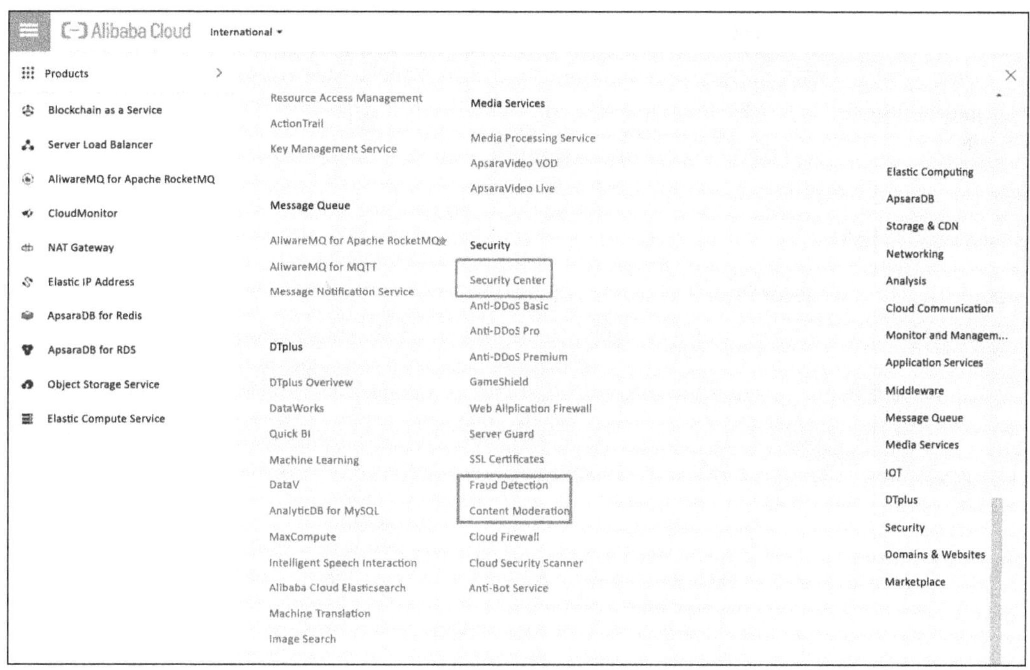

Figure 14.5: Security menu in Alibaba Cloud

There are three levels in Security Center, including a free tier, but then services are limited to setting policies for abnormal login, cloud platform configuration assessment, and vulnerability checks. The advanced tier is charged at approximately 10 USD per server per month, but it does provide sophisticated protection against viruses, ransomware, and cryptomining. Advanced also includes intrusion detection. The enterprise level is the most complete offering, with AccessKey leak detection, attack awareness, and attack tracking among other services. Enterprises are recommended to contact Alibaba Cloud for financial offerings.

Naturally, there is a CIS benchmark (version 1.0.0) for Alibaba Cloud too, which we recommended reviewing.

Implementing security policies in OCI

The best place to start to understand how security policies work in OCI is the **Consensus Assessment Initiative Questionnaire** (**CAIQ**) for OCI, available to read at `https://www.oracle.com/a/ocom/docs/oci-corporate-caiq.pdf`. This is an extensive document that lists the security practices in OCI. It lists all the control domains and how Oracle addresses security issues in each domain. We picked the first topic in the list as an example, asking if Oracle adheres to industry standards such as the OWASP Software Assurance Maturity Model to build security in systems and the Software Development Life Cycle. Oracle's answer to that one is **Oracle Software Security Assurance** (**OSSA**).

Obviously, this is not an implementation of the security policies for our specific environment. The CAIQ is a good starting point to define these policies, adhering to the best practices of OCI.

In OCI, we define policies in a security zone. Creating a security zone is done through a recipe, a collection of the policies in that zone. There's already a predefined recipe called the Maximum Security Recipe, managed by Oracle, which can't be modified. It's shown in *Figure 14.6*:

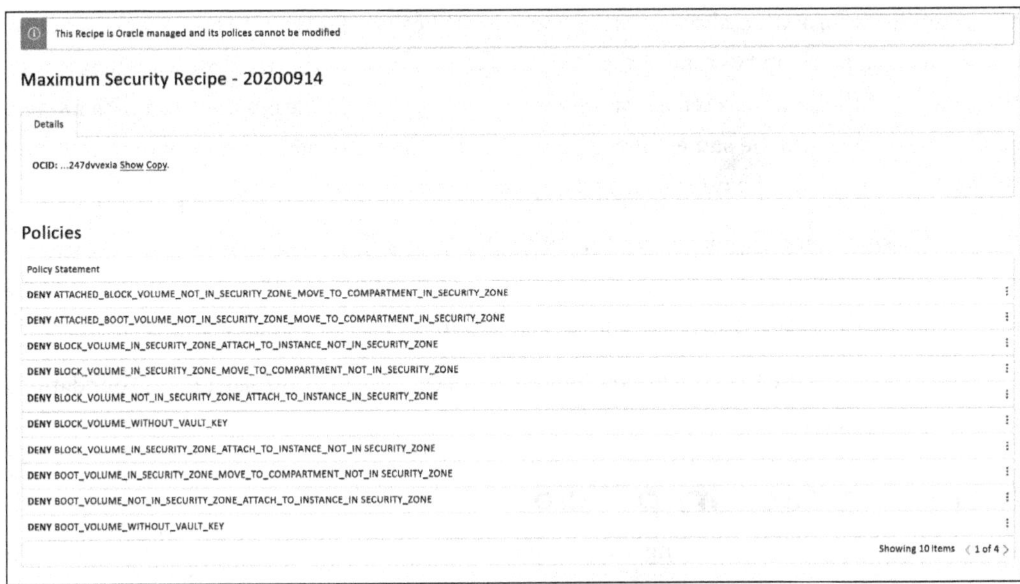

Figure 14.6: Maximum Security Recipe in OCI

We can, however, add recipes with specific security policies, following a categorization. These categories are:

- Restrict resource movement, preventing resources to be moved outside of the zone
- Restrict resource association, ensuring that all components of a specific resource are in the same zone
- Deny public access
- Require encryption
- Ensure data durability concerning automatic backups
- Ensure data security, preventing data from being copied outside the zone
- Use only configurations approved by Oracle

More detailed information about security zones in OCI can be found at `https://docs.oracle.com/en-us/iaas/security-zone/using/security-zone-policies.htm`. Be aware that a security zone is not the same as a tenant or a compartment in OCI; a zone is associated with a single compartment that has the same name as the zone. The compartment holds the resources, and the security zone holds the policies that the resources have to comply with.

How do we keep track of policies in OCI? With Cloud Guard. This tool detects misconfigured resources and activity across tenants that are considered suspicious. Cloud Guard can be easily launched from the **Security** menu in the console. Recipes can now be entered in Cloud Guard.

The latest version of the CIS benchmark for OCI is CIS Oracle Cloud Infrastructure Foundations Benchmark version 1.2.0.

Managing security policies

It doesn't stop with implementing security policies. We need to have governance in place to manage the policies. Governance is required at two levels:

1. The security policies themselves, auditing these against the compliance frameworks that a business has to adhere to.
2. The technical implementation of the security policies, keeping the monitoring up to date, and making sure that all assets are indeed tracked against the policies.

The first level is the domain of people concerned with the security governance in a business, typically, a **Chief Information Security Officer** (**CISO**) or **Chief Information Officer** (**CIO**). They need to set directions for security policies and make sure that the business is compliant with the security strategy, industry, and company frameworks. The CISO or CIO is also responsible for assurance from internal and external auditing.

Level two is more about security management, concerning how to deal with security risks in the IT landscape, including the cloud environments. To make it simple—security governance is about making policies; security management is about (technically) implementing and enforcing policies. So, security engineers should worry about the management of security monitoring tools that were covered in this chapter. They will need to understand how to implement rule sets in Microsoft Defender for Cloud, AWS Security Hub, and Google Security Command Center. They will also need to know what to do in the event of an alert being raised in these systems, who should follow up, and what actions need to be taken. Those will be technical actions, such as isolating an environment when it's breached. The configuration of rules in the security suites is also in their hands.

However, the security policies themselves need to be defined from a higher level in a business. The CISO or CIO will hardly ever completely understand how to program a security console, but they will know what needs to be protected from a business perspective. Obviously, the strategic level of CISO/CIO can't do anything without input from the tactical level—the security architects and engineers. They will all have to work closely together.

Manage risks with Cloud Security Posture Management

We discussed methodologies to implement security policies in the various clouds. Now, we also have to make sure that these policies are followed to ensure that our environments stay compliant. That's the key function of CSPM:

- Detect cloud misconfigurations
- Remediate cloud misconfigurations, preferably through automation
- Manage best practices for different cloud configurations and services
- Check the cloud health status against a security control framework and compliance standards
- Monitor cloud services, including storage solutions, encryption, and account permissions

CSPM is designed to detect and remediate risks that might be caused by bad configurations of cloud services. Since we work in multi-cloud, we have to find tools that can scan multiple environments in clouds. Some of these tools are also able to check against regulatory frameworks such as the HIPAA, alert, and even automatically remediate issues. Palo Alto Networks' Prisma and Trend Micro's Cloud One Hybrid Cloud Security are two such tools, but there are many more. Since CSPM is an evolving market, more providers are expected to enter this domain.

The best practices are:

- Understanding the risk profile of an organization; different organizations have different risk profiles, based on factors such as the type of data they handle, the regulatory environment they operate in, and their overall security posture. Understanding your organization's risk profile will help you prioritize your efforts and allocate resources appropriately.
- Implementing a strong governance framework by establishing clear policies, procedures, and controls for your cloud environment. This should include things like access controls, data classification, and incident response plans.
- Implementing CSPM, which provides continuous monitoring and assessment of your cloud environment and enables fast identification of potential risks. Other tools, such as **cloud access security brokers** (**CASBs**) and SIEM systems, can also help you identify and mitigate risks.

- An obvious point that can't be stressed enough: make sure to keep your cloud environment, including all hardware and software, up to date with the latest patches and security updates. This can help you protect against known vulnerabilities and reduce the risk of attacks.
- Monitoring and reviewing your risk posture regularly. This can involve conducting regular risk assessments, reviewing security logs, and staying up to date with the latest threats and vulnerabilities.

One of the crucial elements of keeping environments secure is only granting access to authorized users and services. Good IAM is key. That's the central topic of the next chapter.

Summary

In this chapter, we discussed the basics of security frameworks as a starting point to define policies for cloud environments. We learned that there are different frameworks and that it depends on the industry to determine the compliance requirements of a business. Then, we must decide which security controls to set to ensure that our cloud environments are compliant too.

One framework that is globally accepted and commonly used for clouds is CIS. We learned that the CIS benchmarks for these cloud platforms not only greatly overlap but also have specific settings that need to be implemented in the respective security suites—Microsoft Defender for Cloud, AWS Security Hub, Google's Security Command Center, Alibaba Cloud, and OCI's Security Zones.

In the last section, we learned how we can keep control of security policies and our environments compliant by implementing CSPM and studying some best practices.

In the next chapter, we will dive into identity and access management, since that's where security typically starts—who is allowed to do what, how, and maybe even when in our cloud environments? In *Chapter 17, Implementing and Integrating Security Monitoring,* we will further explore the use of the monitoring tools that we discussed briefly in this chapter.

Questions

1. We've discussed the CIA principle. What does it stand for?
2. Name the attributes that are included as a minimum in a policy.
3. Where do we implement security policies in OCI?
4. What does a CSPM tool do?

Further reading

You can refer to the following links for more information on the topics covered in this chapter:

- The CIS framework: https://www.cisecurity.org/
- The CIS Benchmark for Azure: https://learn.microsoft.com/en-us/compliance/regulatory/offering-cis-benchmark
- The CIS Benchmark for AWS: https://d0.awsstatic.com/whitepapers/compliance/AWS_CIS_Foundations_Benchmark.pdf
- The CIS Benchmark for GCP: https://www.cisecurity.org/benchmark/google_cloud_computing_platform/
- The CIS Benchmark for Alibaba: https://www.cisecurity.org/benchmark/alibaba_cloud
- The CIS Benchmark for OCI: https://www.cisecurity.org/benchmark/oracle_cloud
- Link to the OWASP community pages: https://owasp.org/www-project-top-ten/#:~:text=The%20OWASP%20Top%2010%20is%20the%20reference%20standard,software%20development%20culture%20focused%20on%20producing%20secure%20code
- *Enterprise Cloud Security and Governance*, by Zeal Vora, Packt Publishing

15

Implementing Identity and Access Management

The core principle of identity and access management in the cloud is that everyone and everything in it is an identity. In this chapter, we will learn how we can manage identities and control their behavior by granting them specific roles, allowing them to perform only those activities that are related to the primary job of an administrator. We will see that **Role-Based Access Control (RBAC)** is very important to keep our cloud environments secure. We will learn about authenticating and authorizing identities, how to deal with least privileged accounts, what eligible accounts are, and why a central depository is needed. We will learn how we can federate with Active Directory from the various public clouds.

After this chapter, you will have a good understanding of technologies such as federation, single sign-on, multi-factor authentication, privileged access management, and **Identity as a Service (IDaaS)**.

In this chapter, we're going to cover the following main topics:

- Understanding identity and access management
- Using a central identity store with Active Directory
- Designing access management across multi-cloud
- Exploring **Privileged Access Management (PAM)**
- Enabling account federation in multi-cloud
- Working with IDaaS

Understanding identity and access management

Identity and access management (**IAM**) is all about controlling access to IT systems that are critical to a business. A key element of IAM is **Role-Based Access Control**, or **RBAC** for short. In an RBAC model, we define who is allowed to have access to systems, what their role is, and what they are allowed to do according to that role. An important principle of RBAC is **least privilege**, meaning that a system administrator will only get the rights assigned that are required to perform the job. For example, a database administrator needs access to the database, but it's not very likely that they will need access to network switches too.

In this chapter, we will discuss concepts such as **Single Sign-On** (**SSO**), **Multi-Factor Authentication** (**MFA**), and **Privileged Access Management** (**PAM**). Before we go into those, let's have a look at the basics of IAM. There are three layers that we have to consider in our architecture:

- **Managed identities**: In this book, we've written a number of times that in cloud environments, everything should be perceived as an identity. Identities must be known: users, systems, APIs—everything that communicates with components in your cloud environment and with people or systems in the outside world.
- **Managed roles**: Roles must be defined in our cloud environments and assigned to identities. This includes the process of adding, removing, and updating roles. This is not only valid for people but also devices and systems. A system is an identity and has a specific role, for instance, a domain controller or application server. Thus, system authorizations must be defined and access rights must be given to resources.
- **Managed access**: This is the definition of who and what are given appropriate levels of access. To use the example of the database administrator once more, a database administrator needs access to the database and not to a network switch. If the database resides in a specific virtual network within the cloud environment, the administrator may need access to that network as well. However, that access should be limited; it is only needed to get to the database. That must be defined in the role.

On all three layers, the principle of least privilege is valid and must be followed through to achieve maximum protection of (sensitive) data in systems. The following diagram shows the main principles and related services of IAM:

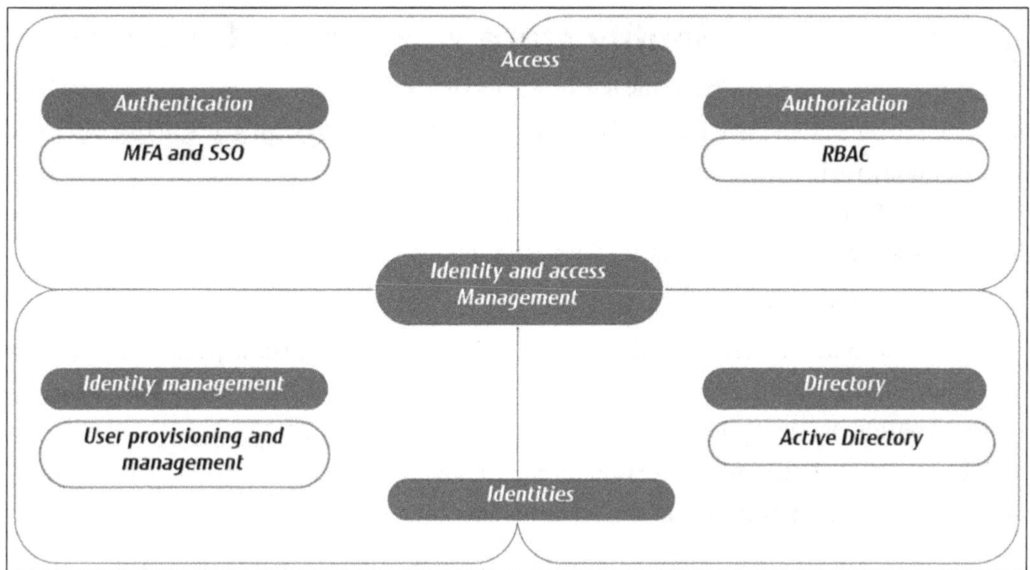

Figure 15.1: Main principles of IAM

The next step is to define what an IAM system needs to do and what sort of tools it provides to control identities and access to our cloud systems.

Primarily, it needs to enable us to control identities. IAM should, therefore, contain a directory—basically, a database that holds all identities. Typically, this directory contains the entities that will access the systems. Almost all enterprises use **Active Directory (AD)** as their central directory. AD uses objects to represent an entity; an entity can be a user, a device, or a system. It also defines which domain an object belongs to.

Next, the IAM systems must be enabled to grant entities roles and the associated access. If a user is added and a role is assigned to that user, then the IAM system makes sure that access rights are provisioned to that user. Typically, roles and groups will be defined in the directory so that all that remains is to assign a user or object to that role or group. Access rights on the appropriate level are then automatically enabled. IAM should also facilitate a review process; only a few admins should have the right to add users to or remove users and objects from a directory. A user can request specific access, but they will always need a review and approval before rights are actually assigned. PAM and **Privileged Identity Management (PIM)** are tools to define that process. We will explore these concepts in the next sections.

Using a central identity store with Active Directory

One of the most used identity stores is still Active Directory. Before we get into AD itself, it's important to understand that it should definitively not be confused with Azure Active Directory. The key difference is that Azure AD is a cloud-native IDaaS solution whereas AD is a traditional **Lightweight Directory Access Protocol (LDAP)**, a network protocol that determines how information is exchanged from directory services using, for instance, TCP/IP.

Understanding AD is not easy, but basic knowledge is necessary when talking about IAM. An enterprise should only have one central directory. Identities should only be kept in one place. That also comes with a risk—if a directory gets breached, an attacker will have access to all identities that exist within the enterprise. It's crucial that the directory and the IAM system are very secure and that directory data is extremely well protected. This is an area where tools such as Saviynt and CyberArk come in; they add an extra security layer on top of IAM.

Both Saviynt and CyberArk offer solutions that are deployed on top of IAM, providing vaults and a way to secure access to systems, for instance, by hashing passwords in encrypted vaults so that users actually don't see passwords, getting them instead from the tools. These tools can also record sessions or system logins to enable maximum visibility of activity in an environment, often referred to as an audit trail.

Let's get back to the identity store and AD. The term is very much associated with Microsoft, as it was developed by that company for Windows domain networks. In the meantime, it has become a widely accepted term for the concept itself. AD comprises basically two major components that are both relevant in cloud environments. The first component is the directory itself; the second component is the domain services.

Domain services comprise a domain controller that authorizes and authenticates objects—users and computers—in a network. That network can be in a public cloud. It can also be a standalone network, but more often, the internal network of the enterprise is extended to a cloud. Extended may not be the right word, though. The enterprise on-premises network and cloud network(s) are merely connected or, to put it a better way, we connect the domains.

To be able to do that, domain controllers are needed in the public cloud. The domain controller makes sure that a specific part of the public cloud is now within our domain. For all of this, AD Federation Services can be used to federate the domain in the cloud with the directory that enterprises already have, commonly on-premises. The following diagram shows **My Company X** with the AD on-premises. There is an environment in a public cloud as well. That environment federates with the on-premises AD:

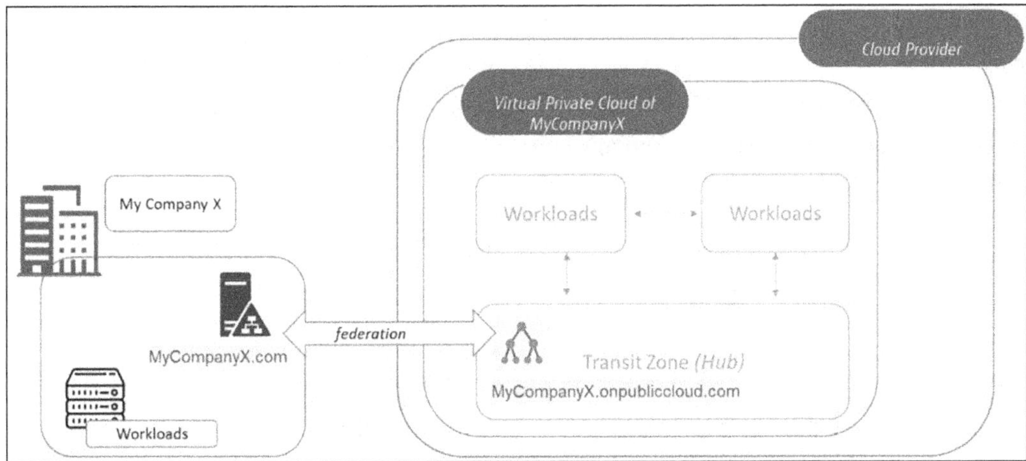

Figure 15.2: Conceptual overview of AD federation

Microsoft AD uses LDAP, Kerberos, and Domain Name Services for these services. LDAP enables the authentication and storing of data about objects and also applications. Kerberos is used to prove the identity of an object to another object that it communicates with. DNS enables the translation of IP addresses to domain names and vice versa.

This concludes the section about AD and how it's used as an identity store. The next section will explain how access to clouds is controlled.

Designing access management across multi-cloud

In the previous section, we learned that we need to have federation with AD in our public cloud environment. The next question is, *how do we do that?* Azure uses **Azure Active Directory (AAD)**. Just as a reminder—AAD is not the same as AD. AAD is an authentication service in Azure, using AD as the directory. The primary function of AAD is to synchronize identities to the cloud. For the synchronization, it uses Azure AD Connect.

With AAD, enterprises will have a system that provides employees of these enterprises with a mechanism to log in and access resources on different platforms. That can be resources in Azure itself or resources such as applications hosted on systems in the corporate network.

But AAD also provides access to SaaS solutions such as Microsoft 365 and applications that can integrate with Azure. AAD makes sure that users only have to log in once using SSO. It's secured by MFA, meaning that when a user logs in by typing in a password, it is not enough. A second validation is needed to prove their identity. This can be a PIN code through a text message or an authenticator app on a mobile device, but also a fingerprint. If the user is authenticated, access is granted to federated services.

The federation between the domains in the corporate network and the corporate domain in Azure cloud is done with **Active Directory Federation Services (ADFS)**. Strictly speaking, there's no hard requirement to use ADFS since AAD integrates with AD natively with hybrid entities, using password hash-sync or passthrough authentication. For third-party MFA you will still need ADFS though.

In the cloud, a corporate cloud domain is situated in the domain of the public cloud itself. In Azure, that is defined by `onmicrosoft.com`; this domain name address signifies that an environment resides in Azure.

Now, if we take company X, which has its domain specified as `companyx.com` and wants to have an environment in Azure, the domain in Azure would probably be `companyx.onmicrosoft.com`. Next, trust must be established between the corporate domain and the domain in Azure with ADFS. AD federation in Azure is shown in *Figure 15.2*.

In AWS, ADFS can be enabled as a component of the AWS Federated Authentication service. With ADFS, a user is authenticated against the central identity store, AD. After authentication, ADFS returns an assertion—a statement—that permits login to AWS using AWS's **Security Token Service (STS)**. STS returns temporary credentials based on `AssumeRoleWithSAML`, which then allows access to the AWS Management Console of the enterprise environments in AWS. The following diagram shows the concept:

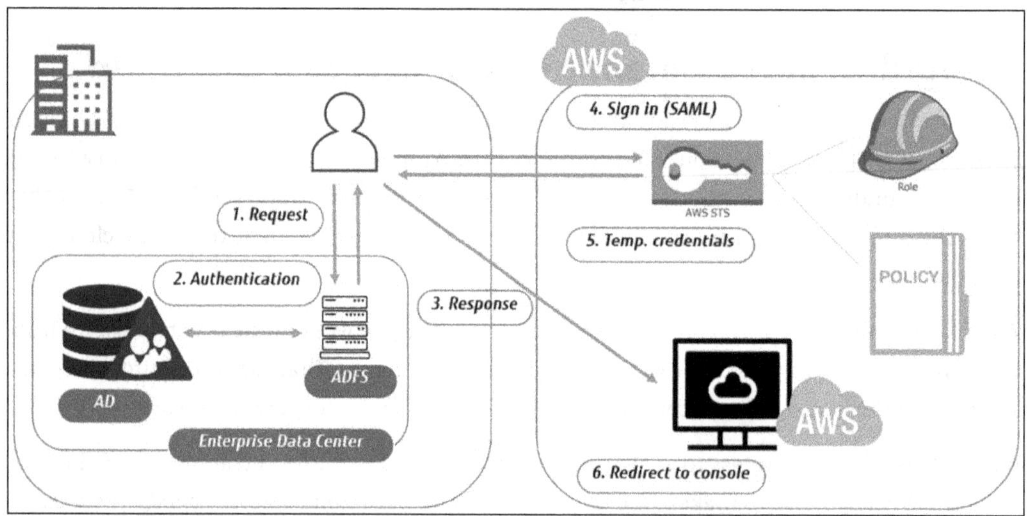

Figure 15.3: Concept of AWS federated authentication

`AssumeRoleWithSAML` is something specific to AWS. This function in STS provides a way to authenticate against the identity store with role-based AWS access. It uses **Security Assertion Markup Language (SAML)**, an open standard for exchanging authentication and authorization data between parties, such as the identity store at a corporate level and the cloud provider. Yes, it's comparable to LDAP, but SAML is more commonly used in the cloud.

A lot of companies still use ADFS, but it's no longer a hard requirement to use AD in AWS or GCP. It is possible to integrate AD directly into other clouds. Refer to `https://learn.microsoft.com/en-us/azure/active-directory/saas-apps/aws-multi-accounts-tutorial` and `https://cloud.google.com/architecture/identity/federating-gcp-with-azure-active-directory`.

Also, GCP embraces SAML to do AD federation. At GCP, it starts with Google Cloud Identity, the service that GCP uses for IAM. But Google also understands that enterprises typically already have an identity store with AD. We can set up federation between GCP's Cloud Identity or G Suite and enable user provisioning from AD, including SSO. SSO is done through SAML.

For the actual federation, we use Google Cloud Directory Sync, a free service from Google. The concept is shown in the following diagram:

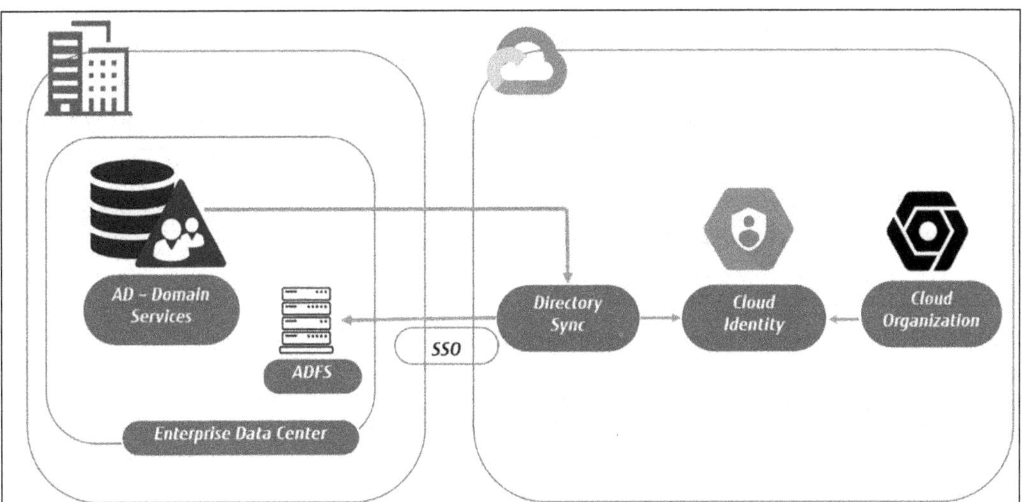

Figure 15.4: Concept of Google Directory Sync

Of course, AD can be used in OCI to provide centralized authentication and authorization for OCI resources as well. We need to take a number of steps to set this up, as we did with the other clouds. First, we need to create a **virtual cloud network** (**VCN**) in OCI and configure a VCN and subnet for AD. In the VCN, we can now create an AD controller by using an AD server image. The best practice is to use an image from the marketplace. We must configure the AD domain controller by setting up the AD forest and domain and adding users and groups. Now we can use the AD server as the identity provider and start adding groups and users to the AD server. Be aware that we also must create a policy that allows access to OCI resources based on the AD groups and users. In the OCI console, we configure the AD identity provider as the default identity provider for our tenancy and configure the **single sign-on** (**SSO**) settings.

It's important to note that this is a general overview of the process, and more detailed steps and specific configurations may vary depending on the specific use case. In all of these concepts, it's also important to understand how the corporate AD is set up. This setup needs to be mapped to the IAM policies in the cloud platform. AD has a logical division with forests, trees, domains, and organizational units. Forests are the top-level segment of an AD and contain the root-level domain. Objects such as computers and users are grouped into domains. A group of domains forms a **tree**. Domains and trees form a forest – simply put, a collection of groups with users and systems in one domain that trust each other.

In a public cloud, this division of forests, trees, and domains might not map by default to the structure that a public cloud has. Using GCP as an example, organizations are the container boundaries that hold the resources within GCP. Organizations contain all the projects that can be hierarchically subdivided into folders. These structures have to be mapped to the AD structure; otherwise, federation will fail, leading to objects and users that can't authenticate or have access to specific resources in cloud environments.

It's beyond the scope of this book to do a deep dive into AD, but in the *Further reading* section, we've listed literature that provides more in-depth insights.

Working with least-privilege access

Least privilege is an important principle in IAM that ensures that users are only given the minimum level of access needed to perform their job functions. This principle is based on the idea that by limiting the access of users, organizations can reduce the risk of security breaches and data loss.

In an ideal scenario, a user should only have access to the specific resources and data that are necessary for them to perform their job. **Just in Time (JIT)** grants them access to do their job only, no longer than is strictly required.

This reduces the risk of unauthorized access to sensitive information and prevents users from accidentally or intentionally causing damage to a system. By implementing least privilege, organizations can ensure that users are only able to access the resources they need to perform their job, while preventing them from accessing or modifying sensitive information.

Least privilege also helps organizations to comply with various regulatory requirements, such as the EU regulations under the **General Data Protection Regulation** (**GDPR**) and the **Health Insurance Portability and Accountability Act** (**HIPAA**). These regulations require organizations to implement strict controls on access to sensitive data and to ensure that only authorized users have access to it.

Least privilege can even become a significant way of lowering costs. By limiting the access of users, organizations can reduce the need for expensive security measures, such as firewalls and intrusion detection systems, and also reduce the risk of system downtime caused by unauthorized access. Having said that, least privilege access doesn't mean that organizations do not need to implement other security measures.

What do we need to do if we implement least privilege access? The following steps are best practices:

1. **Identify and classify sensitive data**. This includes identifying data that is sensitive from a regulatory, legal, or business perspective.
2. **Create roles and responsibilities**. Once the sensitive data has been identified, we should create roles and responsibilities for each employee. This includes determining what access each employee needs to perform their job and what access they should be denied.
3. **Implement access controls**. These controls are required to ensure that users are only able to access the resources they need to perform their job. This includes implementing authentication, authorization, and access control mechanisms such as **Role-Based Access Control** (**RBAC**), but also **attribute-based access control** (**ABAC**).
4. **Regularly review and update access rights**. This should include regularly reviewing user accounts, role assignments, and access rights to ensure that they are still valid and necessary. A good addition to this is Microsoft Entra and, specifically, Microsoft Entra Permissions Management. This is a **Cloud Infrastructure Entitlement Management** (**CIEM**) product that provides visibility and control over permissions for any identity and resources in Azure, AWS, and GCP.
5. **Monitor and audit access**. This includes implementing logging, monitoring, and reporting mechanisms to track user activity and detect any unusual activity.

This is not a one-time activity. We must evaluate these steps every time we implement new systems or applications to ensure that access is restricted only to the necessary parties.

Exploring Privileged Access Management (PAM)

In previous sections, the principle of least privilege was introduced—users only get the minimum set of rights to the systems that they are authorized for/require.

PAM uses the principle of least privilege to grant users only the access they need to perform their job functions. This is achieved by using roles and policies to limit access to specific resources and actions, and by using MFA to ensure the identity of users. PAM also allows you to monitor and audit access to your resources, so you can detect and respond to any unauthorized access. Additionally, PAM provides a feature called Session-Based Authentication, enabling customers to create short-lived access credentials for users that need privileged access to resources, which last for a predefined duration.

Least privilege works with non-privileged accounts or **least-privileged user accounts** (**LUAs**). Typically, there are two types of LUA:

- Standard user accounts
- Guest user accounts

Both types of accounts are very limited in terms of user rights.

There are situations where these accounts simply aren't sufficient and inhibit people from trying to do their job. The user would then need elevated rights—rights that are temporarily assigned so that the user can continue with their work. An account with such elevated rights is called a **privileged account**. Examples of privileged accounts are the following:

- **Domain administrative accounts**: Accessing all resources in the domain
- **AD accounts**: Accessing AD with rights to, for example, add or remove identities
- **Application accounts**: Accessing applications and databases to run, for example, batch jobs or execute scripts

A special category would be break glass accounts, sometimes referred to as emergency accounts. These are accounts that function as a last resort when users are completely locked out of an environment. The break glass account is an account that has access to all resources and has all the rights to literally unlock the environments again.

The issue with these accounts is that they form a much bigger risk than standard, non-privileged accounts. If privileged accounts get breached, a hacker can have control over critical systems and functions. PAM is a solution that mitigates these risks. In short, PAM makes sure that elevated rights can only be used by specified accounts for specific systems at a specified time and for specified reasons.

The principle behind this is called JIT and just enough administration. In this principle, an administrator can decide that specific users need certain privileges to perform tasks on systems. But these users will not get these privileges permanently; that would be a violation of the least privilege principle. These users will get eligible accounts, meaning that when a user needs to perform a certain task, the rights to do so will be elevated. To enable the eligible account, the user will need permission that expires after a pre-set time window.

So, the user requests permission to enable the elevated rights from the privileged account. Permission is granted for one, two, or the number of hours that are needed to perform the tasks. After the time has expired, the rights are automatically withdrawn.

PAM on cloud platforms

How does this work on cloud platforms? PAM only works if the principle of least privilege is applied—what privileged accounts are needed, and what roles they will have. The cloud platforms all have an extensive role-based model that can be applied, enabling execution at a granular access level with a separation of duties for resources.

Cloud providers work with only a few built-in general role types: roles that can do everything in the cloud tenant, roles that can do specific things in certain areas of the tenant, and roles that can only view things. In addition to those, there's often a role for the purpose of adding users and roles in the tenant. That's not sufficient for a role-based access model that requires more granularity. Cloud providers provide that and have roles specifically for network administrators, database administrators, or even very particular roles just for managing the backups of specific websites.

With a clear overview of our accounts and the roles that these accounts have, PAM can be configured in cloud environments. Azure offers PIM as their solution to identify and set eligible accounts. Be aware that this requires a premium license for AD in Azure. PIM sets eligible accounts, activates JIT, and configures MFA.

In AWS, PAM features are included in the IAM solution. Like Azure, AWS offers a role-based access model and the possibility of having privileged accounts, using elevated rights, SSO, and MFA. The logic starts with requests being denied by default, except for the root account, which has full access. There must be an explicit `allow` in an identity or resource policy that overrides the default policy. If these policies exist and are validated, then access is allowed. AWS IAM checks every policy that is connected to the request.

As with many other services on their platform, AWS allows third parties to provide solutions on top of the native technology—in this case, AWS IAM. Both Saviynt and CyberArk are two vendors among the third parties that have developed PAM solutions for AWS.

One solution that is worthwhile mentioning here is IAM Access Analyzer, a solution that helps to identify resources in an organization and accounts that are shared with an external entity. This tool is able to provide recommendations for modifications to IAM policies to ensure that access is following least privilege. It will present findings, showing the resources that have access permissions, how the resources are shared through, for instance, S3 bucket policies or access control lists, and the level of access.

GCP offers cloud identities and, like other clouds, a model for role-based access. One particular feature that needs mentioning here is Recommender. This feature provides usage recommendations to optimize GCP environments, but it also comprises tools to manage IAM in GCP. IAM Recommender automatically detects identities that may lead to security risks and can even remove unwanted access to resources. It uses smart access control.

From the Google Cloud console, the IAM page lists all accounts and the permissions that these accounts have. IAM Recommender displays the number of unused permissions over the past 90 days and makes recommendations, such as replacing a role with an account with a predefined role in Cloud Identity, or creating a custom role with the appropriate rights. By doing so, we enforce least privilege in GCP.

We have studied Azure, AWS, and GCP so far. It's fair to say that the basic principles in these clouds to set up IAM are quite similar. Let's go over the steps to implement IAM, this time using OCI. We can use the following services:

- **Identity**: creates and manages users, groups, and policies.
- **Policy**: defines specific permissions for users, groups, and resources.

- **Compartment**: organizes resources into compartments and then uses policies to control access to those compartments. This is comparable to setting up Resource Groups in Azure and AWS and projects in GCP.
- **Users**: creates and manages users and their associated credentials.
- **Groups**: creates and manages groups of users, making it easier to grant or revoke access for multiple users at once.
- **Roles**: assigns permissions to users and groups for specific tasks or job functions.
- **Multi-Factor Authentication (MFA)**: provides an additional layer of security by requiring users to provide a second form of authentication, such as a fingerprint or a one-time passcode, in addition to their password.
- **Audit and Governance**: monitors and audits access to resources, so we can detect and respond to any unauthorized access.
- **Session-based Authentication**: creates short-lived access credentials for users that need privileged access to OCI resources that last for a predefined duration.

This concludes the comparison of IAM services between various clouds, but what if we really are multi-cloud and want to federate IAM across the clouds? That's the topic of the next section.

Enabling account federation in multi-cloud

Businesses are shifting more and more from software to services by adopting SaaS solutions, for instance. Typically, a user would have to log in to separate SaaS solutions, since these are provisioned from a service provider. The risk is that users create new passwords to log in to SaaS solutions. It's easy to lose control of who has access to what. This can be solved through SSO, but the directories of SaaS solutions or web applications need to be federated in that case.

In the field of account federation, Okta has become an increasingly popular IAM solution in recent years. To avoid confusion, it's not an alternative to AD. AD is typically the primary, central directory; Okta is a solution that utilizes AD and takes care of the federation to web applications using SSO. That's what Okta does—it enables IAM with SSO on top of AD, delivering IDaaS.

IDaaS is a cloud-based service that provides organizations with a way to manage and authenticate users' identities across multiple cloud environments. As more and more businesses adopt a multi-cloud strategy, IDaaS has become an essential tool to ensure secure access to cloud-based resources.

One of the main benefits of IDaaS is its ability to centralize identity management. Instead of managing identities across multiple cloud environments individually, IDaaS allows organizations to manage all identities in one central location. This makes it easier to control access to resources, enforce security policies, and track user activity. Additionally, IDaaS can be integrated with existing on-premises identity management systems, allowing for the seamless management of both on-premises and cloud-based resources. The basic principle of IDaaS is shown in *Figure 15.5*:

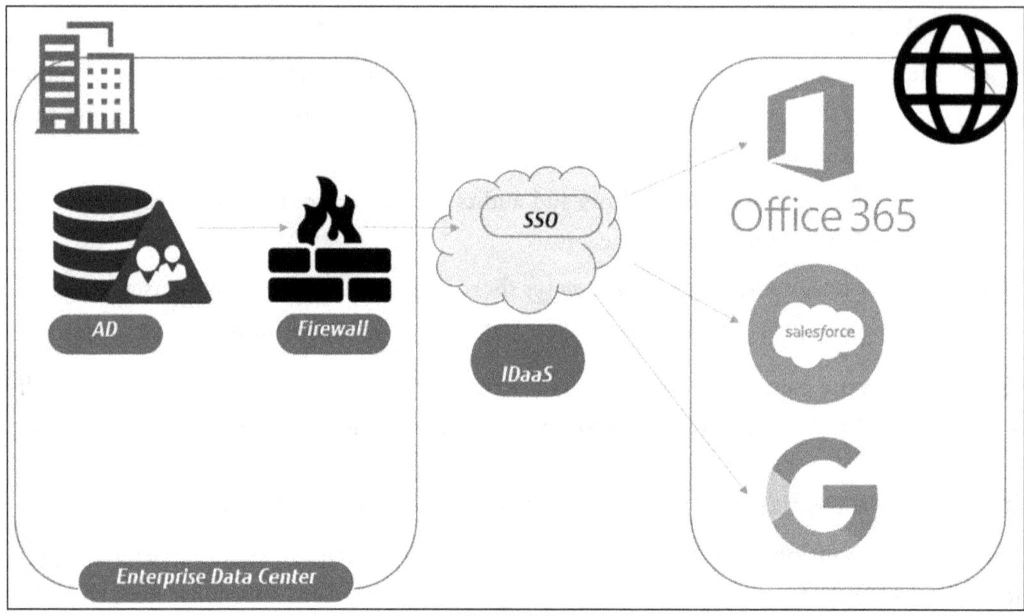

Figure 15.5: High-level concept of IDaaS

The other benefits of IDaaS are the capabilities to provide MFA and SSO, but the most important feature concerns the centralization of the management and monitoring of user access to resources. IDaaS provides granular access controls, allowing organizations to specify exactly who can access which resources and under what conditions. Additionally, IDaaS can provide real-time monitoring of user activity, allowing organizations to quickly identify and respond to security breaches.

To be clear, doing so will move all identities fully to the cloud. It's really a decision for the **Chief Information Security Officer** (**CISO**) or **Chief Information Officer** (**CIO**) to adopt this shift.

Enterprises will need federation using cloud solutions, though. Companies tend to have complex IT environments, comprising on-premises systems, IaaS, PaaS, and SaaS. To be able to connect everything, as it were, all in one domain, federation is required. An IDaaS solution can be a valuable option to connect an existing AD to all these different solutions and enable secure access management to them.

Summary

Security starts with IAM—making sure that we have control over who's accessing our environments and what they are allowed to do in systems. In this chapter, we have learned what identities are and that we need a central identity store. From this identity store, we have to federate between the different cloud solutions that an enterprise has. We have learned how we can set up federation and how IDaaS can be a good solution for this.

We've studied concepts of authorization and authentication in the major cloud platforms. An important concept is least privilege. After this chapter, you should be able to make a distinction between standard accounts and privileged accounts. Lastly, we learned what benefits PAM has in securing access to our clouds.

The reason to have our cloud environments maximally secured is to protect our data. We have studied identities, access management, and security policies to protect our infrastructure. In the next chapter, we will learn how we should define security policies to keep our data safe.

Questions

1. In IAM, we have three layers that we must consider to identify identities. Can you name them?
2. Both AWS and GCP use a specific protocol for authentication. What's that protocol?
3. If a standard account isn't sufficient, we can "promote" users temporarily with another account that holds more rights. How do we name these accounts?

Further reading

- *Mastering Active Directory*, by Dishan Francis, Packt Publishing

16

Defining Security Policies for Data

Data is an important asset of any company. Enterprises store their data more and more in multi-cloud setups. How do they secure data? All cloud platforms have technologies to encrypt data but differ in how they apply encryption and store and handle keys. But data will move from one cloud to another or to user devices, so it needs to be secured in transit, next to data at rest. This is done with encryption, using encryption keys. These keys need to be secured as well, preventing non-authorized users from accessing the keys and encrypted data.

Before we discuss data protection itself, we will briefly talk about data models and how we can classify data. We will explore the different storage solutions the major clouds offer. Next, we will learn how data can be protected by defining policies for **data loss prevention** (**DLP**), labeling information to control access, and using encryption.

In this chapter, we're going to cover the following main topics:

- Storing data in multi-cloud concepts
- Exploring storage solutions
- Understanding data encryption
- Using encryption in Azure, AWS, GCP, OCI, and Alibaba Cloud
- Securing access, encryption, and storage keys
- Securing raw data for big data modeling

Storing data in multi-cloud concepts

If you ask a **chief information officer** (**CIO**) what the most important asset of the business is, the answer will very likely be *data*. The data architecture is therefore a critical part of the entire business and IT architecture. It's probably also the hardest part of business and IT architecture. In this section, we will briefly discuss the generic principles of data architecture and how this drives data security in the cloud.

Data architecture consists of three layers—or **data architecture processes**—in enterprise architecture:

- **Conceptual**: A conceptual model describes the relationship between business entities. Both products and customers can be entities. A conceptual model connects these two entities: there's a relationship between a product and the customer. That relationship can be a sale: the business selling a product to a customer. Conceptual data models describe the dependencies between business processes and the entities that are related to these processes.
- **Logical**: The logical model holds more detail than the conceptual model. An enterprise will likely have more than one customer and more than one product. The conceptual model only tells us that there is a relation between the entity customer and the entity product. The next step would be to define the relationship between a specific customer and a product. Customers can be segregated by adding, for instance, a customer number. The conceptual model only holds the structure; the logical model adds information about the customer entity, such that customer X has a specific relationship with product Y within the entity product.
- **Physical**: Neither conceptual nor logical models say anything about the real implementation of a data model in a database. The physical layer holds the blueprint for a specific database, including the architecture for location, data storage, or database technology.

The following diagram shows the relationship between conceptual data modeling and the actual data—data requirements are set on a business level, leading to technical requirements for the storage of the data and, eventually, the data entry itself:

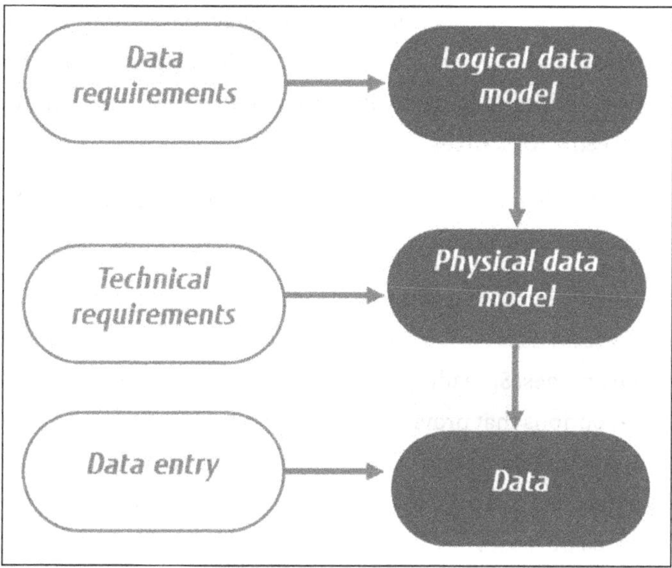

Figure 16.1: Concept of data modeling

Data modeling is about structuring data. It doesn't say anything about the data types or data security, for that matter. Those are the next steps.

Each data type needs to be supported by a data model. Each data type drives the choice of the technology used to store and access data. Common data types are integers (numeric), strings, characters, and Booleans. The latter might be better known as *true/false statements* since a Boolean can only have two values. Alongside these, there are abstract types, such as stacks (a data structure where the last entered data is put on top of the stack), lists (countable sequences), and hashes (associative mappings of values).

Data types are not related to the content of the data itself. A numeric type doesn't say whether data is confidential or public. So, after the model and the definition of data types, there's also data classification. The most common labels for classification are public, confidential, sensitive, and personal data. For example, personal data needs to be highly secured. There are national and international rules and laws forcing organizations to protect personal data at the highest possible level, meaning that no one should be able to access this data without reasons justified by legal authorities. This data will be stored in strings, arrays, and records, and will likely have a lot of connections to other data sources.

An architect will have to think about security on different layers to protect the data, including the data itself, the database where the data is stored, and the infrastructure that hosts the database. If the data model is well-architected, there will be a good overview of what the dependencies and relationships are between data sources.

But what about storing data in multi-cloud concepts, using different providers? This approach can provide a number of benefits, including increased availability, disaster recovery, and cost savings.

Let's start with availability. By storing data across multiple cloud providers, enterprises will have the possibility to still access data even if one provider experiences an outage or other disruption. But there are more advantages. Spreading data across cloud providers can increase flexibility, by making use of the best options that providers have for specific use cases. An example might be to have certain applications and data in one environment, using specific PaaS database services. At the same time, the enterprise can have a data lake in another cloud, accessed by data scientists and analysts for research purposes.

Some companies have a multi-cloud strategy for data for reasons of business continuity and disaster recovery. By storing data in multiple locations, enterprises can ensure that their data is protected against a wide range of potential threats, including natural disasters, power outages, and cyber-attacks such as ransomware holding data hostage. This can help to minimize the risk of data loss and ensure that critical business operations can continue even in the event of an emergency.

Lastly, storing data in various clouds and, by doing so, making use of the best propositions of the provider, can be a way to lower costs. This can be achieved by using different pricing models, as we have seen in *Chapter 12, Cost Modeling in the Cloud*. However, this should never be the first goal. Protecting the data must be priority number one.

Since we need storage to hold data, we will discuss the different types of storage in the next section.

Exploring storage technologies

Aside from data modeling and data types, it's also important to consider the storage technologies themselves, including data lakes, queues, and other types of data services. All cloud platforms offer services for the following:

- **Object storage**: Object storage is the most used storage type in the cloud; we can use it for applications, content distribution, archiving, and big data. In Azure, we can use Azure Blob Storage; in AWS, **Simple Storage Services (S3)**; and GCP simply calls it Cloud Storage. The object storage in OCI is Object Storage. Good to know: OCI also supports S3 compatibility, which allows existing S3 applications to work seamlessly with OCI Object Storage. The object storage service in Alibaba Cloud is simply called **Object Storage Service (OSS)**.
- **Virtual disks**: Virtual machines will either be composed of a virtual disk of block storage or ephemeral. Since every component in the cloud is virtualized and defined as code, the virtual disk is also a separate component that must be specified and configured. There are a lot of choices, but the key differentiators are the required performance of a disk. For I/O-intensive read/write operations to disks, it is advisable to use **solid-state drives (SSDs)**.
- **Shared files**: To organize data in logical folders that can be shared for usage, filesystems are created. The cloud platforms offer separate services for filesystems—in Azure, it's simply called **Files**; in AWS, **Elastic File System (EFS)**; and in GCP, the service is called **Filestore**. GCP does suggest using persistent disks as the most common option. In Alibaba Cloud, the suggested service is called EFS too. OCI provides OCI File Storage to create and manage file systems and mount them as NFS or SMB shares. These protocols are also supported by Azure Files. These shares are then accessible from on-premises or other cloud environments. Alibaba Cloud also provides a file storage service called Apsara File Storage using **NAS (network attached storage)** technology.
- **Archiving**: The final storage tier is archiving. For archiving, high-performance SSDs are not required. To lower storage costs, the platforms offer specific solutions to store data that is not frequently accessed but that needs to be stored for a longer period of time, referred to as a retention period. Be aware that the storage costs might be low but the cost of retrieving data from archive vaults will typically be higher than in other storage solutions. In Azure, there's a storage archive access tier, whereas AWS offers S3 Glacier and Glacier Deep Archive. In GCP, there are Nearline, Coldline, and Archive—basically, different tiers of archive storage. OCI offers Archiving Storage. The archiving service that is recommended in Alibaba Cloud is Cold Storage.

The following diagram shows the relationship between the data owner, the actual storage of data in different solutions, and the data user:

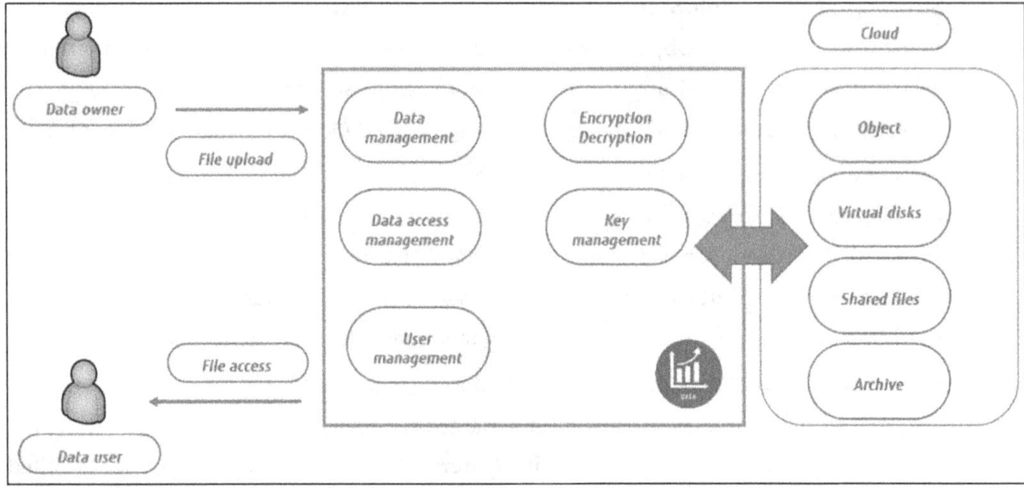

Figure 16.2: Conceptualized data model showing the relationship between the data owner, data usage, and the data user

All mentioned solutions are ways to store data in cloud environments. In the next section, the principles of data protection are discussed.

Understanding data protection in the cloud

In a more traditional data center setup, an enterprise would probably have physical machines hosting the databases of a business. Over the last two decades, we've seen tremendous growth in data, up to the point where it has almost become impossible to store this data in on-premises environments—something that is often referred to as **big data**. That's where public clouds entered the market.

With storing data in external cloud environments, businesses were confronted with new challenges to protect this data. First of all, the data was no longer in on-premises systems but in systems that were all handled by third-party companies. This means that data security has become a shared responsibility—the cloud provider needs to offer tools and technologies to be able to protect the data on their systems, but the companies themselves still need to define what data they need to protect and to what extent. It's still largely the responsibility of the enterprise to adhere to compliance standards, laws, and other regulations.

There's more to consider about data than its current or live state. An enterprise should be equally concerned about historical data that is archived. Too often, data protection is limited to live data in systems, but should also focus on archived data. Security policies for data must include live and historical data. In the architecture, there must be a mapping of data classification and there must be policies in place for DLP.

Data classification enables companies to apply labels to data. DLP prevents sensitive data from being transferred outside an organization. For that, it uses business rules and data classification. DLP software prevents classified data from being accessed and transferred outside the organization. To set these rules, data is usually grouped, based on the classification. Next, definitions of how data may be accessed and by whom are established. This is particularly important for data that can be accessed through APIs, for instance, by applications that connect to business environments. Business data in a **customer relationship management** (**CRM**) system might be accessed by an application that is also used for the sales staff of a company, but the company wouldn't want the data to be accessed by Twitter. A company needs to prevent business data from being leaked to other platforms and users than those authorized.

To establish a policy for data protection, companies need to execute a **data protection impact analysis** (**DPIA**). In a DPIA, an enterprise assesses what data it has, what the purpose of that data is, and what the risk is when the data is breached. The outcome of the DPIA will determine how data is handled, who or what should be able to access it, and how it must be protected. This can be translated into DLP policies. The following table shows an example of a very simple DLP matrix. It shows that business data may be accessed by a business application, but not from an email client. Communication with social media—in this example, Twitter—is blocked in all cases. In a full matrix, this needs to be detailed:

Connection	Data Source		
	Business Data	Email	Twitter
Business Application	Allowed	Denied	Blocked
Email	Denied	Allowed	Blocked
Twitter	Blocked	Blocked	N/A

Table 16.1: Example of data mapping

Labeling and DLP are about policies: they define what must be protected and to what extent. The next consideration is the technologies to protect the data—the how.

Understanding data encryption

One of the first, if not *the* first, encryption devices to be created was the Enigma machine. It was invented in the 1920s and was mostly known for its usage in World War II to encrypt messages. The British scientist Alan Turing and his team managed to crack the encryption code after 6 months of hard work.

The encryption that Enigma used, in those days, was very advanced. The principle is still the same: we translate data into something that can't be read without knowing how the data was translated. To be able to read the data, we need a way to decipher or decrypt the data. There are two ways to encrypt data—asymmetric, or public key, and symmetric. In the next section, we will briefly explain these encryption technologies, before diving into the services that the leading cloud providers offer in terms of securing data.

First, let's get into two forms of encryption: at rest and in transit. Data that is stored on a (virtual) device or in a database can be protected using encryption at rest. If the data is accessed by unauthorized users, these users won't be able to read the data unless they have the appropriate decryption key. A typical method of encryption is **full disk encryption** (**FDE**), commonly implemented on end user devices using, for instance, BitLocker. This works the same in the cloud. For example, FDE can be implemented on VMs in Azure using Azure Disk Encryption, and in AWS, with the encryption feature of EBS.

Another method is the use of database encryption, which encrypts the data stored within a database. This can be done through the use of built-in encryption features of the database management system or through the use of third-party encryption software.

Encryption in transit, on the other hand, refers to the process of protecting data as it is transmitted over a network. This type of encryption is used to protect data from being intercepted and read by an attacker while it is being transmitted from one device to another. There are multiple ways of implementing encryption in transit. **Transport Layer Security** (**TLS**) is used as a protocol to encrypt data between a web server and a browser. Another method is using VPNs.

There's one really important note that must be made: all efforts to encrypt data are useless if encryption keys are not secured too. Encryption uses an encryption algorithm and an encryption key—symmetric or asymmetric. With **symmetric**, the same key is used for both encrypting and decrypting. The problem with that is that the entity that encrypts a file needs to send the key to the recipient of the file so it can be decrypted. The key needs to be transferred. Since enterprises use a lot of (different types of) data, there will be a massive number of keys. The distribution of keys needs to be managed well and be absolutely secure.

An alternative is **asymmetric** encryption, which uses a private and a public key. In this case, the company only needs to protect the private key, since the public key is commonly available.

Both encryption methods are used. A lot of financial and governmental institutions use **AES**, the **Advanced Encryption Standard.** AES works with data blocks. The encryption of these blocks is performed in rounds. In each round, a unique code is included in the key. The length of the key eventually determines how strong the encryption is. That length can be 128-, 192-, or 256-bit. Recent studies have proven that AES-256 is even quantum-ready. The diagram shows the principle of AES encryption:

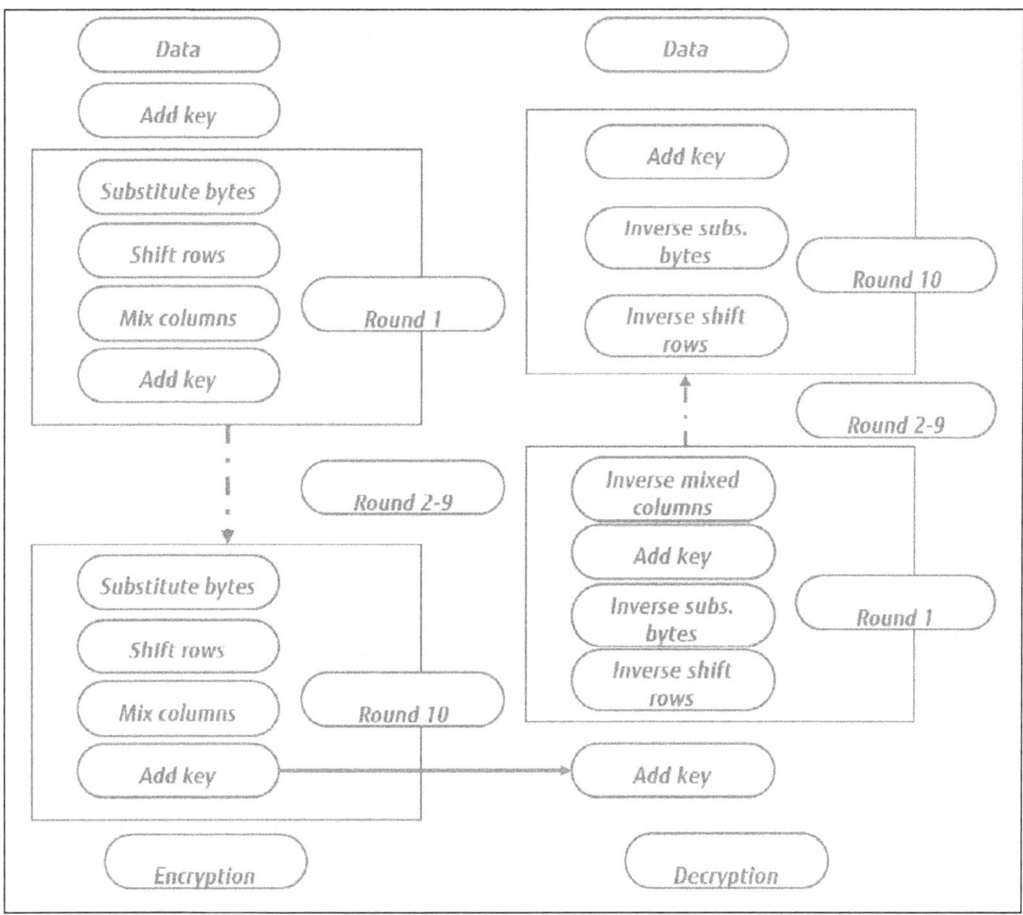

Figure 16.3: Simple representation of AES encryption principle

RSA, named after its inventors, **Rivest, Shamir, and Adleman**, is the most popular asymmetric encryption method. With RSA, the data is treated as one big number that is encrypted with a specific mathematical sequence called **integer factorization**. In RSA, the encryption key is public; decryption is done with a highly secured private key. The principle of RSA encryption is shown in the following diagram:

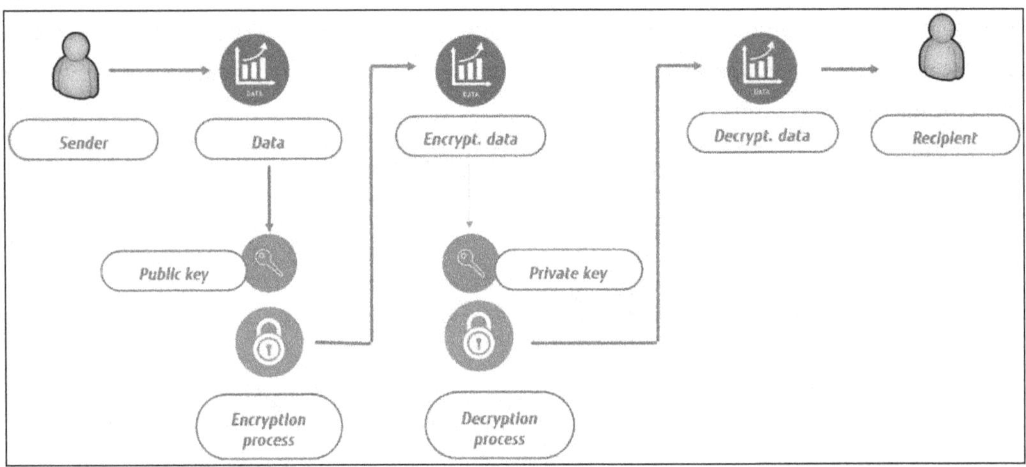

Figure 16.4: Concept of RSA encryption

Both in AES and RSA, the length of the keys is crucial. Even today, the most common way to execute an attack on systems and retrieve data is by brute force, where the attacker will fire random keys on a system until one of the keys matches. With high-performance computers or computer networks executing the attack, the chance of success is still quite high. So, companies have to think about protecting the data itself, but also protecting the keys used to encrypt data.

In the next section, we will explore the different solutions in public clouds for storage and secure keys, and finally, draw our plan and create the data security principles.

Securing access, encryption, and storage keys

Cloud platforms provide customers with technology and tools to protect their assets, including the most important one—data. At the time of writing, there's a lot of debate about who's responsible for protecting data, but generally, the company that is the legal owner of the data has to make sure that it's compliant with (international) laws and standards. In the UK, companies have to adhere to the Data Protection Act, and in the European Union, all companies have to be compliant with the **General Data Protection Regulation** (**GDPR**).

Both the Data Protection Act and GDPR deal with privacy. International standards ISO/IEC 27001:2022 and ISO/IEC 27002:2022 are security frameworks that cover data protection. These standards determine that all data must have an owner so that it's clear who's responsible for protecting the data. In short, the company that stores data on a cloud platform still owns that data and is therefore responsible for data protection.

To secure data on cloud platforms, companies have to focus on two aspects:

- Encryption
- Access, using authentication and authorization

These are just the security concerns. Enterprises also need to be able to ensure reliability. They need to be sure that, for instance, keys are kept in a separate place from the data itself and that even if a key vault is not accessible for technical reasons, there's still a way to access the data in a secure way. An engineer can't simply drive to the data center with a disk or a USB device to retrieve the data. How do Azure, AWS, and GCP take care of this? We will explore this in the next section.

Using encryption and keys in Azure

In Azure, the user writes data to Azure Blob Storage. The storage is protected with a storage key that is automatically generated. Data in a new storage account is encrypted with Microsoft-managed keys by default. You can continue to rely on Microsoft-managed keys for the encryption of your data, or you can manage encryption with your own keys.

The storage keys are kept in a key vault, outside the subnet where the storage is itself. But the key vault does more than just store the keys. It also regenerates keys periodically by rotation, providing **shared access signature (SAS)** tokens to access the storage account. The concept is shown in the following diagram:

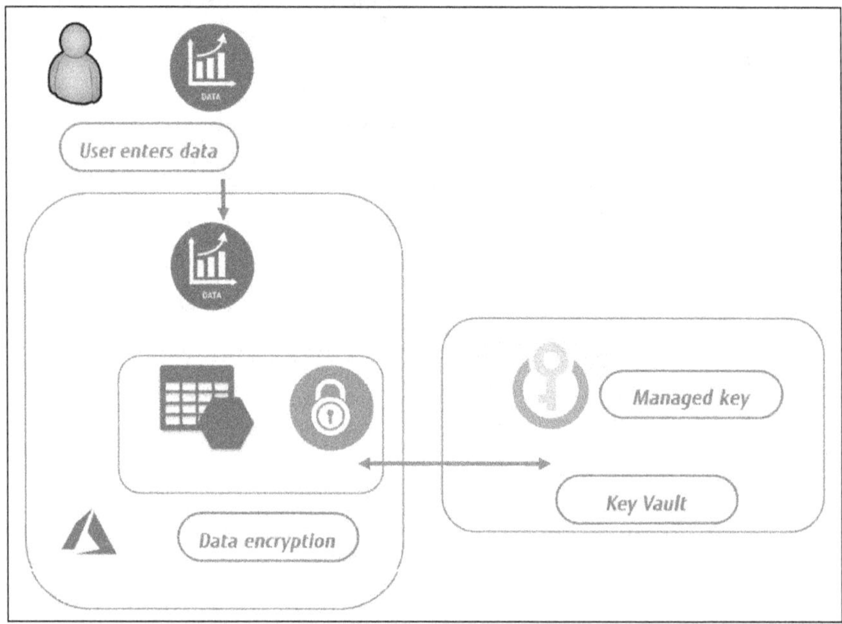

Figure 16.5: Concept of Azure Key Vault

The key vault is highly recommended by Microsoft Azure for managing encryption keys. Encryption is a complex domain in Azure since Microsoft offers a wide variety of encryption services in Azure. Disks in Azure can be encrypted using BitLocker or DM-Crypt for Linux systems. With Azure, **Storage Service Encryption (SSE)** data is automatically encrypted before it's stored in a blob. SSE uses AES-256. For Azure SQL databases, Azure offers encryption for data at rest with **Transparent Data Encryption** (TDE), which also uses AES-256 and **Triple Data Encryption Standard** (3DES or TDES).

Using encryption and keys in AWS

Like Azure, AWS has a key vault solution, called **Key Management Service** (KMS). The principles are also very similar, mainly using server-side encryption. Server-side means that the cloud provider is requested to encrypt the data before it's stored on a solution within that cloud platform. Data is decrypted when a user retrieves the data. The other option is client-side, where the customer takes care of the encryption process before data is stored.

The storage solution in AWS is S3. If a customer uses server-side encryption in S3, AWS provides S3-managed keys (SSE-S3). These are the unique **data encryption keys (DEKs)** that are encrypted themselves with a master key, the **key encryption key (KEK)**. The master key is constantly regenerated. For encryption, AWS uses AES-256.

AWS offers some additional services with **customer master keys (CMKs)**. These keys are also managed in KMS, providing an audit trail to see who has used the key and when. Lastly, there's the option to use **customer-provided keys (SSE-C)**, where the customer manages the key themselves. The concept of KMS using CMKs in AWS is shown in the following diagram:

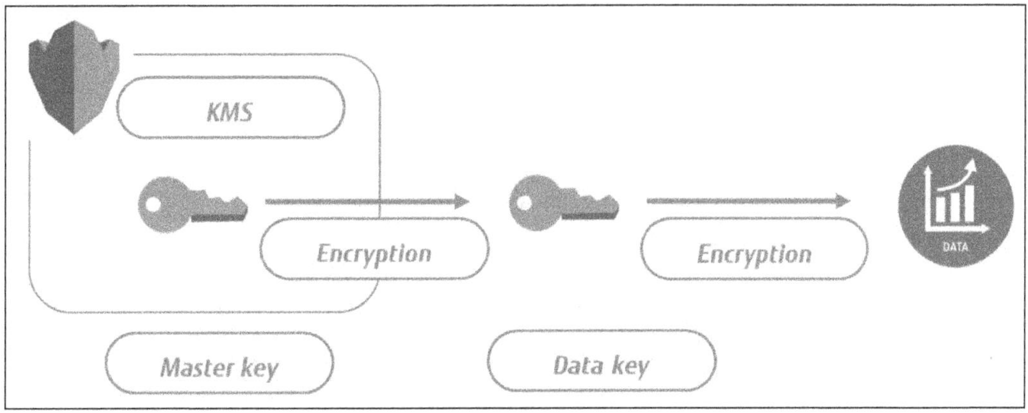

Figure 16.6: Concept of storing CMKs in AWS KMS

Both Azure and AWS have automated a lot in terms of encryption. They use different names for the key services, but the main principles are quite similar. That counts for GCP too, which is discussed in the next section.

Using encryption and keys in GCP

In GCP, all data that is stored in Cloud Storage is encrypted by default. Just like Azure and AWS, GCP offers options to manage keys. These can be supplied and managed by Google or by the customer. Keys are stored in **Cloud Key Management Service**. If the customer chooses to supply and/or manage keys themselves, these will act as an added layer on top of the standard encryption that GCP provides. That is also valid in the case of client-side encryption – the data is sent to GCP in an encrypted format, but GCP will still execute its own encryption process, as with server-side encryption. GCP Cloud Storage encrypts data with AES-256.

The encryption process itself is similar to AWS and Azure and uses DEKs and KEKs. When a customer uploads data to GCP, the data is divided into chunks. Each of these chunks is encrypted with a DEK. These DEKs are sent to the KMS where a master key is generated. The concept is shown in the following diagram:

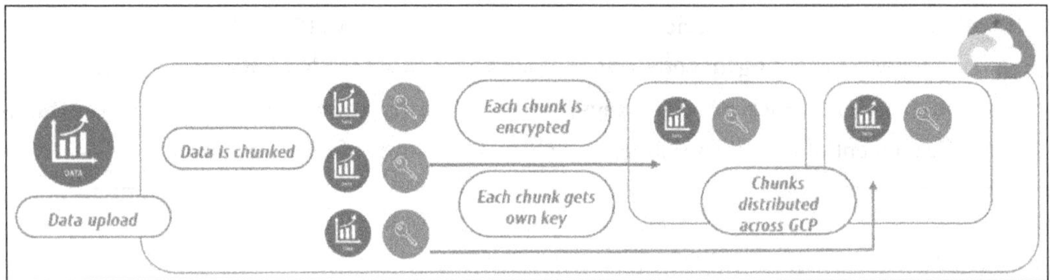

Figure 16.7: Concept of data encryption in GCP

In this book, we also look at OCI and Alibaba Cloud, hence we will touch on encrypting data in these clouds next.

Implementing encryption in OCI and Alibaba Cloud

Like in the named major public clouds, OCI offers services to encrypt data at rest using an encryption key that is stored in OCI's Key Management Service. Companies can use their own keys too, but these must be compliant with the **Key Management Interoperability Protocol** (**KMIP**). The process for encryption is like the other clouds. First, a master encryption key is created that is used to encrypt and decrypt all data encryption keys. Remember, we talked about these KEKs and DEKs before. Keys can be created from the OCI console, the command-line interface, or the SDK.

Data can be encrypted and stored in block storage volumes, object storage, or in databases. Obviously, OCI also recommends the rotation of keys as best practice.

Keys that are generated by customers themselves are kept in the OCI Vault and stored across a resilient cluster.

Storing keys in vaults is done in a **Hardware Security Module** (**HSM**), a physical device that holds and manages secrets such as encryption keys. Most HSMs are compliant with **FIPS** (**Federal Information Processing Standard**) Publication 140-2 Level 3 – a standard that is issued by the American government to safeguard high standards for encrypting data. The standard includes tamper-evident mechanisms, hardening, and automatic deletion of plaintext keys.

The steps are the same for Alibaba Cloud. From the console or CLI, keys are created and then encrypted data is stored in one of the storage services that Alibaba Cloud provides: OSS, Apsara File Storage, or databases.

So far, we have been looking at data itself, the storage of data, the encryption of that data, and securing access to data. One of the tasks of an architect is to translate this into principles. This is a task that needs to be performed together with the CIO or the **chief information security officer (CISO)**. The minimal principles to be set are as follows:

- Encrypt all data in transit (end-to-end)
- Encrypt all business-critical or sensitive data at rest
- Apply DLP and have a matrix that shows clearly what critical and sensitive data is and to what extent it needs to be protected

In the *Further reading* section, some good articles are listed on encryption in the cloud and best practices for securing data.

Finally, develop use cases and test the data protection scenarios. After creating the data model, defining the DLP matrix, and applying the data protection controls, an organization has to test whether a user can create and upload data and what other authorized users can do with that data—read, modify, or delete it. That does not only apply to users but also to data usage in other systems and applications. Data integration tests are therefore also a must.

Data policies and encryption are important, but one thing should not be neglected: encryption does not protect companies from misplaced **identity and access management** (**IAM**) credentials. Thinking that data is fully protected because it's encrypted and stored safely gives a false sense of security. Security really starts with authentication and authorization.

Securing raw data for big data modeling

One of the big advantages of the public cloud is the huge capacity that these platforms offer. Together with the increasing popularity of public clouds, the industry saw another major development in the possibilities to gather and analyze vast amounts of data, without the need to build infrastructure in on-premises data centers to host the data. With public clouds, companies can have enormous data lakes at their disposal. Data analysts program their analytical models to these data lakes. This is what is referred to as **big data**. Big data modeling is about four Vs:

- **Volume**: The quantity of data
- **Variety**: The different types of data
- **Veracity**: The quality of data
- **Velocity**: The speed of processing data

Data analysts often add a fifth V to these four, and that's **value.** The original model does not have it, but many analysts feel that value is likely the most important driver. Big data gets value when data is analyzed and processed in such a way that it actually means something. The four-Vs model is shown in the following diagram:

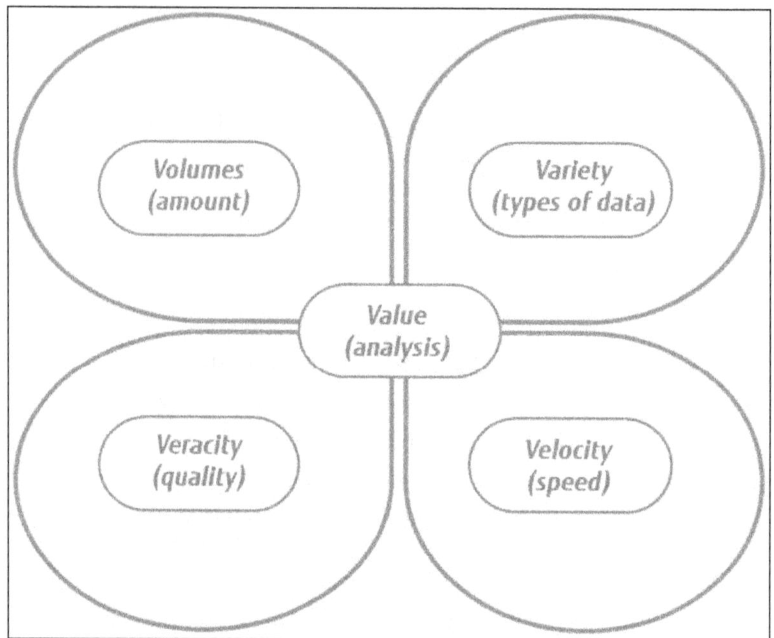

Figure 16.8: The four Vs of big data

Processing and enriching the data is something that is done in the data modeling stage. Cloud providers offer a variety of solutions for data mining and data analytics: Azure Synapse Analytics, Redshift in AWS, BigQuery in GCP, Oracle's BigData, and AnalyticDB in Alibaba Cloud. These solutions require a different view of data security.

As with all data, the encryption of data at rest is required and, in almost every case, is enabled by default on any big data platform or data warehouse solution that is scalable up to petabytes. Examples are Azure Data Lake, AWS Redshift, and Google's BigQuery. These solutions are designed to hold any kind of unstructured data in one single repository.

To use Azure Data Lake as an example, as soon as the user sets up an account in Data Lake, Azure encryption is turned on and the keys are managed by Azure, although there's an option for companies to manage keys themselves. In Data Lake, the user will have three different keys: the master encryption key, the data encryption key, and the block encryption key. The latter is necessary since data in Data Lake is divided into blocks.

AWS Redshift works in clusters to store data. These clusters can be encrypted so that all data created by users in tables is encrypted. This data can be extracted to SQL database clients. In that case, data in transit is encrypted using **Secure Sockets Layer** (**SSL**). Finally, the Redshift cluster will sit in a **virtual private cloud** (**VPC**) in AWS; access to the environment is controlled at the VPC level.

Google's BigQuery is a fully managed service, yet users have a ton of choices for how to treat data in BigQuery. The service comprises over 100 predefined detectors to scan and identify data. Next, GCP offers a variety of tools to execute DLP policies, such as data masking and the pseudonymization of data. Scanning data in BigQuery is easy through the GCP cloud console. This is also the place where the user can enable the DLP API. As with all data in GCP, it will be encrypted upon entry by default. BigQuery doesn't check whether data is already encrypted; it runs the encryption process at all times.

Data entered in OCI is encrypted by default. This is true for all data services in OCI: block, object, or file services storage, or in one of Oracle's platform solutions such as Oracle Analytics Cloud service. In all cases, data encryption at rest is turned on by default. This includes encryption of the database backups. OCI automatically manages the keys, although customers can choose to use their own keys.

For encryption at rest, OCI provides the option to encrypt data stored in Object Storage using the encryption feature of this service. This feature allows for encrypting data using a customer-managed encryption key or a key managed by OCI. For encryption in transit, OCI provides the option to encrypt data transmitted over the network using the Oracle Cloud Infrastructure Networking Transport Encryption feature. This feature uses the TLS protocol. Note that encryption options for big data services may be different from the options available for other OCI services.

To store large amounts of unstructured, raw data in Alibaba Cloud, the use of OSS is recommended. Data analytics can be performed using DataWorks, which allows for data cleansing and then loading the data into data warehouses for further analysis. Again, data entered in OSS is not encrypted by default. For encryption at rest, the **Server-Side Encryption (SSE)** feature in OSS must be used. SSE encrypts data with AES-256. As in other clouds, customers can use their own keys with SSE-C. Alibaba Cloud utilizes the TLS protocol for data in transit.

Summary

This chapter was about securing and protecting data in cloud environments. We have learned that when moving data from on-premises systems to the cloud, companies have to set specific controls to protect their data. The owner of the data remains responsible for protecting the data; that doesn't shift to the cloud provider.

We have learned that companies need to think first about data protection policies. What data needs to be protected? Which laws and international frameworks are applicable to be compliant? A best practice is to start thinking about the data model and then draw a matrix, showing what the policy should be for critical and sensitive data. We've also studied the principles of DLP using data classification and labeling.

This chapter also explored the different options a company has to store data in cloud environments and how we can protect data from a technological point of view. After finishing this chapter, you should have a good understanding of how encryption works and how Azure, AWS, GCP, OCI, and Alibaba Cloud treat data and the encryption of data. Lastly, we've looked at the big data solutions on the cloud platforms and how raw data is protected.

The next chapter is the final one about security operations in multi-cloud. Cloud providers offer native security monitoring solutions, but how can enterprises monitor security in multi-cloud? The next chapter will discuss **integrated security monitoring** in the cloud.

Questions

1. To define the risk of data loss, businesses are advised to conduct an assessment. Please name this assessment methodology.
2. In this chapter, we've studied encryption. Please name two encryption technologies that are commonly used in cloud environments.
3. What's the service in AWS to manage encryption keys?
4. In Azure, companies keep keys in Azure Key Vault. True or false: a key vault is hosted in the same subnet as the storage itself.

Further reading

You can refer to the following links for more information on the topics covered in this chapter:

- Information about the management of storage keys in Azure: `https://learn.microsoft.com/en-gb/azure/security/fundamentals/data-encryption-best-practices`
- Encryption overview in Azure: `https://docs.microsoft.com/en-us/azure/security/fundamentals/encryption-overview`
- Data protection in AWS: `https://docs.aws.amazon.com/AmazonS3/latest/dev/UsingEncryption.html`
- Encryption options in GCP: `https://cloud.google.com/storage/docs/encryption`

- Blog by Kenneth Hui on Cloud Architect Musings about encryption in Azure, AWS, and GCP: `https://cloudarchitectmusings.com/2018/03/09/data-encryption-in-the-cloud-part-4-aws-azure-and-google-cloud/`
- Information about encryption in OCI: `https://blogs.oracle.com/cloudsecurity/post/how-oci-helps-you-protect-data-with-default-encryption#:~:text=Data%20at%2Drest%20encryption,is%20turned%20on%20by%20default`
- Information about encryption in Alibaba Cloud: `https://www.alibabacloud.com/blog/data-encryption-at-storage-on-alibaba-cloud_594581#:~:text=Your%20data%20and%20the%20associated,FIPS)%20140%2D2%20standard`

17

Implementing and Integrating Security Monitoring

Enterprises adopt multi-cloud and use cloud services from different cloud providers. These solutions will be securee, but enterprises want an integrated view of the security status on all of their platforms and solutions. This is what solutions such as **Security Information and Event Management** (**SIEM**) and **Security Orchestration, Automation, and Response** (**SOAR**) do.

In this chapter, we will learn why these systems are a necessity in multi-cloud. First, we will discuss the differences between the various systems, and then we will explore the various solutions that are available on the market today. The big question we're going to answer in this chapter is, *how do we make a choice and, more importantly, how do we implement these complicated solutions?*

We're going to cover the following main topics in this chapter:

- Understanding SIEM and SOAR
- Setting up a Security Operations Center
- Setting up the requirements for integrated security
- Implementing a security model
- Exploring multi-cloud monitoring suites

Understanding SIEM and SOAR

All cloud providers offer native services for security monitoring, such as Microsoft Defender for Cloud, AWS Security Hub, and Security Command Center in Google Cloud. However, companies are going multi-cloud using IaaS, PaaS, and SaaS from different providers. Enterprises want an integrated view of their security in all these solutions. If an enterprise is truly multi-cloud, it will need an integrated security solution with SIEM and SOAR.

Next, the enterprise needs a unit that is able to handle and analyze all the data coming from SIEM and SOAR systems and trigger the appropriate actions in case of security events. Most enterprises have a **Security Operations Center** (**SOC**) to take care of this. In the next section, we will explain what the differences are between SIEM and SOAR, why an enterprise needs these systems in multi-cloud, and what the role of the SOC is.

Differentiating SIEM and SOAR

Let's start with SIEM. Imagine that workloads—systems and applications—are deployed in Azure and AWS, and the enterprise also uses a number of SaaS services, such as Microsoft 365 and Salesforce. All these environments are protected with firewalls in both Azure and AWS, along with on-premises data centers. Traffic is routed through virtual network devices, routing tables, and load balancers. The enterprise might also have implemented intrusion detection and prevention to protect systems in the public clouds and on-premises data centers. All these security systems will produce a vast amount of information on the security status of the enterprise environments.

A SIEM system collects, aggregates, and analyzes this information to identify possible threats. Since it collects data from all environments, it's able to correlate the data and recognize patterns that might hint toward attacks. For this, SIEM uses machine learning and analytics software. It recognizes abnormal behavior in systems with anomaly detection. A simple example is, if user A logs in from an office in London at 9.00 AM and again logs in at 9.30 AM from Singapore, a SIEM system would know that this is impossible and raise an event or alert. The architecture of a SIEM system is shown in *Figure 17.1*:

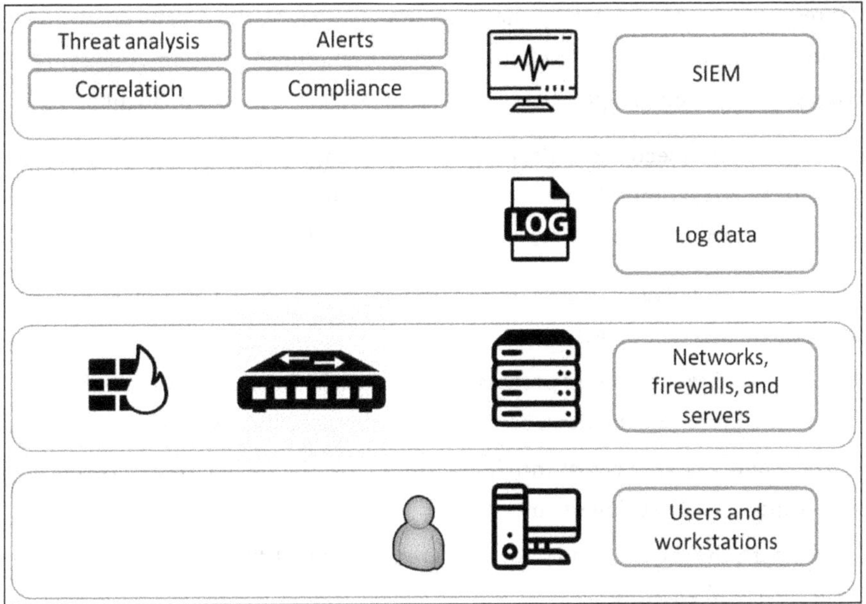

Figure 17.1: The conceptualized architecture of a SIEM system

SOAR goes beyond SIEM. Like a SIEM system, SOAR collects and analyzes data that it gathers from a lot of different sources, such as public cloud environments. But the added value of SOAR is in orchestration and automation. In SOAR systems, companies can define automated responses to events, using playbooks that integrate with security solutions in the platforms. If a SOAR system detects a threat in a system, it can immediately remediate it by taking actions such as closing communication ports, blocking IP addresses, or putting systems into quarantine. It does that fully automated, including logging and raising tickets to service management systems. This gives security professionals time to investigate the actual threat, without needing to worry about remediation first. That task is taken care of preemptively for them by SOAR.

Initiating a Security Operations Center

Since the world is moving to digital, companies are increasingly threatened by digital threats. It also seems that hackers are always one step ahead of the companies themselves in protecting their digital assets. It takes a lot of skills to keep up and counter these attacks. Therefore, enterprises rely more and more on specialized groups where security expertise is bundled—a SOC. Enterprises can have these in-house or outsourced to specialized companies.

The SOC is responsible for monitoring and analyzing the security state of an enterprise on a 24/7 basis. A team of security engineers will use different technology solutions, including SIEM and SOAR, to detect, assess, and respond as quickly as possible to security incidents.

A SOC plays a vital role in securing multi-cloud environments by providing centralized monitoring, incident response, threat intelligence, vulnerability management, compliance and regulation, and security automation. Let's work that out in more detail. The SOC:

- Monitors and analyzes security events from multiple cloud platforms to detect potential security threats and breaches
- Leads the incident response process when a security breach or threat is detected, including investigating the incident, containing the breach, and recovering systems and data
- Gathers and analyzes threat intelligence from a variety of sources to better understand and respond to emerging security threats
- Works with the security and operations teams to identify and remediate vulnerabilities in multi-cloud environments
- Helps organizations ensure compliance with regulations and standards, such as ISO, PCI-DSS, and HIPAA, by monitoring and auditing cloud environments
- Uses security automation and orchestration tools to streamline and automate security processes, such as incident response and vulnerability management

In a multi-cloud environment, a SOC is crucial, since organizations will operate a combination of on-premises, public cloud, and private cloud infrastructure to run their applications and store their data. Then it's essential to have maximum observability across these different platforms. But observability is not enough. An enterprise needs to know what its vulnerabilities are and where it is at risk so that it can develop and implement plans to avoid, deter, or mitigate risks when they materialize.

In the next section, we will explain how an enterprise can set up a SOC. In the last section of this chapter, *Exploring multi-cloud monitoring suites*, we will explore some major SIEM and SOAR solutions that companies can use to protect their systems in multi-cloud environments.

Setting up the requirements for integrated security

Before a company gets into buying licenses for all sorts of security tools, security architects will need to gather requirements. That is done in the following four stages that a security team needs to cover:

1. **Detect**: Most security tools focus on detecting vulnerabilities and actual attacks or attempts to breach systems. Some examples are endpoint protection, such as virus scanners and malware detection, and **Network Traffic Analyzers (NTAs)**. In multi-cloud, architects need to make sure that detection systems can operate on all platforms and preferably send information to one integrated dashboard.
2. **Analyze**: This is the next phase. Detection systems will send a lot of data, including false positives. Ideally, security monitoring does a first analysis of events, checking them against known patterns and behavior of systems and users. This is the first filter. The second phase in the analysis is prioritization, which is done by skilled security staff. They have access to knowledge base repositories of providers and security authorities. They have the information that enables them to give priority to potential threats, based on relevant context. Remember one thing—where there's smoke, there's usually a fire. The question is, how big is the fire?
3. **Respond**: After a threat is detected and prioritized, the security team needs to respond. First of all, they need to make sure that the attack is stopped and exploited vulnerabilities are identified. The next step is remediation— preventing systems from enacting (further) damage or data breaching. The final step in response is recovery—restoring systems and making sure that the data is safe. Be sure that processes for following up security events are crystal clear. Who needs to be informed, who's mandated to take decisions, and what is the escalation path?
4. **Prevent**: SIEM and SOAR systems can do a lot in detecting, analyzing, and responding to security events. However, security starts by preventing vulnerabilities from being exploited in the first place. Security teams need to have continuous visibility on all the platforms that an enterprise uses and must have access to security reports, assessments, and threat detection scans from the providers. It's also essential that recommendations from Azure, AWS, GCP, VMware, or any other provider are followed up. These providers issue security updates on a regular basis and give recommendations to improve the state of security of environments that are deployed on their platforms. These recommendations should be followed.

Market analyst Gartner predicts that by 2024, 80 percent of all SOCs will have invested in tools using artificial intelligence and machine learning. However, Gartner analysts also conclude that these investments will not necessarily bring down the amount of time that security teams have to spend on investigating security events. So, what would be wise investments in terms of security tools and systems?

First, leverage what providers already have. Azure, AWS, GCP, and OCI all have security suites that gather a lot of information on the health and integrity of systems. In almost all cases, it's a matter of ticking the box to enable these security systems, although security engineers will have to set a baseline to which the tools monitor the systems. This was discussed in *Chapter 14, Defining Security Policies*.

Implementing the security model

A lot of companies already have a multi-cloud setup. For example, they use AWS to host websites and have Microsoft 365 from Microsoft, a SaaS solution. In AWS, security teams will work with AWS Inspector and GuardDuty for monitoring security. In Office 365, they might use **Microsoft Defender**, for example. The challenge for security teams is to have an integrated vision of the full IT environment. How do companies get there?

Security models often start with the concepts of zero trust and security by design. What is meant by that? We must realize that with cloud and multi-cloud, the need for security is growing. Organizations collect a lot of data and aggregate it in the cloud, often in vast data lakes that hold petabytes of raw data, making it very attractive for criminals to try to get hold of. That data is constantly exchanged, being used across multiple applications, likely operating on different platforms or at least communicating between different platforms. Multi-cloud environments are more complex and dynamic, making it harder to secure all the various components. As a result, organizations are turning to zero trust and security by design concepts to help ensure that their multi-cloud environments are secure.

Zero trust assumes that any user, device, or network component may be compromised and, therefore, requires multi-factor authentication and continuous verification before access is granted. This approach assumes that the trust level for all access requests is set to zero and that security measures must be in place to protect the organization's assets from malicious actors.

Security by design incorporates security considerations into the design and development of systems, applications, and infrastructure. This approach ensures that security is built into the fabric of the technology and that security is a consideration at every stage of the development process. In multi-cloud architectures, we combine the two concepts, which is the best practice. By doing that, we address security at the foundation of the technology. Enterprises can implement a defence-in-depth approach that includes multi-factor authentication, encryption, and continuous monitoring. Security by design ensures that security is built into the development process and that security considerations are integrated into all aspects of the technology.

Let's give an example: we can use **identity and access management (IAM)** solutions to enforce multi-factor authentication and implement a least privilege access model. This helps to ensure that only authorized users have access to an organization's cloud assets and that their access is monitored and audited. Next, we can encrypt all data by default. All clouds do that already: Azure, AWS, GCP, and OCI encrypt data at entry. We should also use secured networking such as VPNs to ensure that all traffic is encrypted and protected from eavesdropping.

Automation and orchestration tools will help to automate security processes, such as incident response and vulnerability management, and provide a centralized view of security events and alerts from across the cloud environment.

By assuming that all access requests are not trustworthy and incorporating security considerations into the design and development of technology, organizations can help ensure the security of their multi-cloud environments and protect their assets from malicious actors. What steps do we need to take to define and implement a security strategy and associated model?

- **Define a target operating model**: What does the entire environment look like and who's responsible for all or some of it? Companies must have a clear demarcation model on roles and responsibilities in the management of cloud platforms, services, and systems. The target operating model describes the landscape of components and the owners of these components. Security is a component that the security officer is responsible for.
- **Define workflows and escalation procedures**: This defines the workflow when security events occur. What is the procedure in the case of high-priority events, medium-rated events, and low-risk events? When a high-priority event is detected, it should be raised to the security officer. The security officer decides who needs to be informed and what actions must be taken. These are operational tasks. They may report to the **Chief Security Officer** (**CSO**) or the **Chief Information and Security Officer** (**CISO**). The CSO or CISO is responsible for strategic security decisions.
- **Analyze the capabilities of security tools that are already in place**: Evaluate the tools that are in place already. What do these tools cover? How are APIs configured, and can they communicate with overlaying systems? What are the default baselines that these tools use?
- **Gap analysis**: There will always be blind spots. A common example in batch jobs is whether these are monitored as well from a security perspective. What happens when jobs are stopped? Is communication between systems then halted, and is the integrity of systems still safeguarded? In cloud-native environments, companies should also have a good understanding of how containers and serverless solutions are monitored. Not all monitoring tools can handle these native environments yet.

- **Make a strategic plan**: This is what the CSO or CISO must be concerned with. The first question that must be covered in a strategic plan is the maturity goal of the enterprise. The next question is what the major security concerns are for the enterprise—what are the biggest risks and threats? Hint: it is not always about the loss of money. Reputational damage goes far beyond revenue loss when systems are breached. Finally, the company must be able to identify whether existing tools, processes, and expertise are sufficient and what needs to be done to get to the desired maturity goal.

The following diagram shows a maturity model for security:

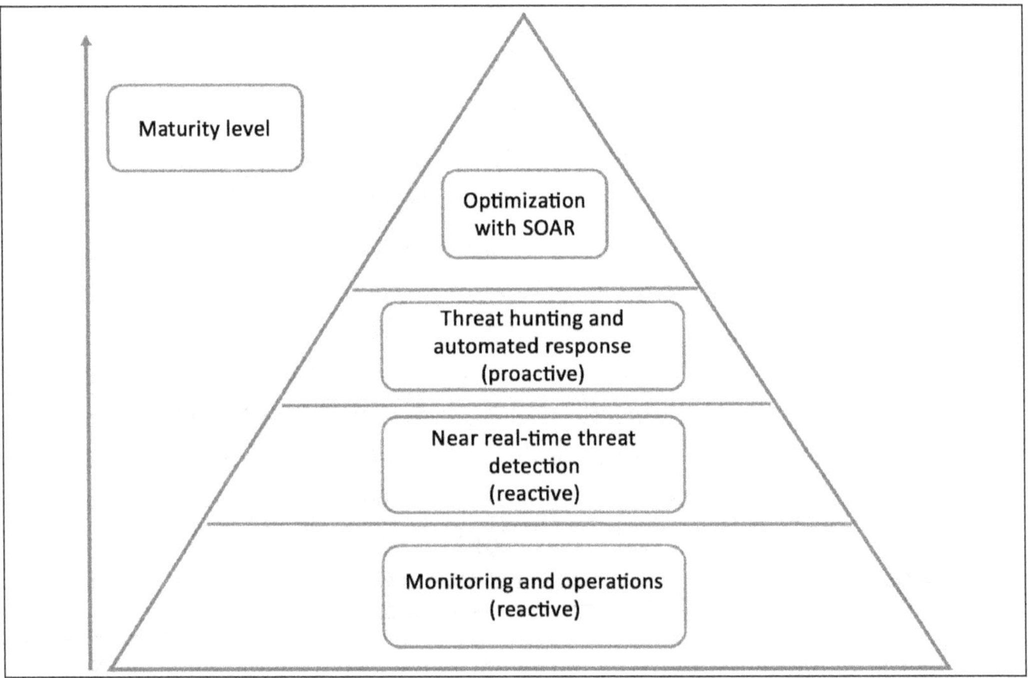

Figure 17.2: Security maturity model, from reactive monitoring to proactive threat hunting

It's strongly advised to set up a security team or SOC. It's not realistic to have one or two security engineers watching over multi-cloud environments. The difficult part is how to get there. The best practice is to plan the setup in three stages:

- **Stage 1: Get visibility for the business**: In this stage, we gather the security policies and align the security processes between the business and IT.

- **Stage 2: Integrate IT security operations with business security**: This is the stage where security operations enable security monitoring and onboard the security baselines—as defined with the business at stage 1—in the monitoring systems. Part of this stage is a risk assessment of the platforms. It's recommended to do an assessment of the security baselines of the cloud providers and analyze whether these baselines concur with the security principles of the enterprise.
- **Stage 3: Optimize**: This is the stage where a truly integrated view is created, using one dashboard that covers the entire security state of the IT landscape.

The stages are shown in the following diagram:

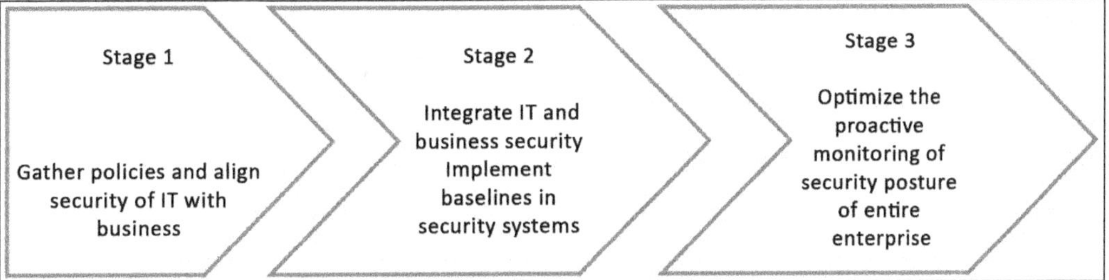

Figure 17.3: Three stages of security onboarding

Integrated security means that a company has a clear model of processes, tools, and expertise. In multi-cloud, this also means that cloud providers are part of these processes, tools, and certainly, expertise. The security architect will have the task of getting this defined, designed, and modeled. SIEM and SOAR tools can help to get an integrated view of the entire security state—or posture—of an enterprise. In the next section, we will discuss popular solutions for this in multi-cloud.

Exploring multi-cloud monitoring suites

Companies have a wide variety of choices when they're looking for a SIEM solution This is a fast-growing market. Each year, market analyst Gartner publishes a list of leading solutions in different IT domains. For a number of years, Splunk, LogRhythm, and Rapid7 have been named as leading products for SIEM by Gartner.

Splunk is a log management and analysis platform that can collect, analyze, and visualize data from a variety of sources, including all major cloud providers.

LogRhythm is a SIEM platform that can collect and analyze log data from cloud environments. LogRhythm provides a set of pre-built connectors and integrations for all clouds, allowing us to collect and centralize log data from these platforms and use LogRhythm's threat detection and incident response capabilities to detect and respond to security threats.

Rapid7 is a security analytics platform that provides threat detection and incident response capabilities for cloud environments. Rapid7's security analytics and incident response capabilities are used to detect and respond to security threats.

To summarize, all of these solutions can work with all major cloud providers using REST APIs. **REST** stands for **REpresentational State Transfer**. A REST API is a programmable interface that connects to a service in the cloud and enables data from that service to be captured and sent to an application. In this case, the SIEM suite uses an API to get security data, such as alerts from the cloud, and transfers it to the dashboard of the SIEM solution.

Splunk, LogRhythm, and Rapid7 have APIs for Azure, AWS, GCP, and OCI. Splunk and LogRhythm integrate with Azure Monitoring and Azure Event Hubs to export logs, coupled with Azure connectors from SIEM vendors that enable the collection of these logs into the SIEM product. In AWS, these tools work with AWS Config, CloudTrail, and CloudWatch to collect data. Splunk can use other cloud-native logging solutions like operations suite for GCP and OCI Logging to collect logs and metadata, which can then be analyzed and visualized in Splunk.

The market for SIEM and SOAR is rapidly growing, also attracting companies that didn't have security as their main focus but, since 2020, have invested heavily in developing or acquiring security products. It's a logical move when you realize how fast cybercrime is growing. The Cloud Security Alliance published the *Pandemic Eleven* in 2022, including the top threats in the cloud:

1. Insufficient identity credentials, access, and key management
2. Insecure interfaces and APIs
3. Misconfiguration of resources and inadequate change control
4. Lack of cloud security architecture
5. Unsecure third-party resources
6. System vulnerabilities
7. Accidental cloud data disclosure
8. Organized crime and hackers
9. Cloud storage data exfiltration
10. Misconfiguration and exploitation of containers and serverless workloads
11. Insecure software development

The last two deserve a more detailed discussion, which we will do in the next chapter about DevSecOps.

Good examples of companies that made big investments in security are VMware and ServiceNow.

VMware transformed itself from a company that virtualized server environments into a company that can perform a central role in managing multi-cloud. In 2019, it introduced Intrinsic Security, which consists of several products, including VMware Secure State. It analyzes misconfigurations of systems and threats, detecting changes that are applied to systems. It calculates the security risk of these systems and is able to automate remedial actions when systems are at risk. In order to do so, security engineers need to load baselines into Secure State, this tool measures the compliance of systems. Secure State is multi-cloud and can be used as a single tool on top of Azure, AWS, GCP, OCI, and hybrid platforms that hold both public and private clouds. The latter does not necessarily have to be built with VMware and can also run, for example, Hyper-V or OpenStack.

In ServiceNow, enterprises can configure the same functionality using SecOps and **Governance, Risk, and Compliance** (**GRC**). GRC can be seen as the repository that holds the security policies and compliance baselines of an enterprise. Next, GRC continuously monitors the compliance of systems, analyzes the business impact of risks, and collects audit data. SecOps is the SOAR module of the ServiceNow suite; it continuously monitors the security posture of the entire IT environment and can automatically mitigate security issues, based on security incident response scenarios that are defined as workflows in SecOps.

A workflow can, for example, include a system being suspended when SecOps detects that software has not been checked for patches in more than 3 months. If the enterprise has a compliance rule that states that software needs to be checked for patches at least once every 3 months, an automated workflow could trigger the action to suspend the use of the software.

One final product that we will review here is Azure Sentinel, the native SIEM and SOAR solution for Azure. Sentinel does what all SIEM and SOAR solutions do: collect data, check it against compliance baselines that have been defined in Azure, and respond to threats and vulnerabilities with automated workflows. It also uses artificial intelligence to detect and analyze possible attacks, by learning the behavior of systems and users. With Sentinel, Microsoft has a very extensive suite of security solutions in the cloud with Defender, Cloud App Security, and Microsoft Defender for Cloud. Although Sentinel is based in Azure, enterprises can also connect, for instance, AWS CloudTrail and GCP to Sentinel using the Microsoft Defender for Cloud Apps, part of Microsoft 365 Defender. This product is a **Cloud Access Security Broker** (**CASB**). Over the years, Sentinel has matured into a more agnostic SIEM, as recognized by market analysts.

This list of tools and suites is, of course, not exhaustive. Enterprise architects and security specialists should, together, start gathering requirements from a business, define the needed security level of systems against compliance frameworks, agree to the security processes between the business and IT, and then decide what sort of security tools would best fit the requirements. SIEM and SOAR solutions are complex. These solutions can add a lot of value to safeguard the security posture of an IT environment, but careful consideration and an evaluation of the business case are needed.

Summary

Enterprises use a wide and growing variety of cloud solutions. Cloud platforms, systems, software, and data need to be protected from threats and attacks. Likely, a company will also have a variety of security solutions. To create one integrated view of the security of the entire IT environment, companies will have to implement security tooling that enables this single point of view. In this chapter, we looked at SIEM and SOAR systems, tools that can collect data from many different sources and analyze it against security baselines. Ideally, these tools will also trigger automated responses to threats, after calculating the risks and the business impact.

The functionality and differences between SIEM and SOAR have been explained. After reading this chapter, you should have a good understanding of how these systems can integrate with cloud platforms.

In the last section of this chapter, leading SIEM and SOAR solutions were discussed. The chapter concludes this section of our book about security operations, or SecOps. In the next chapter, we will learn how to integrate SecOps in software and system development using DevOps and how we can mature organizations, using the DevSecOps maturity model.

Questions

1. What does SOC stand for?
2. What is a common technology to integrate SIEM and SOAR systems into cloud platforms?
3. Monitoring and operations are the first level in the security maturity model. Rate the following statement true or false: the reason for this is that monitoring and operations are reactive.

Further reading

- *Enterprise Cloud Security and Governance*, by Zeal Vora, Packt Publishing

18

Developing for Multi-Cloud with DevOps and DevSecOps

The typical reason why most enterprises adopt the cloud is to accelerate application development. Applications are constantly evaluated and changed to add new features. Since everything is codified in the cloud, these new features need to be tested on the infrastructure of the target cloud. The final step in the life cycle of applications is the actual deployment of applications to the cloud and the handover to operations so that developers have their hands free to develop new features again, based on business requirements.

To speed up this process, organizations work in DevOps cycles, using release cycles for applications with continuous development and the possibility to test, debug, and deploy code multiple times per week, or even per day, so that these applications are constantly improved. Consistency is crucial: the source code needs to be under strict version control. That is what CI/CD pipelines are for: continuous integration and continuous delivery and deployment.

We will study the principles of DevOps, how CI/CD pipelines work with push and pull mechanisms, and how pipelines are designed so that they fit multi-cloud environments. Next, we will discuss how we must secure our DevOps processes using the principles of the DevSecOps Maturity Model and the most common security frameworks.

In this chapter, we're going to cover the following main topics:

- Introducing DevOps and CI/CD
- Using push and pull principles in CI
- Designing the multi-cloud pipeline

- Using the DevSecOps Maturity Model
- Automating security best practices using frameworks

Introducing DevOps and CI/CD

Before we get into the principles of DevSecOps, we need to have a good understanding of DevOps. There are a lot of views on DevOps, but this book sticks to the definition and principles as defined by the **DevOps Agile Skills Association** (**DASA**). It defines a DevOps framework based on six principles:

- **Customer-centric action**: Develop an application with the customer in mind: what do they need and what does the customer expect in terms of functionality? This is also the goal of another concept, **domain-driven design**, which contains good practices for designing.
- **Create with the end in mind**: How will the application look when it's completely finished?
- **End-to-end responsibility**: Teams need to be motivated and enabled to take responsibility from the start to the finish of the application life cycle. This results in mottos such as *you build it, you run it* and *you break it, you fix it*. One more to add is *you destroy it, you rebuild it better*.
- **Cross-functional autonomous teams**: Teams need to be able and allowed to make decisions themselves in the development process.
- **Continuous improvement**: This must be the goal—to constantly improve the application. But DevOps applies to more than *just* the application: it's also about the processes, the people, and the tools. DevOps, at its core, is a culture, a mindset.
- **Automate as much as possible**: The only way to really gain speed in delivery and deployment is by automating as much as possible. Automation also limits the occurrence of failures, such as misconfigurations.

DevOps has been described in the literature as *culture*, a new way of working. It's a new way of thinking about developing and operating IT systems based on the idea of a feedback loop. Since cloud platforms are code-based, engineers can apply changes to systems relatively easily. Systems are code, and code can be changed, as long as changes are applied in a structured and highly controlled way. That's the purpose of **CI/CD pipelines**.

Continuous Integration (**CI**) is built on the principle of a shared repository, where code is frequently updated and shared across teams that work in the cloud environment. CI allows developers to work together on the same code at the same time. The changes in the code are directly integrated and ready to be fully tested in different test environments.

Continuous Delivery and Deployment (**CD**) focuses on the automated transfer of software to test environments. The ultimate goal of CD is to bring software to production in a fully automated way. Various tests are performed automatically. After deployment, developers immediately receive feedback on the functionality of the code. We have to make a note here: continuous delivery and continuous deployment are not the same thing.

Continuous delivery is putting the artifacts, built in the CI process, in environments, typically development, staging, and production, one after the other, with testing and approvals in between. Continuous deployment means code being put into the production environment from CI in a completely automated way without any human intervention.

The following are some key points to differentiate both practices:

- Continuous delivery usually takes time from development to production; weeks or days in the best case.
- The kinds of changes applied under continuous deployment can go to production several times per day.

To learn more, we refer you to the following blog by Martin Fowler: `https://www.martinfowler.com/bliki/ContinuousDelivery.html`

CI/CD enables the DevOps cycle. Combined with CI/CD, all responsibilities, from planning to management, lie with the team, and changes can reach the customer much faster through an automated and robust development process. *Figure 18.1* shows the DevOps cycle with CI/CD:

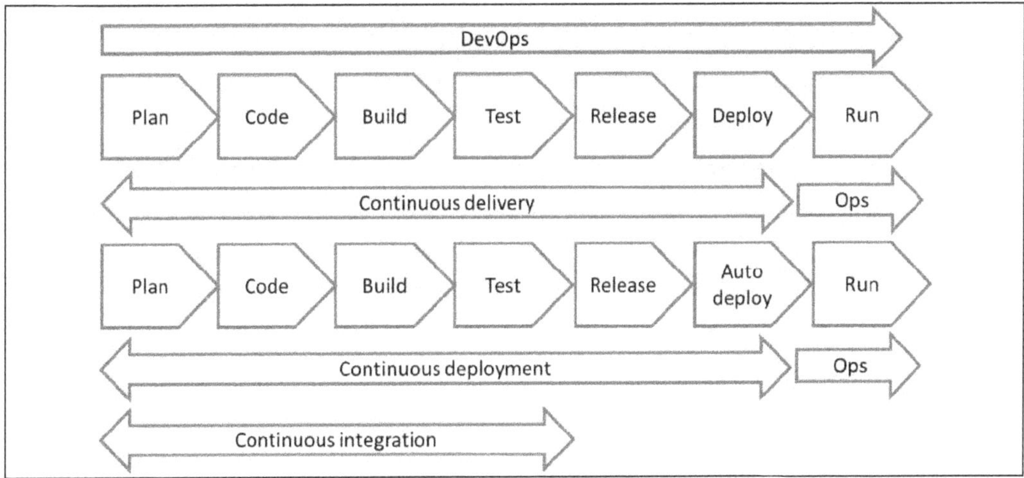

Figure 18.1: DevOps cycle with CI/CD

In the next section, we will study how to get started with CI/CD.

Getting started with CI/CD

CI/CD is widely adopted by enterprises, but a lot of projects fail. This section explains how enterprises can successfully implement CI/CD and how they can avoid pitfalls. The major takeaway should be that an implementation starts with consistency. That counts for cloud implementations as well as for CI/CD.

With CI, development teams can change code as often as they want, leading to the continuous improvement of systems. Enterprises will have multiple development teams, working in multi-cloud environments, which makes it necessary to have one way of working. Fully automated processes in CI/CD pipelines can help keep environments consistent. CI/CD and DevOps are, however, not about tools. They're about culture and *sticking to processes.*

To get to a successful implementation of DevOps, an organization is advised to follow these steps:

1. Implementing an effective CI/CD pipeline begins with all stakeholders implementing DevOps processes. One of the key principles in DevOps is autonomous teams that take end-to-end responsibility. It's imperative that the teams are given the authority to make decisions and act on them. Typically, DevOps teams are agile, working in short sprints of 2 to a maximum of 4 weeks. If that time is wasted on getting approval for every single detail in the development process, the team will never get to finish anything in time.
2. Choose the CI/CD system. There are a lot of tools on the market that facilitate CI/CD. Jenkins is a popular one, but a lot of companies that use Azure choose to work in Azure DevOps. Next, GitHub Actions has gained a lot of popularity. Involve the people who have to work with the system daily and enable them to take a *test drive* with the application. Then, make a decision on the CI/CD system and ensure all teams work with that system. Again, it's about consistency.
3. It's advised to perform a proof of concept. An important element of CI/CD is the automation of testing, so the first step is to create an automated process pipeline. Enterprises often already have quality and test plans, possibly laid down in a **Generic Test Agreement (GTA)**. This describes what and how tests must be executed before systems are pushed to production. This is a good starting point, but in DevOps, organizations work with a **Definition of Done (DoD)**:

- The DoD describes the conditions and the acceptance criteria a system must meet before it's deployed to production. The DoD is the standard of quality for the end product, application, or IT system that needs to be delivered. In DevOps, teams work with user stories. An example of a user story is: "as a responsible business owner for an online retail store, I want to have multiple payment methods so that more customers can buy our products online." This sets requirements for the development of applications and systems. The DoD is met when the user story is fulfilled, meaning that unit testing is done, the code has been reviewed, acceptance criteria are signed off, and all functional and technical tests have passed.
- *Figure 18.2* shows the concept of implementing a build and release pipeline with various test stages. The code is developed in the build pipeline and then sent to a release pipeline where the code is configured and released for production. During the release stages, the full build is tested in a test or **Quality and Assurance** (**Q&A**) Assurance environment. In Q&A, the build is accepted and released for deployment into production:

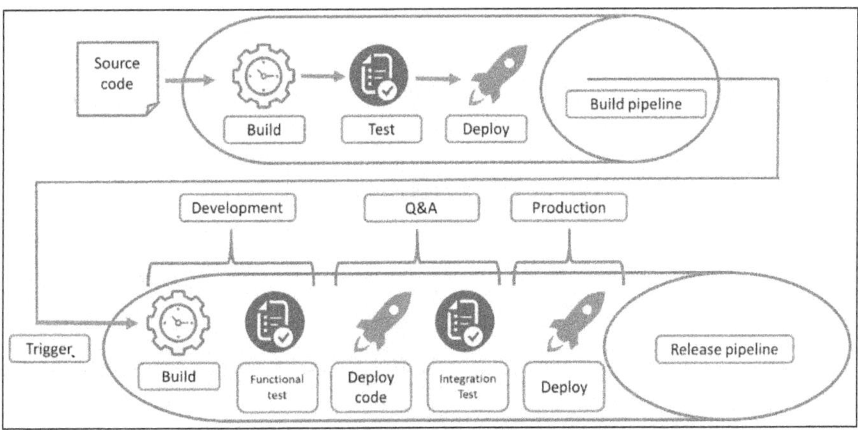

Figure 18.2: Conceptual diagram of a build and release pipeline

4. Automate as much as possible, as one of the principles of DevOps states. This means that enterprises will have to adopt working in code, including **Infrastructure as Code (IaC)**. In CI/CD, teams work from one repository, and this means that the application code and the infrastructure code are in the same repository, so that all teams can access both whenever they need to.

If all these steps are followed, organizations are able to start working in DevOps teams using CI/CD. In the next sections, CI/CD is explained in more detail, starting with version control, and then discussing the functionality of commits and push and pull mechanisms in the pipeline.

Working under version control

By working from one code repository with different teams, version control becomes crucial in CI/CD. Git and Subversion are popular version control systems that enable teams to organize their files that form the source code, test scripts, deployment scripts, and configuration scripts used for applications and infrastructure components. Everything is code, which means that systems consist of a number of code packages: the code for the VM, the code for how the VM should be configured based on policies, and the application code itself. A version control system also enables teams to retrieve the historical state of systems, in case a deployment fails or systems are compromised and need to be rebuilt.

Version control systems keep track of changes to files that are stored in the repository. In DevOps, these changes are commonly referred to as commits, something that we'll discuss further in the next section, *Using push and pull principles in CI/CD*. A **commit** comprises the code change itself, along with metadata on who made the change and the rationale behind the code change. This ensures that code is kept consistent and, with that, repeatable and predictable. It also means that teams are forced to document everything in the repository and bring it under version control.

This list contains many of the items that need to be under version control:

- Application code
- API scripts and references (what is the API for?)
- IaC components such as components such as VMs, network devices, storage, images for operating systems, DNS files, and firewall configuration rules
- Infrastructure configuration packages
- Cloud configuration templates, such as AWS CloudFormation, **Desired State Configuration (DSC)** in Azure, and Terraform files
- Code definitions for containers, such as Docker files
- Container orchestration files, such as Kubernetes and Docker Swarm files
- Test scripts

Once companies have implemented this, they need to maintain it. This is not a one-time exercise. Teams should confirm that version control is applied to application code, systems configuration, and automation scripts that are used in the CI/CD pipeline. Only if this is applied and used in a consistent way will enterprises be able to deploy new applications and systems rapidly, yet securely and reliably.

Using push and pull principles in CI

CI/CD pipelines work with branches, although other terms can be used for this. The main branch is sometimes referred to as a mainline or, when teams work in GCP, as a trunk. The most important principle to remember is that a development team has one main branch or mainline. Next, we will see two ways of pushing new code to that main branch in the following sections.

Pushing the code directly to the main branch

In this method, the developers work directly in the main code; they change small pieces of the code and merge these directly back into the main branch. Pushing code back to the main branch is called a **commit**. These commits are done several times per day, or at least as soon as possible. Working in this way ensures that releases can be done very frequently, as opposed to working in code forks that result in separate or feature branches, which are described in the second method. *Figure 18.3* shows how direct pushes to the main branch work:

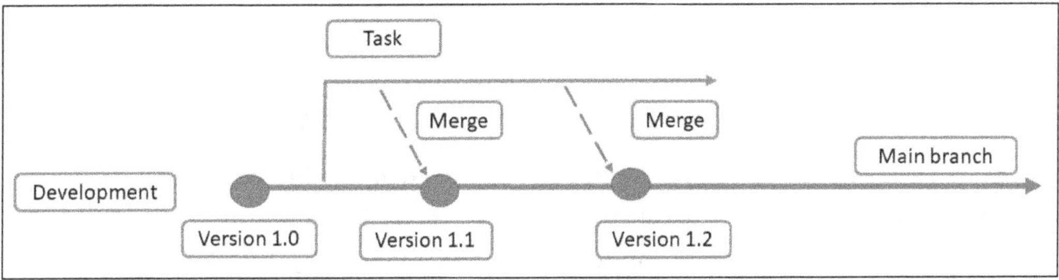

Figure 18.3: Developers merging code directly to the main branch

The idea behind CI is that companies get rid of long, complex integrations. Developers work in small batches of the code that they frequently commit to the main. The big advantage is that developers immediately see whether the change is done correctly, with the possibility to revert the change without having a huge impact on the main as a whole. This is DevOps—the developers are responsible for the build, the commit, and the result: you break it, you fix it. Automated tests that are executed after the code commit are crucial to ensure that systems keep running without failures.

Pushing code to forks of the main

In this method, teams copy code from the main and create a separate or feature branch. This is also referred to as **forking**: developers create a feature branch by taking a copy from the source code on the main branch. They do their development on this **forked code**. In GCP, this is not trunk-based development, or better said: this is referred to as feature-driven development.

This method is often used for major developments, creating new features. Developers can work in isolation on the forked code, and when they're done, commit the code back to the main branch, merging the new features or builds with it. The downside is that this can lead to complex integrations. This can't be done on a frequent basis as intensive testing is required before the merging takes place. *Figure 18.4* shows how feature branches operate in a workflow:

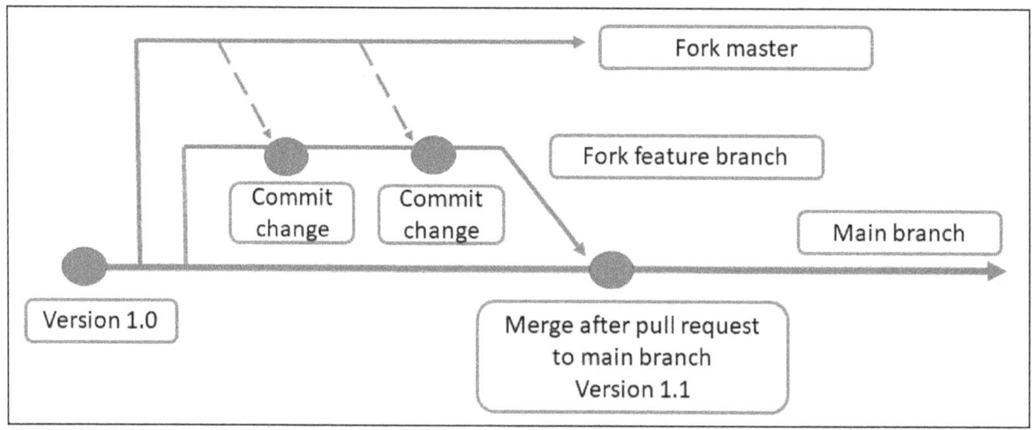

Figure 18.4: Developers working in a feature branch before merging to the main branch

In both methods, code gets pushed to the repository in, for example, GitHub. As soon as a developer has committed their code, they execute a pull request. This is the stage where the new, changed code is reviewed before the changes are actually merged into the main branch.

Best practices for working with CI/CD

There are a few best practices to remember when working with CI/CD. One of the biggest pitfalls is that code reviews can often be too extensive, meaning that developers have to get approval from different stakeholders before they can push the code to production. This will cost a lot of time and effectively slow down the DevOps process. Companies that adopt DevOps should have two principles implemented:

- The **four-eyes principle**: Have code reviewed while it's being written by working in developer pairs, where the second developer reviews the code of the first developer. This is also referred to as **pair programming**. Peer review is another method: here, the authors of the code each review at least one other developer's work, typically at the end of the development process.

- Running automated test scripts is most important. These scripts must be executed before code is actually committed to the main branch to make sure that systems keep functioning after the code commit.

By following these principles, we are already applying the first principles of DevSecOps. But it's not enough: we must really integrate security into DevOps. To do that, we can work with the DevSecOps Maturity Model. We will study this in the next section.

Using the DevSecOps Maturity Model

Security is not a sauce that we put on top of products when they are finished. Security policies have to be applied from the first moment of development, all the way up to deployment to production. That's where **DevSecOps** comes in. The position of security in the DevOps cycle is shown in the following diagram:

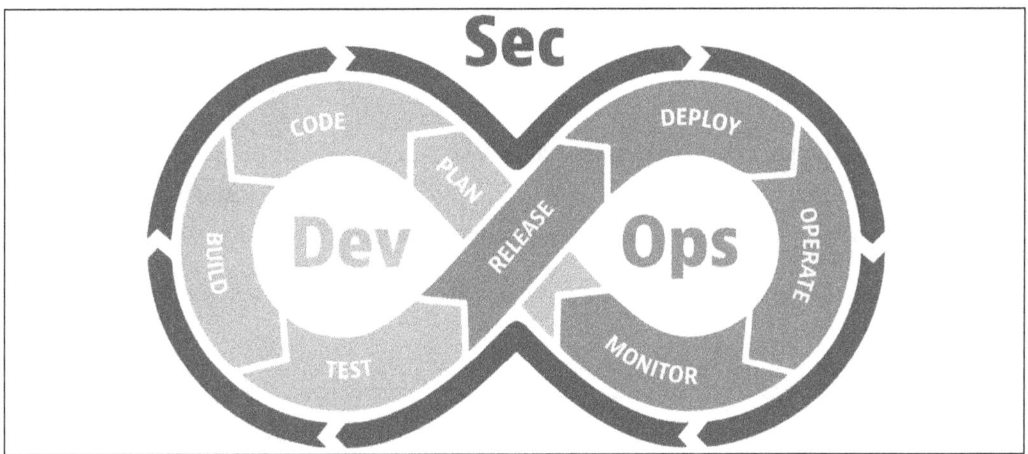

Figure 18.5: The DevSecOps cycle

The **DevSecOps Maturity Model** of the **Open Web Application Security Project (OWASP)** is a framework that helps organizations assess and improve their software development and delivery practices. The model aims to integrate security practices into the DevOps process that we described in the previous sections. By using this model, businesses can improve the security of their software products and reduce the risk of data breaches and cyber-attacks.

The DevSecOps Maturity Model can be found at `https://owasp.org/www-project-devsecops-maturity-model/`.

The OWASP **DevSecOps Maturity Model (DSOMM)** is divided into five levels, each representing a different level of maturity in terms of security integration into the DevOps process. These five levels represent where companies stand in terms of the maturity of their DevSecOps implementation:

- **Level 0—No security culture**: There is no formalized security program or processes, and security is not a consideration in the software development process.
- **Level 1—Siloed security**: Security is taken into account but is limited to a separate team or department. Security is not integrated into the software development process, and there is no collaboration between the security and development teams.
- **Level 2—Integrated security**: Security is integrated into the software development process and there is collaboration between security and development teams. However, the security process is not yet automated, and security testing is not performed in every stage of the development process.
- **Level 3—Continuous security**: Security is integrated into every stage of the software development process. Security testing is automated, and security checks are performed continuously throughout the development process.
- **Level 4—Continuous improvement**: The DevOps process is continuously monitored and improved to ensure that security is always a top priority. The organization has a culture of security, and security is considered as part of every decision and action taken.

Now, let's look at how businesses can use this maturity model to improve their DevSecOps practices:

1. **Assess current security practices**: The first step in using the DevSecOps maturity model is to assess the current security practices of the organization. During the assessment, we identify the current level of security integration into the DevOps process and consider the gaps and weaknesses in the process.
2. **Set goals for security integration**: The goals should be **Specific, Measurable, Achievable, Relevant, and Time-Constrained (SMART)**. For example, the organization may set a goal to automate security testing in every stage of the development process by the end of a specified period.
3. **Create a roadmap for improvement**: The roadmap should include specific actions and milestones that need to be achieved to reach the desired level of maturity. The roadmap should also include timelines and responsibilities for each action item.

4. **Implement security automation tools**: We must automate security testing in every stage of the development process. This can be achieved by implementing security automation tools, such as **Static Application Security Testing** (**SAST**) and **Dynamic Application Security Testing** (**DAST**) tools. These tools can identify vulnerabilities and weaknesses in the software code and provide developers with feedback on how to fix them. With this, we reach Level 3 of the DSOMM.
5. **Integrate security into the DevOps process**: Security aspects must be considered in the design, development, testing, deployment, and operations phases of the process. This is driven by culture: the mindsets of all of those involved have to be focused on developing and deploying software that is secure. To achieve this, the organization needs to create a culture of security, where everyone in the organization is responsible for the security of the software. This is the target of Level 4 of DSOMM.

We have been talking about software in this section, but DevSecOps also includes the underlying infrastructure in the clouds that we use to run our software. In major clouds such as Azure, AWS, and GCP, infrastructure is provided through a variety of services, including virtual machines, containers, and serverless computing, as we have learned throughout this book. These infrastructure services can be configured and managed using IaC tools such as AWS CloudFormation, **Azure Resource Manager** (**ARM**), and Google Cloud Deployment Manager. IaC tools allow infrastructure to be treated as code, enabling developers and operations teams to version, test, and automate infrastructure changes just like they would with software changes.

DevSecOps teams can leverage the infrastructure services and IaC tools provided by the cloud providers to implement security controls throughout the software development life cycle. For example, they can use **Identity and Access Management** (**IAM**) policies to control access to resources, implement network security controls to protect against attacks, and use logging and monitoring to detect and respond to security incidents. For that reason, cloud infrastructure must also be included in the DevSecOps practices.

One of the first topics that we must address in security and DevSecOps is observability: we need to be able to see what is happening in our pipelines and detect issues in a timely manner. That's the topic of the next section.

Manage traceability and auditability

DevSecOps starts with observability in order to enable the management of traceability and auditability. This is becoming increasingly relevant in today's complex cloud-native environments where companies execute multiple releases of their software per month, week, or even day. And in the context of multi-cloud models and workflows, they might release software across various clouds and use services from different providers.

Observability is essential for maintaining the security and stability of modern software systems. By prioritizing traceability and auditability, organizations can achieve a higher level of observability, enabling them to identify potential security threats and respond to them quickly and effectively. This will help them to reduce the risk of security incidents, improve the reliability and performance of software systems, and ensure compliance with regulatory requirements.

At its core, observability refers to the ability to collect and analyze data from across the entire software stack, including infrastructure, applications, and user interactions. This can be in one cloud, but also across stacks that are deployed in multi-cloud settings. Here, we certainly need to know what service is hosted where and how services are interconnected.

Two critical components of observability are traceability and auditability. Traceability refers to the ability to track and trace events and activities across the entire software system. This means that DevSecOps teams must be able to follow the flow of data and code through the system, identifying potential points of vulnerability and ensuring that security controls are in place at each stage of the process.

Auditability is closely related to traceability, but it focuses more on the ability to review and verify the actions taken by various stakeholders in the software development and deployment process. This includes developers, security analysts, operations teams, and other stakeholders who play a role in ensuring the security and stability of the software system. By maintaining a complete and accurate audit trail of all actions taken, organizations can ensure accountability and transparency while also facilitating more effective incident response and forensic analysis in the event of a security incident.

To achieve this level of observability, we must invest in the right tools and processes. This includes leveraging automation tools to collect and analyze data from across the software stack, the cloud infrastructure that we have deployed and manage with IaC, implementing robust logging and monitoring capabilities, and integrating security and compliance checks into every stage of the software development and deployment process. In multi-cloud systems, we therefore need "agnostic" tools that can work with various clouds and software stacks. Popular examples of such tools are:

- **Prometheus**: An open-source monitoring system that can be used to collect and query metrics from multiple sources, including different cloud providers. It supports a variety of data sources and can be integrated with many different tools.
- **Grafana**: Like Prometheus, Grafana is open source. It helps to visualize and analyze processes to create dashboards and alerts based on data from multiple sources, including various cloud stacks.
- **Fluentd**: Fluentd is an open-source data collector that can be used to collect, process, and forward log data from multiple sources, including different cloud providers. It supports a variety of outputs and can be configured to integrate with many different tools, making it a flexible choice for multi-cloud observability.
- **Jaeger**: Jaeger is an open-source distributed tracing system that can be used to trace requests across multiple services and cloud providers. It supports various tracing protocols and can be integrated with many different tools, making it a good choice for multi-cloud observability.
- **OpenTelemetry**: OpenTelemetry is a vendor-neutral observability framework that provides a standard way to collect, process, and export telemetry data from multiple sources, including different cloud providers. It supports various programming languages and can be integrated with many different tools, making it a popular choice for multi-cloud observability.

Choosing the right tools is one thing, but what do we monitor against? The answer to that question is: we implement security guardrails and guidelines to be secure and compliant. We can use security frameworks to help us in setting the appropriate levels when securing our platforms. This is the topic of the final section of this chapter.

Automating security best practices using frameworks

The hardest part in getting security to the appropriate level in organizations is to define when the organization is compliant, and environments are "secure enough"—if such a thing exists. The problem with security in any IT environment is that just like cloud technology itself, the tactics, techniques, and processes used to attack environments are also evolving fast. Hackers will constantly find new ways to compromise environments. That's why every team member in a DevOps team must be fully aware of security risks. Every choice that a team makes comes with a consequence that must be thought through in terms of security. Are we introducing a vulnerability or other risk by developing and deploying software or by using a specific cloud service? What do we need to do to protect the data, application, underlying infrastructure, connectivity, and ultimately, the user?

Frameworks such as OWASP, CIS, and MITRE ATT&CK can help in defining the level of risk and determining what security measures must be applied. The next step is to automate this. Automating security practices is an essential aspect of modern DevSecOps practices. By automating security practices, developers and operations teams can ensure that security controls are implemented consistently and effectively across their software and infrastructure.

Let's have a more detailed look at the most commonly used frameworks—OWASP, the CIS Controls, and MITRE ATT&CK:

- The OWASP Top 10 is a list of the most critical security risks to web applications. To automate security practices based on the OWASP Top 10, organizations can use automated tools to scan their applications for vulnerabilities, such as SQL injection or **Cross-Site Scripting** (**XSS**). These tools can be integrated into the CI/CD pipeline to automatically test code changes for vulnerabilities before they are deployed. Additionally, organizations can use automated code analysis tools to identify vulnerabilities in code as it is being written.
- The CIS Controls are a set of guidelines for securing information systems. To automate security practices based on the CIS Controls, organizations can use configuration management tools such as Chef, Puppet, or Ansible to enforce security policies on infrastructure. These tools can be used to automatically configure security settings on servers, networks, and applications, ensuring that security controls are consistent across the organization. CIS has issued controls for each separate cloud, including Azure, AWS, GCP, Alibaba Cloud, and OCI.
- Lastly, the MITRE ATT&CK framework is a knowledge base of **Tactics, Techniques, and Processes** (**TTPs**). To automate security practices based on the MITRE ATT&CK framework, organizations can use tools that monitor network and system activity for signs of attack. For example, they can use **Security Information and Event Management** (**SIEM**) tools to detect suspicious activity and use automated incident response tools to respond to security incidents as they occur. Using the framework to define what specific vulnerabilities are in an environment is not an easy task. Some adjacent tools might be handy when working with MITRE ATT&CK. One example is DeTTECT, originally invented for the Dutch bank Rabobank, one of the biggest financial organizations in the Netherlands. It assists in using ATT&CK to score and compare data log source quality, visibility coverage, detection coverage, and threat actor behavior.

Mitre ATT&CK references can be found at `https://attack.mitre.org/`.

DeTTECT has a GitHub repository that can be found at `https://github.com/rabobank-cdc/DeTTECT`.

Automation can and must be used to implement security best practices, such as vulnerability management, identity and access management, and data protection. By automating security practices, we can achieve more effective and efficient security controls, reducing the risk of security incidents and improving an organizations, overall security posture.

This concludes our chapter on DevOps and DevSecOps. We have two more Ops-related concepts that we have to discuss, both of which are up and coming in the world of multi-cloud. The next chapter covers AIOps and GreenOps.

Summary

After completing this chapter, you should have a good understanding of the DevOps way of working and the use of CI/CD pipelines in cloud environments. Everything is code in the cloud, from the application to the infrastructure and the configuration. Code needs to be stored in a central repository and brought under version control. That's where the CI/CD pipeline starts. Next, the DevOps team defines the phases of the pipeline, typically build, test, and deploy. Actions in these phases are automated as much as possible.

We discussed the concepts of push and pull in CI/CD pipelines using main and feature branches, describing the different methodologies to push and commit code to branches. If teams work consistently from one repository using a unified way of working, they can deploy code to different clouds.

Security must be intrinsic, from the first moment developers pull or start developing new code, all the way up to deployment to production. Adopting DevSecOps enables the integration of security practices into the DevOps cycle. To help us in achieving this, we can work with the DevSecOps Maturity Model from OWASP, which sets guidelines to improve security practices in the development of cloud environments.

In the next chapter, the final concepts of operations will be discussed: AIOps and GreenOps.

Questions

1. Systems must meet the acceptance criteria before they can be signed off as ready. In DevOps, a specific term is used for this sign-off—what is that term?
2. What's the term that we use when we push code back to the main branch?
3. Name the three security frameworks that we discussed in this chapter and that help organizations in setting up security practices.
4. What are Prometheus and Grafana?

Further reading

You can refer to the following links for more information on the topics covered in this chapter:

- **DevOps Agile Skills Association (DASA)**: `https://www.DevOpsagileskills.org/`
- *Enterprise DevOps for Architects*, by Jeroen Mulder, Packt Publishing

19

Introducing AIOps and GreenOps in Multi-Cloud

AIOps stands for **Artificial Intelligence for Operations**, but what does it really mean? AIOps is still a rather new concept but can help to optimize your multi-cloud platform. It analyzes the health and behavior of workloads end to end—that is, right from the application's code all the way down to the underlying infrastructure. AIOps tooling will help in discovering issues, thereby providing advice for optimization. The best part is that good AIOps tools do this cross-platform since they operate from the perspective of the application and even the business chain.

This chapter is an introduction to the concept of AIOps. The components of AIOps will be discussed, including data analytics, automation, and Machine Learning (ML). After completing this chapter, you will have a good understanding of how AIOps can help in optimizing cloud environments and how enterprises can get started with implementing AIOps.

We will also discuss a new concept in cloud operations, which is focusing on sustainability. With the enormous growth of the cloud, we have to be aware that the cloud uses data centers all over the world. These have an impact on our environment. GreenOps makes us conscious of the usage of the cloud in the most environmentally friendly way.

In this chapter, we're going to cover the following main topics:

- Understanding the concept of AIOps
- Optimizing cloud environments using AIOps
- Exploring AIOps tools for multi-cloud
- Introducing GreenOps

Understanding the concept of AIOps

AIOps combines analytics of big data and ML to automatically investigate and remediate incidents that occur in the IT environment. AIOps systems learn how to correlate incidents across the various components in the environment by continuously analyzing all logging sources and the performance of assets within the entire IT landscape of an enterprise. They learn what the dependencies are inside and outside of IT systems.

Especially in the world of multi-cloud, where enterprises have systems in various clouds and still on-premises, gaining visibility over the full landscape is not easy. How would an engineer tell that the bad performance of a website that hosts its frontend in a specific cloud is caused by a bad query in a database that runs from a data lake in a different cloud?

AIOps requires highly sophisticated systems, comprising the following components:

- **Data analytics**: The system gathers data from various sources containing log files, system metrics, monitoring data, and also data from systems outside the actual IT environment, such as posts on forums and social media. A sudden high number of incidents logged into the systems of the service desk may also be a source. AIOps systems will aggregate the data, look for trends and patterns, and compare these to known models. This way, AIOps is able to determine issues quickly and accurately.
- **ML**: AIOps uses algorithms. In the beginning, it will have a baseline that represents the normal behavior of systems, applications, and users. Applications and the usage of data and systems might change over time. AIOps will constantly evaluate these new patterns and learn from them, teaching itself what the new normal behavior is and what events will create alerts. From the algorithm, AIOps will prioritize events and alerts and start remediating actions.
- **Automation**: This is the heart of AIOps. If the system detects issues, unexpected changes, or abnormalities in behavior, it will prioritize and start remediation. It can only do that when the system is highly automated. From the analytics output and the algorithm, AIOps systems can determine what the best solution is to solve an issue. If a system runs out of memory because of peak usage, it can automatically increase the size of memory. Some AIOps systems may even be capable of predicting the peak usage and already start increasing the memory before the actual usage occurs, without any human intervention. Be aware that cloud engineers will have to allow this automated scaling in the cloud systems themselves.

- **Visualization**: Although AIOps is fully automated and self-learning, engineers will want to have visibility of the system and its actions. For this, AIOps offers real-time dashboards and extensive ways of creating reports that will help in improving the architecture of systems. That's the only thing AIOps will not do: it will not change the architecture. Enterprises will still need cloud architects for that. The next section discusses how AIOps can help in improving cloud environments.

AIOps is a good extension of DevOps, where enterprises automate the delivery, deployment, and operations of systems. With AIOps, they can automate operations.

Also, AIOps evolves, for instance, by including security into the concept. A good article about AISecOps can be found at `https://www.forbes.com/sites/forbestechcouncil/2022/02/16/coining-the-term-ai-secops-why-your-business-should-consider-aiops-for-cybersecurity/?sh=57406c1f5449`.

Is there something after AIOps? Yes, it's called **NoOps**, or **No IT Operations**, where all operational activities are fully automated. The idea here is that teams can completely concentrate on development. All daily management routines on IT systems are taken over by automated systems, such as system updates, bug fixing, scaling, and security operations. But there is a trade-off here: although it's called NoOps, engineers and some form of governance are still needed to set up the systems and implement the operation's baseline.

Optimizing cloud environments using AIOps

The two major benefits of AIOps are, first, the speed and accuracy in detecting anomalies and responding to them without human intervention. Second, AIOps can be used for capacity optimization. Most cloud providers offer some form of scale-out/-up mechanism driven by metrics, already available natively within the platform. AIOps can optimize this scaling since it knows what thresholds are required to do this, whereas the cloud provider requires engineers to define and hardcode it.

Since the system is learning, it can help in predicting when and what resources are needed. The following diagram shows the evaluation of operations, from descriptive to prescriptive. Most monitoring tools are descriptive, whereas AIOps is predictive:

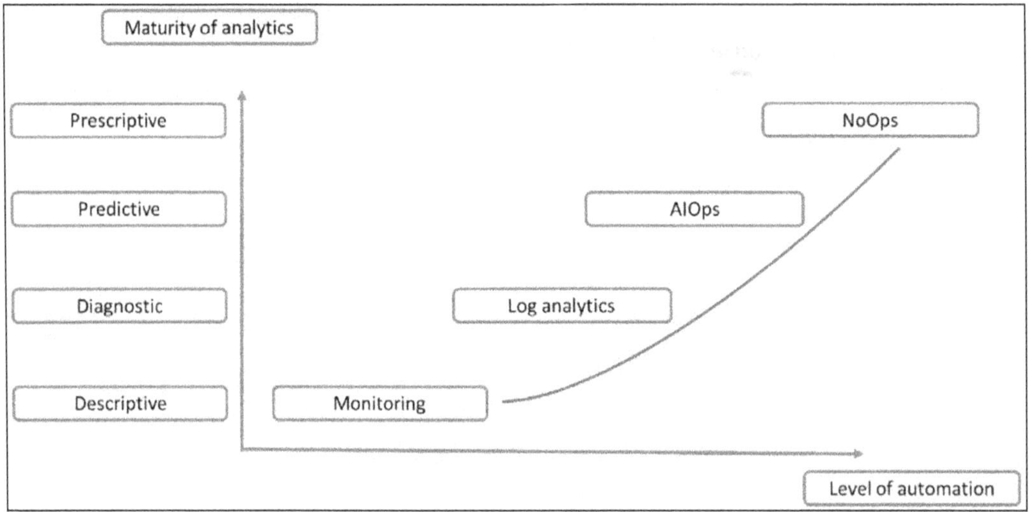

Figure 19.1: Evolution of monitoring to AIOps

Monitoring simply registers what's happening. With log analytics, companies can set a diagnosis of events and take remediation actions based on the outcomes of these analyses This is all reactive, whereas AIOps is proactive and predictive. By analyzing data, it can predict the impact of changes. The last step is systems that are prescriptive, being able to tell what should happen and already preparing systems for events, fully automated. Some very sophisticated AIOps systems can already do that.

So, we can use AIOps to help us manage and optimize the complex and dynamic nature of these environments. But how do these systems do that? AIOps will use ML algorithms to analyze data from cloud infrastructure, applications, and logs to identify patterns and anomalies that could lead to issues. This enables IT teams to predict and prevent problems before they occur, reducing downtime and improving availability.

Next, with AIOps, we can **automate** the response to common issues in cloud environments, such as scaling resources up or down based on demand, restarting failed services, or reconfiguring network settings. This reduces the need for manual intervention, saving time and reducing the risk of human error.

One of the most common areas where AIOps shows value is **root cause analysis**. When issues occur in cloud environments, AIOps can help identify the root cause by analyzing data from multiple sources and correlating events. This will help to quickly diagnose and resolve problems, reducing downtime and improving service levels. Automation and root cause analysis are ways to continuously optimize our cloud environments. Tools will continuously monitor cloud environments and recommend optimizations based on usage patterns and performance metrics. This includes optimizing resource allocation, identifying opportunities to reduce costs, and recommending improvements to application performance.

Enterprises are discovering AIOps because it helps them in optimizing their IT infrastructure. But how do companies start with AIOps? The following guidelines are recommended to successfully implement an AIOps strategy:

- **AIOps systems are learning systems**: Enterprises will have to learn how to work with and interpret analysis from these systems as well to get the best out of it. So, don't try to get the entire IT environment under AIOps in one go, but start with a small pilot and iterate from there.
- **Data is essential in AIOps**: This should not only be data that comes from IT systems but also business data. After all, the great benefit of AIOps is that it can take actions that are based on business data. If AIOps knows that certain products sell better at specific times of the year—which is business-driven data—it can take actions to optimize IT systems for that peak period. Also, if it turns out that systems are not used as expected, AIOps will be able to analyze the usage and correlate it with other events. In that way, AIOps can be a fantastic source for the business in becoming a truly data-driven organization. Businesses, therefore, absolutely need to be involved in the implementation of AIOps.
- **Most important in a successful implementation is to standardize**: Throughout this book, it has been stressed that multi-cloud environments need to be implemented in a consistent way, meaning that infrastructure must be defined and configured as code so that it can be deployed in a consistent manner to various cloud platforms. The code must be centrally managed from one repository, as much as possible. This will ensure that AIOps systems will learn quickly how systems look and how they should behave so that anomalies can be detected quickly.

AIOps can also help in testing against real-life scenarios and take much more into consideration in terms of testing. As such, AIOps is a great extension to DevOps and CI/CD pipelines executing frequent releases to applications. AIOps will know which systems will be impacted when changes are applied to a certain system, and also vice versa: which systems will respond to changes in terms of performance and stability. These can be systems that are hosted in different clouds or platforms; they can be part of the application chain.

This problem of the coexistence of applications and systems that disproportionately consume resources is referred to as **noisy neighbor**. AIOps will identify the neighbors, warn them of upcoming changes, and even take proactive measures to avoid the applications and systems from running into trouble. This goes beyond the unit and integration tests that are triggered by a CI/CD pipeline.

Today's multi-cloud environments are complex, with servers and services running in various clouds. Systems are connected over network backbones of different cloud platforms, routing data over the enterprise's gateways, yet continuously checking whether users and systems are still compliant with applied security frameworks. There's a good chance something may be missed when distributing applications across these environments.

AIOps can be used to improve the overall architecture. Architects will have much better insight into the environment and all the connections between applications and systems; this includes not only servers but also network and security devices. Next, AIOps will help in the distribution of applications across platforms and the scaling of infrastructure without impacting the neighbors, even if the neighbors are sitting on a different platform.

Exploring AIOps tools for multi-cloud

AIOps helps enterprises in becoming data-driven organizations. From the first chapter of this book, the message has been that IT—and IT architecture—is driven by business decisions. But business itself is driven by data: how fast does a market develop, where are the customers, what are the demands of these customers, and how can IT prepare for these demands? The agility to adapt to market changes is key in IT, and that's exactly what cloud environments are for—that is, cloud systems can adapt quickly to changes. It becomes even faster when data drives the changes directly, without human interference. Data drives every decision.

That's the promise of AIOps. An organization that adopts the principle of becoming a data-driven enterprise must have access to vast amounts of data from a lot of different sources, inside and outside IT. It needs to embrace automation. But above all, it needs to trust and rely on sophisticated technology with data analytics, AI, and ML. That's a true paradigm shift for a lot of companies. It will only succeed when it's done in small steps. The good news is that companies already have a lot of business and IT data available, which they can feed into AI and ML algorithms. So, they can get started, but first, they will need to select a platform or a tool.

Market analysts expect that the use of AIOps will grow from being worth around 26 billion USD in 2020 to well over 600 billion USD in 2030. This explains why a lot of leading IT companies are investing heavily in AIOps, including big names such as IBM, VMwire, and ServiceNow. Also, cloud providers themselves invest heavily in AIOps capabilities, for instance, in Azure Monitor, which already provides deep insights into the behavior of resources and applications in Azure.

AWS offers CloudWatch and X-Ray. The latter is a distributed tracing service that provides end-to-end visibility into application performance and behavior and allows users to visualize and analyze the flow of requests and responses across distributed systems. X-Ray can identify performance bottlenecks and errors in complex microservices environments.

In GCP, we could identify Cloud Operations—formerly Stackdriver—as an AIOps-enabled tool, since it collects and analyzes metrics, logs, and traces from GCP resources and applications, and provides visibility into operational health and performance. It can also use ML algorithms to detect anomalies and perform root cause analysis.

In OCI, we would look at **Oracle Management Cloud**, a suite of management and monitoring services for OCI. It includes features such as log analytics, application performance monitoring, and infrastructure monitoring, and can be used to automate operational tasks and improve application performance and reliability. It also includes machine learning algorithms to detect anomalies and perform root cause analysis.

How does an enterprise choose the right tool? When an enterprise is working in multi-cloud, it needs an AIOps solution that can handle multi-cloud. These are AIOps platforms that have APIs for the major cloud providers and can integrate with the monitoring solutions of these providers and the third-party tools that enterprises have in the cloud environments. An example of such a platform is Splunk Enterprise, which collects, correlates, and analyzes data from IT infrastructure, applications, and security systems.

In essence, all of these tools work in layers. The layers are depicted in the following diagram:

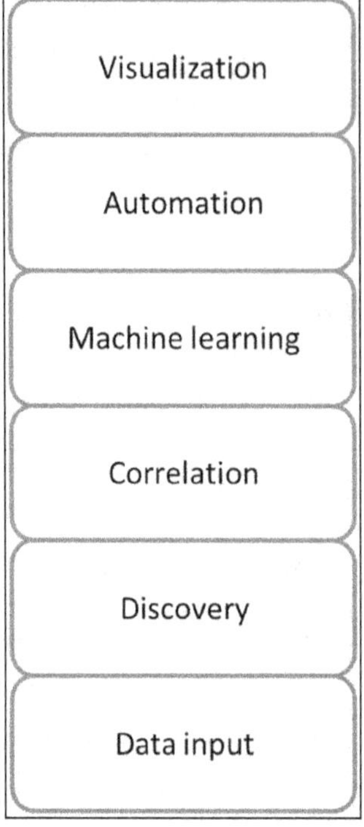

Figure 19.2: Layers of AIOps

Most AIOps systems combine a set of tools in the different layers into an AIOps platform that can handle the various aspects of AIOps.

The market for AIOps is thus rapidly growing. Some examples of more cloud-agnostic AIOps tools are Dynatrace, Splunk, Cisco AppDynamics, Moogsoft, and IBM Cloud Pak for Watson. All these platforms use AI and automation to monitor and analyze cloud environments, including containerized applications and microservices. They provide real-time insights into performance and user experience and can often automatically remediate issues by scaling resources and reconfiguring settings.

These are just a few examples. Keep in mind that each tool has its own unique features and capabilities, and the best tool for a particular organization will depend on its specific needs and requirements.

Key in all these solutions is that they auto-discover any changes in environments in real time and can predict the impact on any other component in the IT environment before events actually occur, as well as from changes that are planned from CI/CD pipelines.

Introducing GreenOps

The concept of GreenOps is becoming increasingly relevant in today's world, where concerns about climate change and sustainability are growing. What do we mean by GreenOps? Before we dive into that, it's relevant to notice that GreenOps and AIOps are actually quite intensively related. Both concepts make use of AI, as we will learn.

In short, we can define GreenOps as the practice of using cloud technology to optimize the environmental sustainability of IT operations, helping organizations to reduce their carbon footprint and operate in a more environmentally friendly manner. The cloud offers a number of benefits when it comes to achieving sustainability goals. One of the most significant advantages of the cloud is that it allows organizations to use shared resources, which can reduce the overall energy consumption of IT operations. By using virtualized infrastructure and shared resources, organizations can improve the efficiency of their IT operations and reduce the number of physical servers they need to run.

Another advantage of the cloud is that it allows organizations to scale their operations up or down quickly and easily, depending on their needs. This can be especially useful for organizations that experience fluctuations in demand, as they can adjust cloud resources in real time to match the level of demand. By avoiding the need to maintain excess capacity, organizations can reduce their energy consumption and save money on their energy bills.

But how does this compare to AIOps? The answer to that question is that GreenOps also involves automation and machine learning to optimize IT operations and, with that, reduce energy consumption. By automating repetitive tasks and using machine learning algorithms to identify areas for optimization, organizations can improve the efficiency of their cloud environments. For example, machine learning algorithms can be used to optimize data center cooling, which can reduce energy consumption and improve the efficiency of IT operations.

Throughout this book, we have discussed the public clouds: Azure, AWS, GCP, OCI, and sometimes Alibaba Cloud. These are all so-called hyperscalers, meaning that they operate at a large scale, serving thousands of customers around the globe. These customers host their environments in the data centers of these providers. How will customers be able to influence and monitor their energy consumption and environmental footprint? Many cloud providers offer energy monitoring and reporting tools that allow organizations to track their energy usage and identify areas for improvement. They can also use cloud-based analytics tools to optimize their energy consumption and reduce their carbon footprint.

Measures to reduce the use of (heavy) resources, implementing automation and efficient automated scaling, will show directly in this energy monitoring. By optimizing IT operations and reducing energy consumption, we can reduce the carbon footprint, save money on energy bills, and—most important of all—help our world to become more sustainable.

Summary

AIOps was a new kid on the block but, since 2020, it has emerged as an almost essential platform for managing complex cloud environments. AIOps systems help organizations in detecting changes and anomalies in their IT environments and already predicting what impact these events might have on other components within their environments. AIOps systems can even predict this from planned changes coming from DevOps systems such as CI/CD pipelines. To be able to do that, AIOps makes use of big data analysis: it has access to a lot of different data sources, inside and outside IT environments. This data is analyzed and fed into algorithms: this is where **AI** comes in, and ML. AIOps systems learn so that they can actually predict future events.

AIOps are complex systems that require vast investments from vendors, and thus from companies that want to start working with AIOps. However, most organizations want to become more and more data-driven, meaning that data is driving all decisions. This makes a company more agile and faster in responding to market changes.

In the last section, we briefly discussed a new concept that becomes increasingly important in using cloud technology: GreenOps. We have to become more aware of the environmental impact of using cloud technology and the data centers of public cloud providers. GreenOps will help us to reduce our carbon footprint and operate our business in the cloud in a more environmentally friendly manner.

After completing this chapter, you should have a good understanding of the benefits as well as the complexity of AIOps. You should also be able to name a few of the market leaders in the field of AIOps. At the end of the day, it's all about being able to respond quickly to changes, but with minimum risk and, preferably, with a low carbon footprint.

Questions

1. AIOps correlates data from a lot of different systems, including IT systems that are not directly in the delivery chain of an application but might be impacted by changes to that chain. What are these systems called in terms of AIOps definitions?
2. Name at least two vendors of cloud-agnostic AIOps systems, recognized as such by market analysts.

3. AIOps works in layers. Rate the following statement as true or false: most AIOps systems have separate solutions for the layers that are combined in an AIOps platform.
4. What is the aim of GreenOps?

Further reading

You can refer to a blog and video on AIOps at `https://searchitoperations.techtarget.com/feature/Just-what-can-AI-in-IT-operations-accomplish`.

Helpful blog posts about GreenOps can be found at `https://www.computerweekly.com/opinion/IT-Sustainability-Think-Tank-Embedding-GreenOps-into-enterprises` and `https://blogs.gartner.com/lydia_leong/2023/03/13/greenops-for-sustainability-must-parallel-finops-for-cost/`.

20

Conclusion: The Future of Multi-Cloud

This book has dealt with designing, implementing, and controlling a multi-cloud platform. We talked about five major clouds—Azure, AWS, GCP, Oracle Cloud, and Alibaba Cloud—and discussed strategies to get the best out of these clouds for our businesses. We discovered that building and managing in the cloud can be complex. Yet, the cloud will definitively grow. We will look at the future of the cloud in this final chapter.

The cloud will grow and multi-cloud will grow. The biggest challenge is how organizations can stay in control of their applications in a multi-cloud setting since the cloud can become very complex. Maybe Google has the answer: **Site Reliability Engineering** (**SRE**). SRE incorporates aspects of software engineering and applies them to infrastructure and operations problems. We will also use this chapter to introduce the concept of SRE and its main principles.

In this chapter, we're going to cover the following main topics:

- The growth and adoption of multi-cloud
- Understanding the concept of SRE
- Working with risk analysis in SRE
- Applying monitoring principles in SRE
- Applying principles of SRE to multi-cloud—building and operating distributed systems

The growth and adoption of multi-cloud

In recent years, multi-cloud has emerged as a popular approach for businesses to manage their cloud infrastructure. Let's recap the definition of multi-cloud one more time: we speak about multi-cloud when we use two or more cloud service providers to host and run applications and services. As we look toward the near future, we can expect to see continued developments in multi-cloud as businesses seek to take advantage of its benefits while managing its risks. We'll talk about managing risks later in this chapter when we explore the concept of SRE.

One of the primary reasons that businesses are looking more into multi-cloud is the need for flexibility and agility. Multi-cloud allows businesses to avoid vendor lock-in and take advantage of the unique features and capabilities offered by different cloud providers. This allows them to optimize their applications and services for specific use cases, such as high-performance computing or machine learning, and to quickly respond to changing market conditions or customer needs.

Another benefit of multi-cloud is improved reliability and resilience. By spreading their applications and services across multiple cloud providers, businesses can reduce the risk of downtime or service interruptions due to outages or other issues. This also allows them to implement disaster recovery and business continuity strategies that can help them recover quickly in the event of a major outage or data loss.

However, multi-cloud is not without its risks and pitfalls. One of the biggest challenges with multi-cloud is the complexity of managing multiple cloud environments, as we have seen throughout this book. This can lead to issues with security, compliance, and data governance, as well as increased operational costs and management overhead. Additionally, there may be challenges with data migration and integration across multiple cloud providers, as well as the risk of vendor lock-in with specific tools or services. We discussed these topics when applying BaseOps, FinOps, and DevSecOps.

To mitigate risks, businesses need to carefully plan and implement their multi-cloud strategy. This includes selecting the right cloud providers and services to meet their specific needs, developing a strong governance framework to manage security and compliance, and implementing robust monitoring and management tools to ensure visibility and control across all cloud environments.

Looking toward the future, we can expect to see continued developments in multi-cloud, including the emergence of new technologies and tools to help businesses manage their cloud environments more effectively. We may also see increased collaboration and standardization across cloud providers to make it easier for businesses to integrate and manage their multi-cloud environments.

Multi-cloud is a rapidly evolving area of cloud computing that offers significant benefits for businesses in terms of flexibility, agility, and resilience. Businesses will have to adopt strategies and methodologies to mitigate risks and avoid pitfalls associated with managing multiple cloud environments. A way to overcome the complexity of multi-cloud is by adopting SRE, which we will discuss in the following sections.

Understanding the concept of SRE

Originally, SRE was meant for mission-critical systems, but overall, it can be used to drive the DevOps process in a more efficient way. The goal is to enable developers to deploy infrastructure quickly and without errors. To achieve this, the deployment is fully automated. In this way of working, operators will not be swamped with requests to constantly onboard and manage more systems.

The original description of SRE as invented by Google is well over 400 pages long. In the *Further reading* section, a good book is listed to give you a real deep dive into SRE. This chapter is merely an introduction.

Key terms in SRE are **service-level indicators** (**SLIs**), SLO, and the error budget, or the number of failures that lead to the unavailability of a system. The terms are explained in more detail in the next paragraphs.

SLI and SLO differ from **SLA**, the **service-level agreement**. The SLA is an agreement between the supplier of a service and the end user of that service. SLAs comprise **key performance indicators** (**KPIs**), typically, indicators about the uptime of systems. For example, an SLA may contain a metric concerning an uptime—or **mean time to failure** (**MTTF**)—of 99.9% for a system. It means that the system may be unavailable to the end user for 44 minutes per month, often called downtime. Even the most reliable systems will suffer from failure every now and then, and then KPIs such as **mean time to repair or recovery** (**MTTR**) become important: the average time needed to fix an issue in systems.

Fixing problems can become a large part of the work that operation teams do: in SRE, this is referred to as toil. Explained simply, toil is manual reactive tasks that keep teams away from other proactive tasks and eventually slow down development. SRE is built on the principle that SRE teams have to spend up to 50% of their time improving systems, which means that toil has to be reduced as much as possible so that teams can spend as much time as possible developing.

To enable this, SRE teams set targets, defined in SLIs and SLOs:

- **SLO**: In SRE, this is defined as *how good a system should be*. The SLO is more specific than an SLA, which comprises a lot of different KPIs. One could also state that the SLA comprises a number of SLOs. However, an SLO is an agreement between the developers in the SRE team and the product owner of the service, whereas an SLA is an agreement between the service supplier and the end user.
- The SLO is a target value. For example, the web frontend should be able to handle 100 requests per minute. Don't make it too complex in the beginning. By setting this SLO, the team already has a number of challenges to be able to meet this target, since it will not only involve the frontend but also the throughput on, for instance, the network and involved databases. In other words, by setting this one target, architects and developers will have a lot of work to do to reach that target.
- **SLI**: SLOs are measured by SLIs. In SRE, there are a few indicators that are really important: request latency, system throughput, availability, and error rate. These are the key SLIs. Request latency measures the time before a system returns a response. System throughput is the number of requests per second or minute. Availability is the amount of time a system is usable to the end user. The error rate is the percentage of the total number of requests and the number of requests that are successfully returned.
- **Error budget**: This is probably the most important term in SRE. The SLO also defines the error budget. The budget starts at 100 and is calculated by deducting the SLO. For example, if we have an SLO that says that the availability of a system is 99.9%, then the error budget is 100 - 99.9 = -0.1. This is where the SRE teams have to apply changes without impacting the SLO. It forces developers in the SRE team to either limit the number of changes and releases or to test and automate as much as possible to avoid the disruption of the system and overspending the error budget.
- Remember that SRE is about reducing toil in operations. SRE teams are DevOps teams and they have to make sure that they can spend more time in Dev than in Ops. That starts with the architecture of systems: are these systems fault-tolerant, meaning that systems will still continue to run even if one or more components fail? There might be a reduction in throughput or an increase in latency, but systems should still be available and usable.

To detect failures or performance degradation, monitoring is extremely important in SRE. But before teams get to monitoring, building, and operating with SRE, architects need to define the SLO and SLI. This is done through risk analysis, to be discussed in the next section.

Working with risk analysis in SRE

The basis of SRE is that reliability is something that you can design as part of the architecture of applications and systems. Next to that, reliability is also something that one can measure. According to SRE, reliability is a measurable quality, and that quality can be influenced by design decisions. Engineers can take measures to decrease the detection, response, and repair time, and they can develop systems in such a way that changes can be executed safely without causing any downtime. Architects can design fault-tolerant systems; engineers can develop them.

The major issue is it all comes at a cost, and whether systems really need to be fault-tolerant is a business decision, based on a business case. Already, in *Chapter 1, Introduction to Multi-Cloud*, we've learned that business cases are driven by risks. Let's go over risk management one more time.

The basic rule is that *risk = probability x impact*. Enterprises use risk management to determine the business value of implementing measures that limit either the probability and/or the impact – or, to put it in SRE terminology: risk management is used to determine the value of reliability engineering. Risk management is also used to prioritize reliability measures in the product backlogs of SRE teams. That is done by following the risk matrix, sometimes referred to as PRACT:

- **Prevent**: The risk is avoided completely.
- **Reduce**: The impact or likeliness of the risk occurring is reduced.
- **Accept**: The consequences of the risk are accepted.
- **Contingency**: The measures are planned and executed when the risk occurs.
- **Transfer**: The consequences of the risk are transferred, for instance, to an insurance company.

If the impact of the failure is great, it might be worthwhile looking at a strategy that prevents the risk. This will drive the SLO—how good a system should be. In this case, the availability might be set to 99.99%, leaving only 0.01% for the error budget. This has consequences for the architecture of the system; after all, the risk rating only allows 52 minutes of downtime per year. The diagram shows how business risks drive SLOs in SRE:

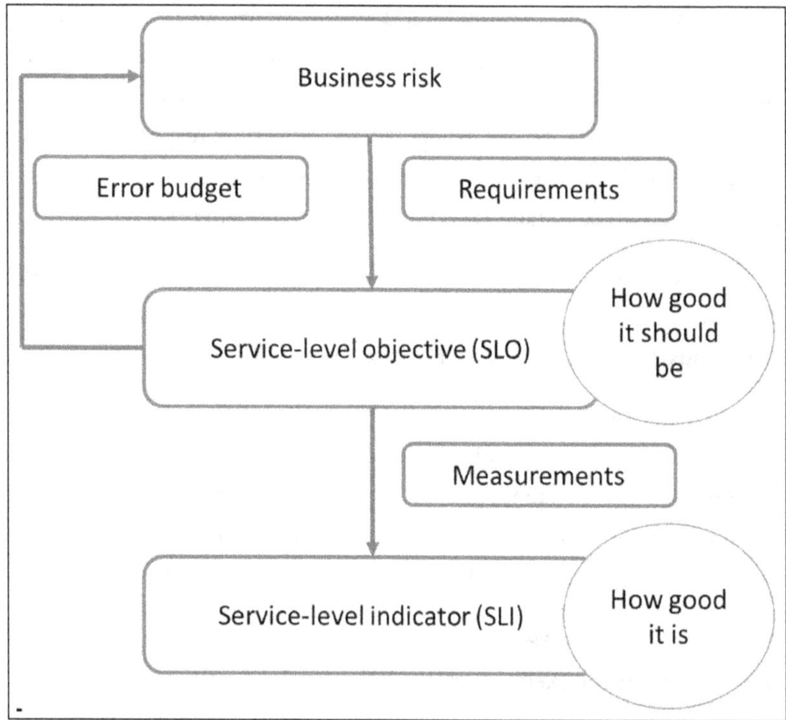

Figure 20.1: The concepts of SLO and SLI

The error budget is used to control the risks and make decisions that don't compromise the SLOs. To calculate the impact on the SLO, the following items have to be taken into consideration:

- **Time to detect** (**TTD**): The time taken to detect an issue in software or a system.
- **Time to resolve** (**TTR**): The time taken to resolve or repair the issue.
- **Frequency/year**: The frequency of errors per year.
- **Users**: The number of users that are impacted by the error.
- **Bad/year**: The number of minutes per year that a system is not usable, or the "bad minutes" per year.

An example will make things clearer. A team deploys bad code to an application. It takes 15 minutes to detect that the application is not performing well with this code and another 15 minutes to resolve the issue by executing a rollback. It's estimated that this will happen at least once every 2 weeks, so around 26 times per year. It will impact 25% of the user population. This will lead to a number of bad minutes per year. If this is higher than the error budget, then the team needs to take measures to reduce the risk. If the SRE team doesn't do that, it will lead to a lot of work in operations. Engineers will have to spend more time fixing problems.

Applying monitoring principles in SRE

Reliability is a measurable quality. To be able to measure the quality of the systems and their reliability, teams need real-time information on the status of these systems. As mentioned in the previous section, the TTD is a crucial driver in calculating risk and, subsequently, determining the SLO. Observability is therefore critical in SRE. However, SRE stands with the principle that monitoring needs to be as simple as possible. It uses the four golden signals:

- **Latency**: The time that a system needs to return a response.
- **Traffic**: The amount of traffic that is placed on the system.
- **Errors**: The number of requests placed on a system that fail completely or partially.
- **Saturation**: The utilization of the maximum load that a system can handle.

Based on these signals, monitoring rules are defined. As the starting point in SRE is avoiding too much work for operations or toil, the monitoring rules follow the same philosophy. Monitoring should not lead to a tsunami of alerts. The basic rules are as follows:

- The rule must detect a condition that is urgent, actionable, and visible to the user. The condition would not be detected without the rule.
- Can the team ignore the alert that is triggered by the rule? What would happen if the alert is ignored?
- If the alert can't be ignored, then how can teams action the alert? For example, if a majority of users are affected by the condition, then the alert can't be ignored, and action must be taken.
- Are there short-term workarounds to improve the condition? This doesn't mean that SRE promotes short-term workarounds, but it does promote actions that ensure that systems are available and usable, even when an error occurs. Remember that SRE is about making systems reliable. And also remember, there's nothing as permanent as a temporary workaround (the origin of this lies in this poem: `https://www.poetryfoundation.org/poetrymagazine/poems/55235/after-a-greek-proverb`). Architects and engineers really should avoid accepting workarounds as solutions.

- Are teams allowed to take action after the alert has been raised?
- Can actions on alerts be automated in a safe manner?

All monitoring rules in SRE must adhere to these principles. In short, monitoring in SRE is about making a good distinction between signals that require action and noise. Monitoring should only do two things in theory: define what is broken in a system and, next, determine why it is broken, getting to the root cause. Most monitoring systems focus on what is broken, the symptoms. It requires more sophisticated monitoring to correlate data and get to the cause of an error.

Especially in multi-cloud environments, the error of a system can find its cause in a system that is hosted on a different platform. Since this is already complex in itself, monitoring rules should be designed in such a way that teams are only alerted when the thresholds of the four golden signals are compromised, making systems unavailable to users.

Applying principles of SRE to multi-cloud—building and operating distributed systems

This book exists because a majority of enterprises are moving or developing systems in cloud environments. Today's enterprises are in a constant transformation mode. This also means a big change in operations. To put it simply, they have to keep up with the speed of change. Traditional operations can't handle this. We need SRE in the future of multi-cloud. SRE teams create reliable systems in cloud environments.

There are a couple of important rules for SRE to enable this:

- **Automate everything**: Automation leads to consistency, but automation also enables scaling. This requires a very well-thought-out architecture. Automation enables issues to be fixed faster since it only has to be fixed in one place: the code. Automation makes sure that the proper code is distributed over all systems involved. With large distributed systems spanning various cloud platforms, this would take days to do manually. SRE was invented by Google, which already had massive services running from cloud services—services that consumers were relying on, such as Gmail. Without automation, these services never would have been as stable as they are today. Without automation, operations would simply be drowned by manual tasks.
- **Eliminate toil**: This is a specific term used in SRE and might be a bit difficult to understand. It's not just work that teams would rather not do; it's every piece of work that keeps teams away from developing.

Toil is manual work, repetitive, and can be automated. But toil is also work that doesn't add value to the product: it's interruptive and slows down the development of services that add value. SRE has a rule for toil: an SRE team should not spend more than 50% of their work on toil. The rationale behind that is that toil can easily consume up to 100% of a team's time.

- How does SRE deal with that? This is where the error budget is important again. If the SRE team needs to spend more than half its time on operations and toil, the error budget is likely exceeded. This calls for engineering, typically meaning that systems need to be refactored by the product team that is responsible for the system. Refactoring aims to improve the design and often also to reduce complexity. The following diagram shows the concept of eliminating toil in SRE.

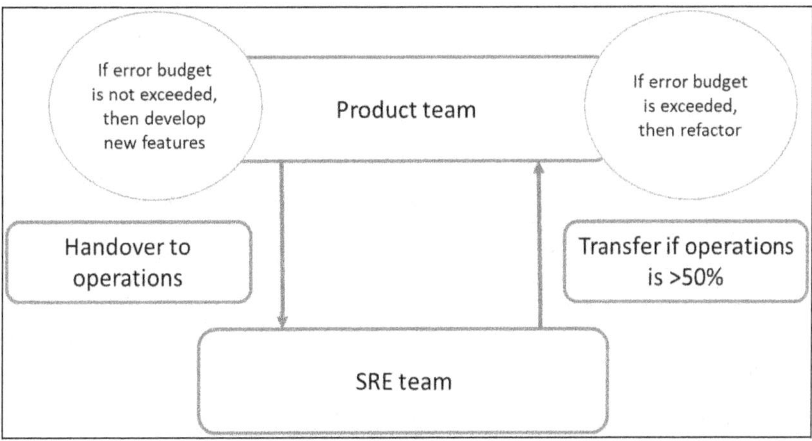

Figure 20.2: Concept of eliminating toil

- **Keep it simple**: Simplicity is a key principle in SRE. Software needs to be simple as a prerequisite to be a stable and reliable system. As a consequence, SRE teams have a strong mandate to push back against product teams when systems are getting too complex. This is often caused by the fact that code for new features is added, but old code is not removed. Code needs to be simple and clean, the use of APIs should be limited, and, if used, APIs should be as simple as possible. SRE lives by the golden rule of less is more.
- **Release engineering**: To keep systems stable and reliable, while changes to developments are applied constantly and at high speed, companies need a rock-solid release process.

Google added release engineers to their teams, specialists that are experts in source code management, software compiling, packaging, configuration builds, and automation. In *Chapter 18, Designing and Implementing CI/CD Pipelines*, the principle of branches was discussed. SRE works with the principle of checking in code, not directly to a master branch but through feature branches.

- Testing is, as in any pipeline, a crucial gate in the release process. Here, another term is introduced that is more or less specific to SRE: the canary test. It refers to the tests that were used in the coal mines to detect whether shafts contained toxic gasses. To determine this, a canary would be sent into the shaft of the mine. If the canary came back alive, it was meant to be safe for miners to go in.
- In SRE, the canary test refers to a subset of servers or services where new code is implemented. These servers are then left for an incubation period. If the servers run fine after this period, all other servers get the new code. If the canary servers fail, then these servers are rolled back to the last known healthy state.

Testing is done against the key values of SRE, that is, latency, traffic, errors, and saturation.

- **Postmortem analysis**: Of course, SRE doesn't mean that mistakes will not happen at all anymore. Multi-cloud environments with distributed systems in different clouds will eventually fail. Systems are getting more complex, mainly because of increasing demands by their users. New features are applied at an ever-increasing speed. Systems are getting more and more intertwined, so they're bound to encounter issues every now and then because of a deployment mistake, bugs, or hardware failure—remember also that in cloud environments, there is some hardware involved—or, indeed, security breaches.
- In SRE, these issues are opportunities to improve systems and software. As soon as an issue has been solved, a postmortem analysis is conducted. It's important to know that these postmortem analyses are blameless. Google itself even talks about a postmortem culture or even a philosophy. Teams register the issue, fix it, document the root causes, and implement the lesson learned, all without finger-pointing, all to grab the opportunity to make systems more resilient.

SRE is about constantly learning. It's about learning by failure and learning by doing. If there's one message that one should remember after completing this book, it is that multi-cloud architecture and governance are also about learning. Azure, AWS, Google, OCI, Alibaba Cloud, and all the other platforms will change constantly.

It's a constant transformation.

Summary

Systems are getting more complex for many reasons: customers constantly demand more functionality in applications. At the same time, systems need to be available 24/7 without interruption. Cloud platforms are very suitable to facilitate development at high speed, and thus we foresee cloud providers growing fast. In other words, the cloud will definitively grow. This comes with challenges for a lot of businesses. Throughout this book, we discovered that building and managing cloud environments can be complex.

The cloud will grow, and likely the complexity of the cloud will grow too. To ensure reliability, especially with systems that are truly multi-cloud and distributed across different platforms, we should adopt the principles of SRE. The most important principles of SRE have been discussed in this chapter. You should have an understanding of the methodology, based on determining the SLO, measuring the SLI, and working with error budgets.

We've learned that these parameters are driven by business risk analysis. We also studied monitoring in SRE and learned how to set monitoring principles. In the last section, some important guidelines of SRE were introduced, covering automated systems, eliminating toil, simplicity, release engineering, and postmortem analysis.

The conclusion of the chapter and this book is to *learn by doing and learn by failure*. The world of multi-cloud is changing rapidly and thus companies will see themselves as being in a constant transformation mode.

Questions

1. Risk analysis is important in SRE. What are the five risk strategies, often referred to as PRACT?
2. SRE mentions four golden signals in applying monitoring rules. Latency and traffic are two of them. Name the remaining two.

3. SRE has a specific term for manual work that is often repetitive and should be avoided. What's that term?
4. Postmortem analysis is a key principle in SRE. True or false: Postmortem analysis is about finding the root cause and finding out who's to blame for the error.

Further reading

For more information on SRE, you can refer to *Practical Site Reliability Engineering* by Pethuru Raj, Packt Publishing.

Join us on Discord!

Read this book alongside other users, cloud experts, authors, and like-minded professionals. Ask questions, provide solutions to other readers, chat with the authors via. Ask Me Anything sessions and much more.

Scan the QR code or visit the link to join the community now.

`https://packt.link/cloudanddevops`

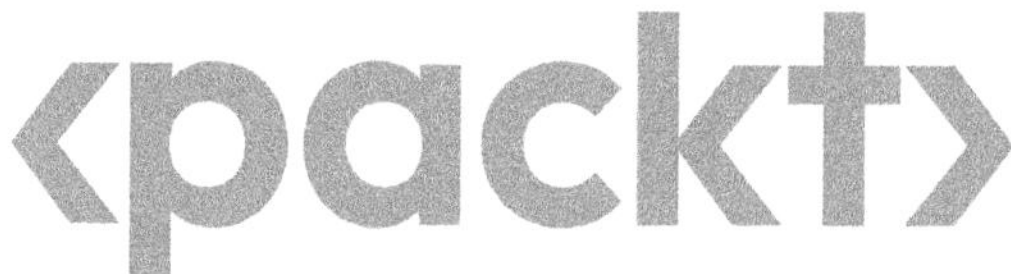

packt.com

Subscribe to our online digital library for full access to over 7,000 books and videos, as well as industry leading tools to help you plan your personal development and advance your career. For more information, please visit our website.

Why subscribe?

- Spend less time learning and more time coding with practical eBooks and Videos from over 4,000 industry professionals
- Improve your learning with Skill Plans built especially for you
- Get a free eBook or video every month
- Fully searchable for easy access to vital information
- Copy and paste, print, and bookmark content

At www.packt.com, you can also read a collection of free technical articles, sign up for a range of free newsletters, and receive exclusive discounts and offers on Packt books and eBooks.

Other Books You May Enjoy

If you enjoyed this book, you may be interested in these other books by Packt:

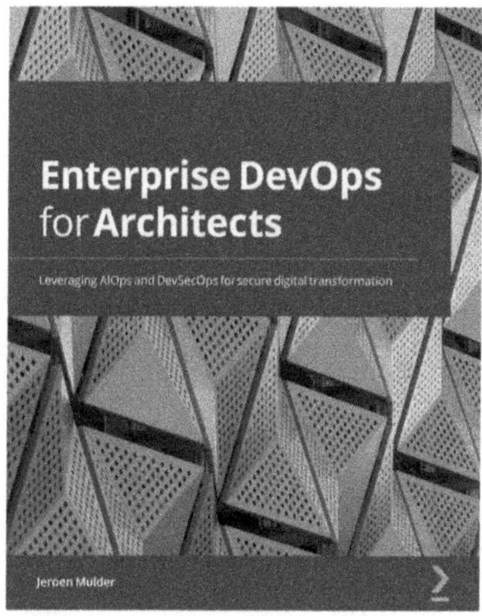

Enterprise DevOps for Architects

Jeroen Mulder

ISBN: 9781801812153

- Create DevOps architecture and integrate it with the enterprise architecture
- Discover how DevOps can add value to the quality of IT delivery
- Explore strategies to scale DevOps for an enterprise
- Architect SRE for an enterprise as next-level DevOps
- Understand AIOps and what value it can bring to an enterprise
- Create your AIOps architecture and integrate it into DevOps
- Create your DevSecOps architecture and integrate it with the existing DevOps setup
- Apply zero-trust principles and industry security frameworks to DevOps

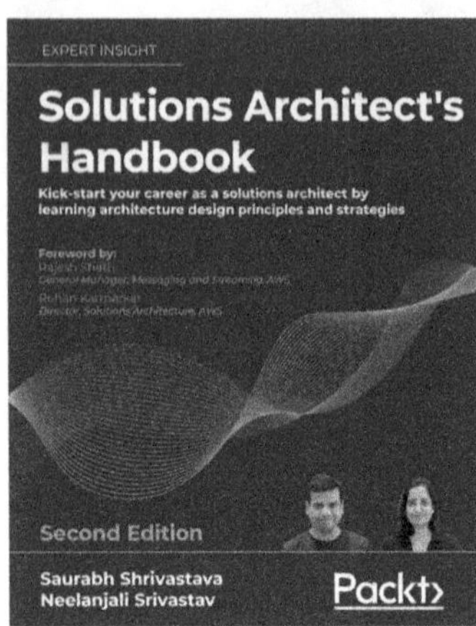

Solutions Architect's Handbook

Saurabh Shrivastava, Neelanjali Srivastav

ISBN: 9781801816618

- Explore the various roles of a solutions architect and their involvement in the enterprise landscape
- Approach big data processing, machine learning, and IoT from an architect's perspective and understand how they fit into modern architecture
- Discover different solution architecture patterns such as event-driven and microservice patterns
- Find ways to keep yourself updated with new technologies and enhance your skills
- Modernize legacy applications with the help of cloud integration
- Get to grips with choosing an appropriate strategy to reduce cost

Packt is searching for authors like you

If you're interested in becoming an author for Packt, please visit `authors.packtpub.com` and apply today. We have worked with thousands of developers and tech professionals, just like you, to help them share their insight with the global tech community. You can make a general application, apply for a specific hot topic that we are recruiting an author for, or submit your own idea.

Share your thoughts

Now you've finished *Multi-Cloud Strategy for Cloud Architects, Second Edition*, we'd love to hear your thoughts! Scan the QR code below to go straight to the Amazon review page for this book and share your feedback or leave a review on the site that you purchased it from.

`https://packt.link/r/1804616737`

Your review is important to us and the tech community and will help us make sure we're delivering excellent quality content.

Index

Symbols

A

C

D

E

F

J

K

L

M

P

Download a free PDF copy of this book

Thanks for purchasing this book!

Do you like to read on the go but are unable to carry your print books everywhere? Is your eBook purchase not compatible with the device of your choice?

Don't worry, now with every Packt book you get a DRM-free PDF version of that book at no cost.

Read anywhere, any place, on any device. Search, copy, and paste code from your favorite technical books directly into your application.

The perks don't stop there, you can get exclusive access to discounts, newsletters, and great free content in your inbox daily

Follow these simple steps to get the benefits:

1. Scan the QR code or visit the link below

`https://packt.link/free-ebook/9781804616734`

2. Submit your proof of purchase
3. That's it! We'll send your free PDF and other benefits to your email directly

www.ingramcontent.com/pod-product-compliance
Lightning Source LLC
Chambersburg PA
CBHW060644060326
40690CB00020B/4506